Fossil Crinoids

HANS HESS
WILLIAM I. AUSICH
CARLTON E. BRETT
MICHAEL J. SIMMS

CAMBRIDGE
UNIVERSITY PRESS

PUBLISHED BY THE PRESS SYNDICATE OF THE UNIVERSITY OF CAMBRIDGE
The Pitt Building, Trumpington Street, Cambridge, United Kingdom

CAMBRIDGE UNIVERSITY PRESS
The Edinburgh Building, Cambridge CB2 2RU, UK
40 West 20th Street, New York, NY 10011-4211, USA
10 Stamford Road, Oakleigh, Melbourne 3166, Australia
Ruiz de Alarcón 13, 28014 Madrid, Spain
Dock House, The Waterfront, Cape Town 8001, South Africa

http://www.cambridge.org

© Cambridge University Press 1999

This book is in copyright. Subject to statutory exception
and to the provisions of relevant collective licensing agreements,
no reproduction of any part may take place without
the written permission of Cambridge University Press.

First published 1999
First paperback edition 2002

Printed in the United States of America

Typeface Goudy 10/13 pt.   *System* DeskTopPro$_{/\text{UX}}$®   [BVC]

A catalog record for this book is available from the British Library

Library of Congress Cataloging in Publication data is available

ISBN 0 521 45024 1 hardback
ISBN 0 521 52440 7 paperback

Fossil Crinoids

Crinoids have graced the oceans for more than 500 million years. Brightly coloured feather stars fascinate divers in coral reefs, and stalked sea lilies are among the most attractive fossils. Crinoids had a key role in the ecology of marine communities through much of the fossil record, and their remains are prominent rock-forming constituents of many limestones.

This is the first comprehensive volume on crinoids to bring together information on their form and function, classification, evolutionary history, occurrence, preservation and ecology. The main part of the book is devoted to assemblages of intact fossil crinoids, which are described in their geological setting, ranging from the Ordovician to the Tertiary. The final chapter deals with living sea lilies and feather stars. The volume is exquisitely illustrated with abundant photographs and line drawings of crinoids from sites around the world.

This authoritative account re-creates a fascinating picture of fossil crinoids for palaeontologists, geologists, evolutionary and marine biologists, ecologists and amateur fossil collectors.

Hans Hess is affiliated with the Basel Natural History Museum.

William I. Ausich is chairman of and professor in the Department of Geological Sciences, The Ohio State University.

Carlton E. Brett is a professor of Geology at the University of Cincinnati, Ohio.

Michael J. Simms is curator of palaeontology at the Ulster Museum, Belfast.

Contents

List of Contributors   page vii
Acknowledgements   ix
Prelude   xi
D. Bradford Macurda, Jr.

Introduction   xiii
Hans Hess and William I. Ausich

GENERAL PART

1. **Crinoid Form and Function**   3
   William I. Ausich, Carlton E. Brett, Hans Hess and Michael J. Simms

2. **Systematics, Phylogeny and Evolutionary History**   31
   Michael J. Simms

3. **Fossil Occurrence**   41
   William I. Ausich, Stephen K. Donovan, Hans Hess and Michael J. Simms

4. **Taphonomy**   50
   William I. Ausich, Carlton E. Brett and Hans Hess

5. **Ecology and Ecological Interactions**   55
   William I. Ausich and Michael J. Simms

ASSEMBLAGES

6. **Middle Ordovician Trenton Group of New York, USA**   63
   Carlton E. Brett

7. **Middle Ordovician of the Lake Simcoe Area of Ontario, Canada**   68
   Carlton E. Brett and Wendy L. Taylor

8. **Upper Ordovician of the Cincinnati, Ohio, Area, USA**   75
   William I. Ausich

9. **Silurian of Gotland, Sweden**   81
   Hans Hess

10. **Middle Silurian Rochester Shale of Western New York, USA, and Southern Ontario, Canada**   87
    Wendy L. Taylor and Carlton E. Brett

11. **Scyphocrinitids from the Silurian–Devonian Boundary of Morocco**   93
    Hans Hess

12. **Lower Devonian Manlius/Coeymans Formation of Central New York, USA**   103
    Carlton E. Brett

13. **Lower Devonian Hunsrück Slate of Germany**   111
    Hans Hess

14. **Middle Devonian Windom Shale of Vincent, New York, USA**   122
    Carlton E. Brett

15. **Middle Devonian Arkona Shale of Ontario, Canada, and Silica Shale of Ohio, USA**   129
    Carlton E. Brett

16. **Lower Mississippian Hampton Formation at LeGrand, Iowa, USA**   135
    William I. Ausich

17. **Lower Mississippian Burlington Limestone along the Mississippi Valley in Iowa, Illinois, and Missouri, USA**   139
    William I. Ausich

18. **Lower Mississippian Edwardsville Formation at Crawfordsville, Indiana, USA** 145
William I. Ausich

19. **Upper Pennsylvanian LaSalle Member, Bond Formation of Central Illinois, USA** 155
William I. Ausich

20. **Permian** 160
Hans Hess

21. **Triassic Muschelkalk of Central Europe** 164
Hans Hagdorn

22. **Pentacrinites from the Lower Jurassic of the Dorset Coast of Southern England** 177
Michael J. Simms

23. **Lower Jurassic Posidonia Shale of Southern Germany** 183
Hans Hess

24. **Middle Jurassic of Southern England** 197
Michael J. Simms

25. **Middle Jurassic of Northern Switzerland** 203
Hans Hess

26. **Upper Jurassic Solnhofen Plattenkalk of Bavaria, Germany** 216
Hans Hess

27. **Uintacrinus Beds of the Upper Cretaceous Niobrara Formation, Kansas, USA** 225
Hans Hess

28. **Tertiary** 233
Hans Hess

29. **Recent** 237
Hans Hess

Appendix I. Geological Time Table with Crinoid Assemblages 245
Hans Hess and William I. Ausich

Appendix II. Glossary of Rocks 247
Hans Hess

Bibliography 249
Hans Hess

General Index 263
Hans Hess

Taxonomic Index 271
Hans Hess

Contributors

William I. Ausich (Introduction; Chapters 1, 3, 4, 5, 8 and 16–19; review of manuscript) was born in 1952 and raised in Galva, Illinois. He earned a bachelor of science degree in geology from the University of Illinois, and master's and doctoral degrees from Indiana University. At Indiana University, he studied under the guidance of Professor N. Gary Lane. His research interests have included Mississippian, Silurian, and Ordovician crinoids and the topics of palaeoecology, constructional morphology, systematics and evolutionary history. He has held faculty positions at Wright State University, Dayton, Ohio, and The Ohio State University, Columbus, where he is presently Professor and Chair.

Carlton E. Brett (Chapters 1, 4, 6, 7, 10, 12, 14, and 15) was born in 1951. He received his doctoral degree in 1978 from the University of Michigan for research on Silurian pelmatozoan echinoderms and their palaeoecology. He taught at the University of Rochester from 1978 to 1998 and is now Professor of Geology at the University of Cincinnati, Ohio. He has authored numerous papers and edited four volumes on various subjects, including Palaeozoic echinoderms, modern and ancient taphonomy, sequence stratigraphy, regional geology of the Ordovician–Devonian rocks in the Appalachian Basin, ancient organism interaction and evolutionary ecology of Palaeozoic marine faunas. He was co-recipient (with William I. Ausich) of the Palaeontological Society's Schuchert Medal in 1990 and is a Fellow of the Geological Society of America.

Stephen K. Donovan (Chapter 3; review of manuscript) was born in 1954 and raised in London. After leaving school he spent five years working in telecommunications before entering the University of Manchester, where he was awarded his bachelor of science degree in 1980. His doctoral degree supervised by Professor Christopher R. C. Paul at the University of Liverpool, was awarded in 1983. From 1986 to 1998 he taught at the University of the West Indies in Kingston, Jamaica, as Professor of Palaeozoology. He is now Keeper of Palaeontology at The Natural History Museum, London. In 1994 he was awarded a bachelor of science degree by the University of Liverpool. His principal research interests centre on the palaeozoology of the Caribbean region.

Hans Hagdorn (Chapter 21) was born in 1949 in Schwäbisch-Hall, Baden-Württemberg. He studied Germanic languages and geography at the University of Tübingen and currently teaches at the Künzelsau Commercial School. Since his school days, he has accumulated a fine collection of Triassic fossils, which is exhibited at the Muschelkalkmuseum in Ingelfingen. His research concentrates on the stratigraphy and palaeoecology of the Germanic Triassic, including the systematics and evolutionary biology of crinoids. He is a member of two commissions on Triassic stratigraphy. In 1988 he received an honorary degree from Tübingen University and in 1995 the Zittel Medal from the Paläontologische Gesellschaft.

Hans Hess (originator and main editor; Introduction and Chapters 1, 3, 4, 9, 11, 13, 20, 23, 25–29; Appendix I and II; Bibliography; Indices) was born in 1930 and raised in Basel, Switzerland, where he studied pharmacy, leading to a career in the chemical industry. He retired in 1991 as the head of Pharmacy Research and Development of Ciba-Geigy Corporation. From his school days he was interested in fossils and specialized in Jurassic echinoderms, an area well suited to a hobby palaeontologist. His work, mainly on asteroids, ophiuroids and crinoids, earned him an honorary degree from Basel University in 1989. He now works part time for the Basel Natural History Museum.

René Kindlimann (all line drawings, redrawn from sources as indicated or new) was born in 1950 and raised in Zürich, where he attended the Industrial School of Art. He works as an independent graphic artist for fairs and exhibitions, more recently also for museums and special exhibitions. As a hobby palaeontologist, he has accumulated an important collection of remains of fossil sharks, on which he has also published.

D. Bradford Macurda, Jr. (Prelude) was born in 1936 and raised in Westchester County, New York. He earned bachelor's and doctoral degrees from the University of Wisconsin, where he studied blastoids under the guidance of Professor L. M. Cline. Subsequently, as a faculty member at the University of Michigan, he focused on Palaeozoic blastoids and living crinoids. He also held positions at Exxon Production Research and with the Energists, in Houston, Texas.

Michael J. Simms (Chapters 1–3, 5, 22, 24) was born in 1960. He first developed an interest in fossils at the age of 6 and by the age of 14 had come to specialize in the Lower Jurassic, in particular its echinoderm fauna. He obtained a first class honours degree in geology with zoology from Bristol University in 1983 and in 1986 completed a doctorate at Birmingham University on Lower Jurassic crinoids. Since then he has held a succession of research fellowships and short-term posts in higher education and in national museums. He has worked on phylogenetic relationships among crinoids. Other research interests include evidence for climate change in the Triassic and karst geomorphology and hydrology. Currently he is serving a two-year appointment as Curator of Palaeontology at the Ulster Museum, Belfast.

Wendy L. Taylor (Chapters 7 and 10) was born in 1962 in Erie, Pennsylvania. She graduated with a bachelor's degree in geosciences and biology from the State University of New York College at Fredonia in 1989. She obtained a master's degree in geology in 1993 at the University of Rochester and completed her doctorate in 1996. She was hired as the Paleontological Research Institute's first collections manager in 1995. Ms. Taylor's master's research describes the taphonomy and palaeoecology of extraordinary fossil assemblages from the Rochester Shale. Her doctoral research is a comparative study of the Palaeozoic echinoderm fossil assemblages in western New York, southern Ontario and selected parts of eastern North America. Her other research interests include problematic animals known as machaeridians and Palaeozoic trace fossils.

Photographs: Source indicated in figure captions.

Acknowledgements

We wish to express our sincere thanks to the following colleagues and specialists, whose support was essential in writing this book, a task that took almost six years and included numerous corrections of working drafts. They are (in alphabetical order): N. Améziane-Cominardi (Paris, France), C. Bartels (Unna-Afferde, Germany), T. K. Baumiller (Ann Arbor, Mich., USA), J. P. Bourseau (Lyon, France), G. Brassel (Flensburg, Germany), G. Dietl (Stuttgart, Germany), W. Etter (Zürich, Switzerland), H. Falk (Linz, Austria), C. Franzén (Stockholm, Sweden), A. Gazdzicki (Warsaw, Poland), U. Gerhard (Bonn, Germany), R. Haude (Göttingen, Germany), R. Hauff (Holzmaden, Germany), T. Heinzeller (Munich, Germany), O. C. Honegger (Zollikerberg, Switzerland), B. Imhof (Trimbach, Switzerland), J. Jackson (Panama City, Panama), M. Jäger (Dotternhausen, Germany), H. Lierl (Linau, Germany), Guanghua Liu (Tübingen, Germany), Hp. Luterbacher (Tübingen, Germany), C. G. Messing (Dania, Fla., USA), D. L. Meyer (Cincinnati, Ohio, USA), C. Neumann (Chapel Hill, N.C., USA), T. Oji (Tokyo, Japan), B. Pabst (Zürich, Switzerland), D. L. Pawson (Washington, D.C., USA), R. J. Prokop (Prague, Czech Republic), M. Röper (Kallmünz, Germany), Hj. Siber (Aathal, Switzerland), A. B. Smith (London, England), J. Todd (London, England), G. Viohl (Eichstätt, Germany), K. Vogel (Frankfurt a.M., Germany), G. D. Webster (Pullman, Wash., USA), A. Wetzel (Basel, Switzerland), and F. Wiedenmayer (Basel, Switzerland). We would also like to thank John Saunders (formerly Natural History Museum, Basel) for his advice during the book's gestation.

We would like to give special thanks to Mary Racine for the meticulous copy editing of the multi-authored manuscript and to production editor Holly Johnson for turning this into an enjoyable book.

We are greatly indebted to the Hans E. Moppert Foundation, Stiftung für Lebensqualität, and the Tobler Fonds of the Natural History Museum, Basel, for making it possible to publish colour photos.

Prelude

D. BRADFORD MACURDA, JR.

When we stand on a wave-swept shore, the physical force of the waves is the dominant stimulus to our imagination. The surface of the ocean is grey and blue, stretching as far as the eye can see, frosted by waves curling and breaking under the wind. The sea appears to be lifeless and empty. No hint of the life teeming within it is evident, except in tide pools along the shore or in the bounty offered by fishermen who have just landed their catch.

During the 1800s, we began to probe the sea in a scientific manner. Our sampling was remote – we used dredges and nets. We were surprised at some of the animals we recovered. Among these were great masses of stalked crinoids, which looked like living fossils. The advent of scuba diving and deep-diving submersibles finally opened the door to direct observation and study. Crinoids were described as being virtually extinct. But dive the Great Barrier Reef in Australia: more than 50 species live there. Free-living crinoids festoon the reef, arms spread in broad fans to filter the water, or they are tucked into crevices, hidden from predators. Dive reefs throughout the western Pacific – in Indonesia, Malaysia, Papua New Guinea and Fiji, and the vertical underwater cliffs at Palau; free-moving crinoids are abundant and numerous. Cross the Indian Ocean to Mauritius, the Seychelles, Tanzania, and the fabled spice island of Zanzibar, and there they are again. Dive the Red Sea at night. By day, the white plates of coralline animals are barren of apparent life. At night they are covered with masses of red crinoids that have come out to feed.

Sail the glittering waters of the Bahamas. Wherever you dive on the reefs, crinoids are to be found. Follow the Antilles south, cross northern South America in Venezuela and Colombia, dive the San Blas Islands of Panama and the islands off Nicaragua, Honduras and Belize. Crinoids are there to greet you. In the shallow waters of the tropics, we have come to learn they are alive, diverse and doing well.

For those who have collected fossil crinoids from Palaeozoic rocks and those of the Triassic and Jurassic, the ultimate time machine is a submersible. Fasten down the hatch, descend below 100 m into the world where the sunlight doesn't shine, and you will find living stalked crinoids, some with stems a metre long. Crinoids with great parabolic fans recurved into the currents have shown their life habits to be different from many reconstructions and have enabled us to better understand their ancestors. Small five-armed crinoids come into view, revealing a set of different strategies as to how high to live above the bottom. In the Straits of Florida,

Hans Hess, William I. Ausich, Carlton E. Brett, and Michael J. Simms, eds., *Fossil Crinoids*. © 1999 Cambridge University Press. All rights reserved. Printed in the United States of America.

Fig. 1. Group of living *Neocrinus decorus* on carbonate hard bottom, with crowns forming parabolic filtration fan. The current runs from the foreground to the background, and the oral surfaces of the crowns face away from the viewer (i.e., downcurrent). Stems are bent proximally at nearly a right angle, orienting the crowns perpendicular to the current. A comatulid (*Stylometra spinifera*) clings to the stem of the specimen third from left. Northeastern Straits of Florida, south of the west end of Grand Bahama Island, at 420 m. (Photograph from the Johnson Sea Link submersible; courtesy C. G. Messing.)

crinoids anchor on the bare rock of the sea floor looking like ghostly sentinels in the light or like a field of sunflowers turned toward the sun (Fig. 1).

We are fortunate indeed to live at a time when we can compare the modern and the ancient first-hand. It is these insights that enable us to better understand the fossil remains of crinoids, to reconstruct their life habits and to learn how they have changed and evolved.

Introduction

HANS HESS AND WILLIAM I. AUSICH

Brightly coloured feather stars fascinate divers in tropical coral reefs, and stalked sea lilies are among the most attractive fossils. Both belong to the class of crinoids, suspension feeders that have graced the oceans for more than 500 million years. Crinoids are common fossils, had a key role in the ecological structuring of marine communities through much of the fossil record and played an important role in rock-building. Crinoids are the topic of this book, which brings together for the first time the essentials of crinoid morphology, preservation, systematics, phylogeny and mode of life in the context of the most important and interesting crinoid fossil assemblages. The book attempts to re-create a picture of fossil crinoids alive in their natural habitats. We hope that some of our enthusiasm for these living and fossil animals will be conveyed to our readers, professionals and amateurs alike.

Crinoids are a class of echinoderms (phylum Echinodermata). These marine animals are characterized by a calcareous endoskeleton of distinct plates or ossicles, radial symmetry[1] and an internal water-vascular system. The echinoderm skeleton is highly porous in a living animal but is easily transformed into solid calcite during fossilization, which explains the richness of crinoidal remains throughout geological history. Crinoids differ from other classes of living echinoderms, such as starfish (Asteroidea), brittle stars (Ophiuroidea), sea urchins (Echinoidea) and sea cucumbers (Holothuroidea), by their morphological organization and their way of feeding. These other echinoderms seek food mostly on the sea floor, so their mouths are directed downward, whereas crinoids filter the water for plankton and direct their mouths upward. Crinoids are placed in the subphylum Pelmatozoa along with other, extinct classes, commonly grouped in the blastozoans.

Crinoids differ fundamentally from other pelmatozoans in the development of true arms. Arms are extensions of the body wall that contain the food-collecting grooves and also the extensions of the haemal, nervous, water-vascular and reproductive systems. Crinoid arms are supported by plates directly continuous with the highest plates of the cup. In contrast, blastozoans had simple armlets or brachioles attached to their cups or thecae that also contained specialized respiratory pores. Like crinoids, blastozoans lived mostly upright. Their oral side was directed upward, and their appendages must have served as food-gathering organs, so that their ambulacra were developed into food grooves. The Lower to Middle Cambrian genus *Gogia* (Fig. 2) provides an example of primitive blastozoans. The oldest known pelmatozoans are from the Lower Cambrian, whereas the first true crinoids did not appear until the Lower Ordovician. Blastozoans lived exclusively during the Palaeozoic Era.

Crinoids flourished during the Palaeozoic Era, when they reached their peak numbers during the Early Carboniferous (Fig. 3). They almost became extinct at the end of the Permian, but recovered to flourish again

Hans Hess, William I. Ausich, Carlton E. Brett, and Michael J. Simms, eds., *Fossil Crinoids*. © 1999 Cambridge University Press. All rights reserved. Printed in the United States of America.

Fig. 2. *Gogia spiralis*. Two individuals of an early pelmatozoan. Wheeler Shale, Middle Cambrian, Antelope Spring, House Range, Utah. The specimens are on the bedding plane of a highly argillaceous, micritic limestone. See Robison (1965) for further details. (Hess Collection; photograph S. Dahint.) ×1. To view this figure in colour, see the colour plate section following page xv.

during the Mesozoic. More than 6,000 fossil species, belonging to more than 800 genera, have been described. Today, approximately 600 living species are known; most are free-living feather stars or comatulids living in the shallow seas. Approximately 80 species of stalked sea lilies are restricted to the deeper water of today's oceans.

Like other echinoderms, crinoids are exclusively marine. Nevertheless, during their long history, they adapted to almost all marine habitats and developed an incredible array of morphologies. Crinoids have populated reefs and hardgrounds as well as muddy bottoms. Many were anchored with a holdfast and others moved sluggishly. The great majority of crinoids were stalked (stemmed). In some, notably the post-Palaeozoic comatulids, the stem was greatly reduced or is absent, but this conferred other advantages, such as the ability to hide from predators, to attain various feeding positions or even to swim. A number of forms became free-floating, including very small, pelagic crinoids; large, free-floating forms; and the largest stalked sea lilies ever found, which were anchored to floating logs.

Crinoids are efficient food collectors, and much of the skeleton is devoted to this purpose. A central cup of calcareous plates contains most of the soft parts, such as the mouth, gut and anus. The stem is composed of disc-shaped columnals, and for most crinoids the stem elevates the animals into a favourable feeding position up in the water column. The cup bears the food-gathering arms with the tube feet, extensions of the water-vascular system. Suspended food is trapped on the tube feet and passed along the food groove to the mouth by tube foot

INTRODUCTION

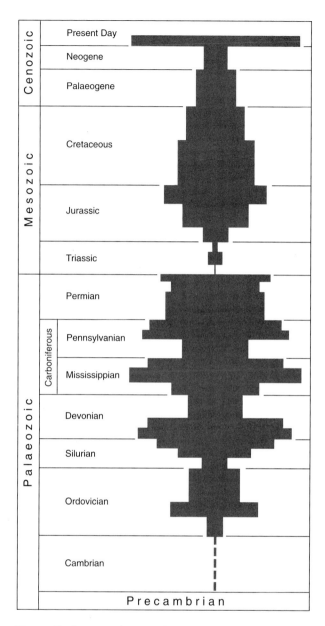

Fig. 3. Evolutionary history of crinoids based on number of genera. (Redrawn from Broadhead & Waters 1980.)

and ciliary action. In living crinoids the arms also contain the gonads, which release a large number of eggs and sperm into the surrounding water during spawning.

The Palaeozoic Era produced the majority of described crinoid species, which are traditionally classified into three subclasses – the Camerata, Flexibilia and Inadunata (Moore & Teichert 1978). During the Mesozoic the subclass Articulata diversified from survivors of the end-Permian extinction. This subclass, which includes all the living crinoids, is named for the development of muscular articulations in the arms, which increases the mobility between ossicles.

Most crinoid ossicles, other than those in the arms, were bound together in life by ligaments. Both ligamentary and muscular articulations produced characteristic patterns that are important not only for classification but also for understanding crinoid function. However, upon death these connective tissues quickly decay, and the great majority of fossil crinoidal remains are disarticulated. The study of disarticulated crinoid ossicles is an essential task in crinoid palaeontology, and characteristics of isolated ossicles are described in the first part of this book.

The introductory chapters discuss the form and function of crinoid morphology, classification, evolutionary history, occurrence, preservation and ecology. In view of the vast fossil record of crinoids, which contains so many unique and extraordinary forms, these chapters cannot be exhaustive. Rather, they prepare the scene for understanding the fossil assemblages. For more detailed information, readers are directed to Part T of the *Treatise on Invertebrate Paleontology* (Moore & Teichert 1978). The primary focus of this book is the complete crinoid organism and assemblages of intact crinoids. Complete specimens occur only rarely but are found repeatedly on bedding planes throughout geological history. Such *Lagerstätten*, treasuries of unique information, are described in their geological setting in the second, and main, part of this book, where we also try to reconstruct the fossils' mode of life. In view of the widespread occurrence of crinoid assemblages, and the corresponding difficulties of gaining access to material and information, this task has been divided among the authors. Even so, the advice of other specialists was essential, and our sincere thanks are expressed to all of these who share our fascination with the crinoids.

NOTE

1. Radial, typically pentameral symmetry is superimposed on basically bilateral symmetry.

Fig. 2. *Gogia spiralis*. Two individuals of an early pelmatozoan. Wheeler Shale, Middle Cambrian, Antelope Spring, House Range, Utah. The specimens are on the bedding plane of a highly argillaceous, micritic limestone. See Robison (1965) for further details. (Hess Collection; photograph S. Dahint.) ×1.

Fig. 6. Close-up of the oral disc (tegmen) of a comasterid comatulid with yellow-tipped oral pinnules; terminal comb teeth show on some of the pinnules. A black ophiuroid is sprawled across the disc, and the crinoid is releasing a bolus of faecal material from the anal tube. (Photograph O. C. Honegger, taken off Manado, northern tip of Sulawesi, depth around 20 m.)

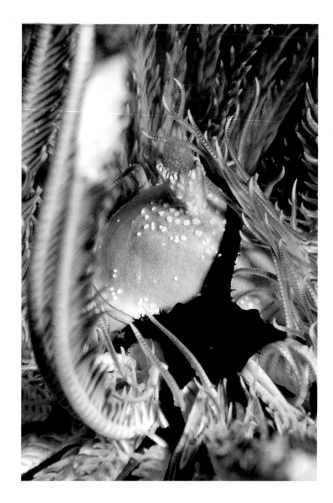

Fig. 62. Root of *Liliocrinus munsterianus*, coloured with fossil organic pigments called Fringelites. The root, which is made up of partly uncoloured calcite, has grown stepwise, by successive accretion, in the muddy sediment. Liesberg Beds (Middle Oxfordian), Liesberg. (National History Museum, Basel; photograph S. Dahint.) ×0.5.

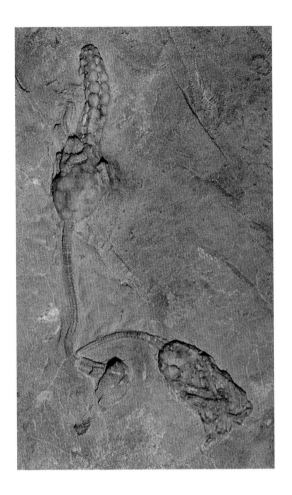

Fig. 81. *Illemocrinus amphiatus*, a small, simple cladid from the Bobcaygeon Limestone near Brechin, Ontario. Note complete short, pentameric stem and small, cemented discoidal holdfast, also stout anal sac. (Royal Ontario Museum.) ×1.5.

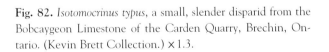

Fig. 82. *Isotomocrinus typus*, a small, slender disparid from the Bobcaygeon Limestone of the Carden Quarry, Brechin, Ontario. (Kevin Brett Collection.) ×1.3.

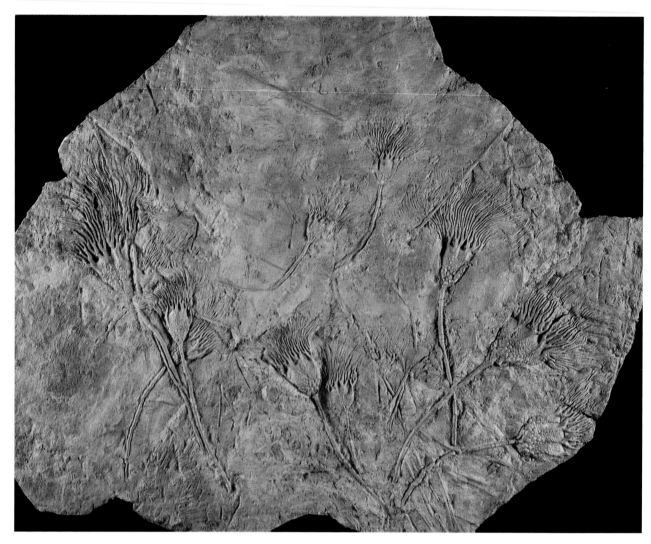

Fig. 109. Lower surface of a slab with *Scyphocrinites* sp. in natural colour. Upper Silurian, from an excavation east of Haroum near wadi of Oued Ziz. (National History Museum, Basel; photograph S. Dahint.) ×0.15.

Fig. 211. Unweathered section with well-developed *Chariocrinus andreae* Beds, Lower Hauptrogenstein, eastern part of quarry at the Lausen railway station. Exposed are two sets of crinoidal limestone beds interbedded with blue marl-clay and separated from oolitic limestone below and above by thicker layers of clay. In the lower set with three distinct beds (arrows), complete crinoids such as those of Fig. 210 occur on the lowest bedding plane, just above the hammer. The upper set passes into marly limestone in its upper part and into oolitic limestone to the right of the picture (not shown). The clay between the two sets has an elevated content of organic carbon (see text). The lower set of three crinoid beds thins to one bed, which wedges out at 3 m to the right of the section shown in the photograph. Clay near the top of the section is deformed through faulting. For detailed profiles in different parts of the quarry see C. A. Meyer (1988). (Photograph 1993 by H. Hess.)

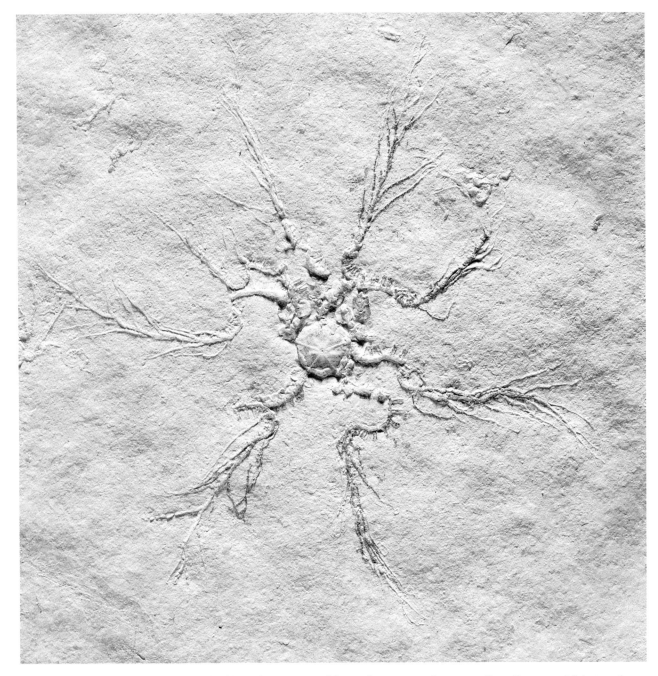

Fig. 219. *Saccocoma tenella* on a lower bedding plane, Langenaltheim. the specimen has unusually well-preserved *Schwimmplatten*, which also occur on the third secundibrachials; on one of the arms the second primibrachial is not axillary. (Hess Collection; photograph S. Dahint.) ×3.

Fig. 233. *Oxycomanthus bennetti*, attached to a sponge, with a trumpet fish hovering among the arms. The fish is not preying on the crinoid but rather is using the crinoid for shelter, waiting to ambush small fish that approach within range (D. L. Meyer, pers. comm., 1996). Off Manado, northern tip of Sulawesi, depth around 20 m. (Photograph O. C. Honegger.)

Fig. 234. *(Below) Oxycomanthus bennetti* showing cirri attached to a sponge; the terminal claws of several cirri are embedded in the surface of the sponge. Some of the cirri that are down have epibionts attached to the distal ends; such cirri might be raised for periodic cleaning by the comb-bearing oral pinnules (see also Fig 6) (D. L. Meyer, pers comm., 1996). Note ophiuroid hiding under the cirri. Off Manado, northern tip of Sulawesi at 20 m. (Photograph O. C. Honegger.)

Fig. 235. Two individuals of the living insocrinid *Cenocrinus asterius* with 40–50 arms, attached to hardground. Crowns are in the wilted-flower posture (Baumiller *et al.* 1991) with the oral side directed upward, indicative of a slack current. Hogsty Reef, Bahamas at 310 m. (Courtesy Harbor Branch Oceanographic Institution, Inc., through D. L. Pawson.)

Fig. 236. Living isocrinid *Endoxocrinus parrae* on coral. Crown with 32 arms in oral view. In this parabolic filtration fan posture (Macurda & Meyer 1974), the stem is bent proximally at a right angle (behind the crown), the oral side of the crown is directed downcurrent and the distal part of the arms recurve aborally into the current. San Salvador, Bahamas, at 692 m. (Courtesy Harbor Branch Oceanographic Institution, Inc., through D. L. Pawson.)

General Part

Crinoid Form and Function

WILLIAM I. AUSICH, CARLTON E. BRETT, HANS HESS AND MICHAEL J. SIMMS

ANCESTORS, ARCHITECTURE AND ADAPTATION

Environmental adaptation accounts for much of the morphological variety within the class Crinoidea, but two other factors also have an important influence on gross morphology. First, crinoid morphology is constrained by the evolutionary history of the group; in other words, much of a crinoid's morphology is inherited from its ancestors. This is particularly evident in the pentaradiate symmetry and calcite endoskeleton that dominate echinoderm morphology. Second, the crinoid skeleton and soft tissues have certain physical properties and limits within which the animal must operate. Some aspects of crinoid morphology are strategies for reducing these architectural constraints rather than being direct adaptations to particular environmental factors. For example, the crinoid skeleton is composed largely of discrete ossicles connected by ligaments and other soft tissue. By adopting this multi-element construction, crinoids overcome the inherent inflexibility of individual calcite ossicles.

To understand the functional morphology of crinoids, fundamental constraints of ancestry, constructional materials and ecology must be considered. Crinoids are the most primitive group among extant echinoderms and, typically, retain at least a vestige of the stem that characterizes the largely extinct pelmatozoans. Like all pelmatozoans, crinoids are largely sessile and exclusively suspension-feeding.

SOFT AND HARD PARTS

The numerous calcareous plates of living crinoids are produced within the body wall, so that they are actually part of an endoskeleton. The bulk of the animal is the skeleton, with only a small percentage of living tissue. Under high magnification, crinoid plates are seen to be highly porous (Fig. 4). In life these pores were filled with tissue. This skeletal microstructure is called stereom, and it is easily recognized in well-preserved fossil ossicles and in thin sections.

The soft parts of crinoids are quite inconspicuous. The digestive tract with mouth, oesophagus, gut, rectum and anus is situated in the aboral cup. The anus and mouth are on the upper surface (Figs. 5, 6), with the anus commonly elevated on a cone or tube (Fig. 6) that is reinforced by platelets (Figs. 37, 38). A system of fluid-filled tubes, called the water-vascular system, is unique and vital for all echinoderms. The central element of the water-vascular system is the ring canal around the oesophagus. Radial canals extend from the ring canal into arms and pinnules, and these extensions underlie the ambulacra (Fig. 7). The water-vascular system canals terminate in the tube feet. This system has a hydrostatic function, as in other echinoderms; it seems to counteract muscular contractions of the tube feet by lengthening them. Tube feet, also called podia or tentacles, are part of the food-gathering ambulacral system, which is made up of ciliated, sensory and mucus-

Hans Hess, William I. Ausich, Carlton E. Brett, and Michael J. Simms, eds., *Fossil Crinoids*. © 1999 Cambridge University Press. All rights reserved. Printed in the United States of America.

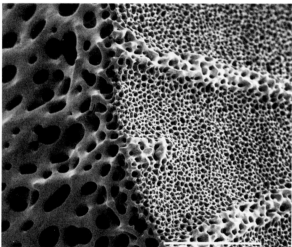

Fig. 4. Scanning electron photomicrographs of a brachial plate from *Promachocrinus kerguelensis* (Recent from the McMurdo Shelf, Antarctica) showing stereomic microstructure. This brachial facet with radiating crenulae is a ligamentary articulation, called syzygy. (Courtesy C. P. Hart and W. I. Ausich.) ×50 and ×300 (scale bars = 100 μm).

living crinoids is situated on specialized pinnules (see the subsection on pinnules). Crinoids also possess a haemal system; this is actually a network of spaces in the connective tissue of the body cavity.

This brief overview would not be complete without mention of the coelom, or body cavity. Adoral coelomic canals underlie the water-vascular and ambulacral systems of the arms and pinnules. The aboral coelomic compartment surrounds the intestine and continues into the arms and pinnules as aboral coelomic canals (Fig. 7). Crinoids do not have special respiratory organs. Respiration commonly occurs on the surface of the tube feet, probably by diffusion of oxygen through the body wall. Oxygen is transported to internal organs through the coelomic fluids rather than by the haemal fluid.

The skeleton of most crinoids is composed of a crown, a stem (also called stalk or column), which elevates the crown above the sea floor, and a holdfast for attachment to the substrate (Fig. 8). The lower part of the crown, the aboral cup (or calyx), contains the bulk of the soft parts, as already described. The food-gathering arms are attached to the cup. The oral (also

secreting cells. In living crinoids, food particles are detected on impact by tube foot sensory cells and secured by mucus secretions of the finger-like extensions (papillae) of the tube feet (Fig. 7). Food particles are then passed along the ciliated food grooves to the mouth. The food grooves are commonly protected by platelets (Fig. 50c), or they may be concealed by enrolling the arms (Figs. 31–34). This ambulacral epidermis is underlain by a layer of nerve cells, to which the sensory cells connect. The nervous system is formed by a ring in the cup with extensions into the stem and cirri, as well as into the arms and pinnules. The reproductive system of

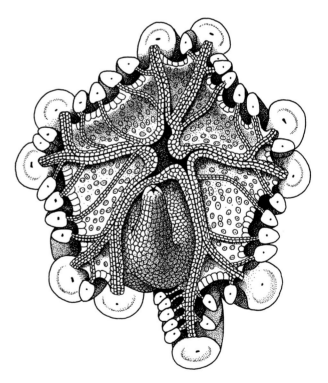

Fig. 5. Oral view of the disc of *Metacrinus angulatus* with tegmen. Food grooves end in the slightly displaced mouth; anal opening at the tip of an eccentric cone. (Redrawn from Carpenter 1884.) ×3.

called ventral) side of the arms is the site of the food grooves and is always directed upward or downcurrent (Fig. 1), whereas the aboral side (dorsal) is directed toward the bottom (or upcurrent). The parts making up the skeleton usually consist of individual plates or ossicles that are more or less firmly joined together. All articulations between ossicles of the stem (called columnals) are bound by ligamentary connective tissue and allow only passive movements. Innervated epithelial cells along cirri (branches off the stem) of certain groups effect slow movement for these stem appendages (Baumiller et al. 1991). Muscular articulations, allowing

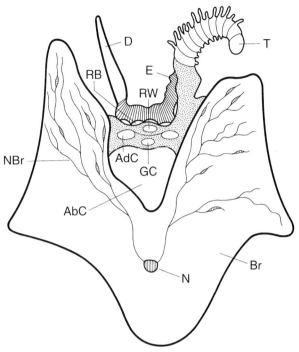

Fig. 7. Section through arm of *Bathycrinus aldrichianus*. Key: T, tentacle or tube foot or podium; D, covering plate of ambulacral groove; E, epidermis of ambulacral groove; Br, brachial ossicle; N, main brachial nerve; NBr, nerve branches; AbC, aboral coelomic canal; AdC, adoral coelomic canal; RW, radial water vessel; RB, radial blood vessel; GC, genital cord. (Redrawn from Carpenter 1884.) ×200.

Fig. 6. Close-up of the oral disc (tegmen) of a comasterid comatulid with yellow-tipped oral pinnules; terminal comb teeth show on some of the pinnules. A black ophiuroid is sprawled across the disc, and the crinoid is releasing a bolus of faecal material from the anal tube. (Photograph O. C. Honegger, taken off Manado, northern tip of Sulawesi, depth around 20 m.) To view this figure in colour, see the colour plate section following page xv.

movements, have apparently developed only between arm ossicles. Ligament fibres penetrate the interior of plates, producing a specific microscopic (galleried) pattern of stereom. Muscles do not extend into the stereom, so that areas of muscular insertion have a more irregular (labyrinthic) stereom. The structure of articular surfaces between ossicles is one of the keys to understanding crinoid function. For orientation of the different parts, we use the terms 'oral' and 'aboral', as well as 'proximal' (towards the base of the cup) and 'distal' (away from the base of the cup) (Figs. 10, 11).

THE STEM AND ITS APPENDAGES

The crinoid stem can serve several functions. The two most important are attachment to the substrate and elevation of the food-gathering system, represented by the arms, above the sea floor. In the majority of non-crinoid pelmatozoans the stem was short and rather weakly developed, suggesting that attachment or an-

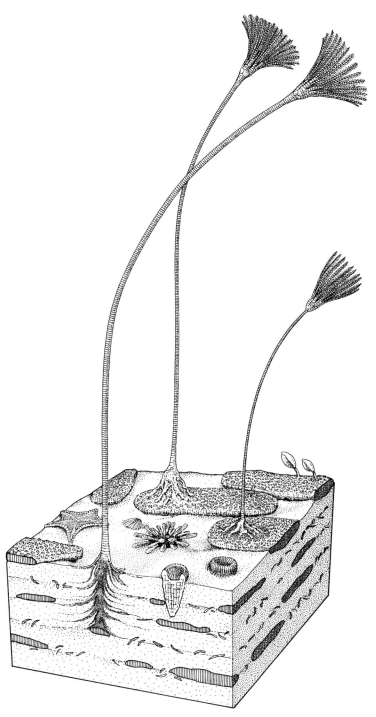

Fig. 8. Reconstruction of the Jurassic sea lily *Liliocrinus munsterianus*. Two individuals attached to dead corals (*Thamnasteria*) by a compact root, another individual anchored in muddy sediment (Liesberg Beds, Middle Oxfordian, Swiss Jura). This environment is comparable to today's lagoon southeast of Nouméa (New Caledonia), where flat corals lie loose on a muddy bottom in 35- to 40-m-deep water (L. Hottinger, pers. comm., 1996). *Liliocrinus munsterianus* had a stem with a length of up to 2 m and a crown with a height of 15 cm. The crinoid was fixed to hard substrates by a massive root; alternatively, it was anchored in the soft bottom by roots that became quite long, growing in step with accumulating sediment (Fig. 62). It must be assumed that such roots first attached to some hard object (piece of coral, shell fragment). Also shown are other parts of the fauna such as the echinoid *Paracidaris florigemma*, the asteroid *Tylasteria*, a pectinid bivalve and two terebratulids; the solitary coral *Montlivaltia* (dead specimens) is partly embedded in the mud.

chorage may have been its primary function. The echinoderm stem appears to have originated from an aboral evagination of the body, leading first to hollow tubes that were reinforced with irregular ossicles. Such primitive stems occur in Middle Cambrian pre-crinoidal pelmatozoans like *Gogia* (Fig. 2). In contrast, even in the earliest known crinoid, *Aethocrinus*, the stem is significantly more robust and longer (Fig. 9), a pattern largely maintained throughout the history of the group. A stem is not required for attachment – the only purpose of a stem is elevation off the bottom so that the animal can escape the benthic boundary layer for better feeding, and perhaps reproduction. It has been suggested that the development of true columns in the Early Ordovician contributed largely to the huge success of crinoids in the Palaeozoic. The comatulids, which flourish today, have become detached, with the potential to climb to a higher position for feeding or to crawl into cavities to avoid predation.

Stem Morphology and Growth

Stems of modern isocrinids will serve as the starting point for our discussion. They are composed of columnals with a central canal, so that the stem contains a central tubular cavity with extensions of the coelom and nervous system. Columnals bearing cirri are nodals or cirrinodals; those without cirri are internodals (Fig. 10). Stems have two distinct regions. In the distal part, away from the cup, the arrangement of the columnals remains constant, and nodals are separated by a nearly constant number of internodals. New nodals are formed just below the cup, so the short proximal region is the immature stem. Near the cup, the developing internodals are completely hidden by the nodals (Fig. 11), but internodals are successively introduced between nodals. Proximal columnals are shorter (thinner) than distal ones, so the stem grows or matures by sequentially adding columnals in the proximal region, first by adding nodals, then by intercalating internodals and finally by increasing the diameter of individual columnals.

Columnals are bound together by two types of elastic ligament fibres or mutable collagenous tissue,[1] which occur in a characteristic pattern (muscles are absent in the stem). Short, intercolumnar ligaments connect each pair of adjacent columnals. Longer, through-going ligaments connect a set of internodals and one associated nodal (Fig. 12). The corresponding articulations are called symplexies and are recognized in lateral profile by their crenulate appearance: interlocking grooves and ridges on adjacent columnals (Figs. 10, 13). The grooves and ridges occur commonly as a petaloid pattern (Fig. 10), which presumably gives the stem a certain flexibility in different directions, preventing twisting and allowing for easy return to the original position. Longer, through-going ligaments are limited to the areola of each interradial petal; each ligament extends all the way through a series of columnals and terminates at the aboral (distal or lower) side of a nodal (Grimmer *et al.* 1985). Thus, longer ligament fibres are lacking between a nodal and the internodal immediately below. At this

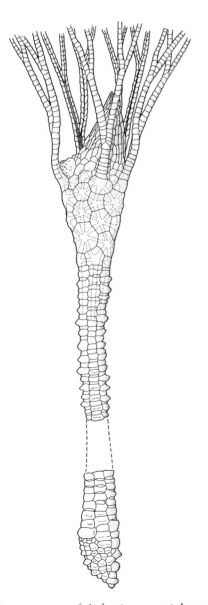

Fig. 9. Reconstruction of *Aethocrinus moorei*. Lower Ordovician, Montagne Noire, France. (Redrawn from Ubaghs 1969.) ×1.

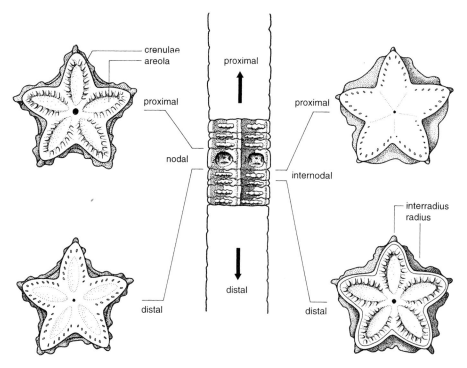

Fig. 10. Part of the stem of the isocrinid *Metacrinus angulatus*, with nodal and internodals, showing the different articular facets. (Redrawn from Carpenter 1884.) ×3.

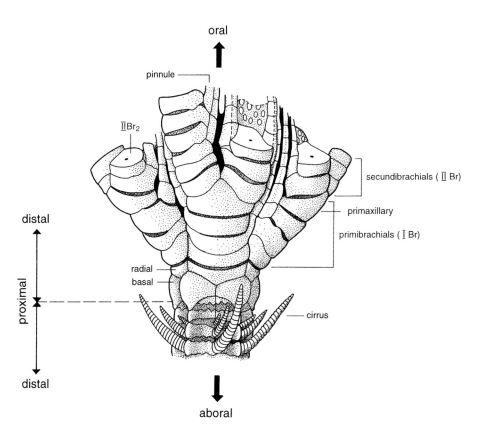

Fig. 11. Proximal stem, cup and base of arms of *Metacrinus angulatus*. (Redrawn from Carpenter 1884.) ×3.

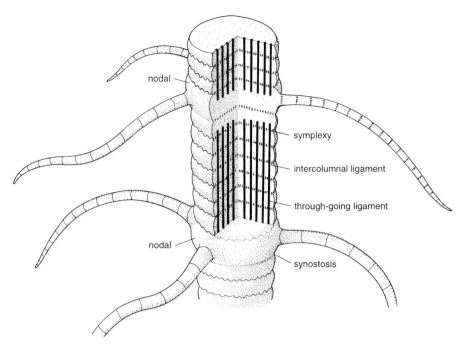

Fig. 12. Ligaments between nodal and internodals of an isocrinid stem. (Adapted from Baumiller & Ausich 1992.)

place, a tight junction, called synostosis, or more properly cryptosymplexy (hidden symplexy), is developed (Fig. 10). Synostoses and cryptosymplexies are easily recognized from the outside by a straight suture between nodal and distal internodal (Figs. 10, 12). These articulations have a simple low-relief topography and are held together only by the short intercolumnar ligaments. Breaking of the stem at this point guarantees that stem segments always end with a whirl of cirri for better attachment. It has been suggested by Hagdorn (1983) that this articulation developed among Middle Triassic isocrinids as a result of a habitat change from hard to soft bottoms. However, in contrast to Middle Jurassic isocrinids that thrived on soft bottoms (see Chapter 25), extant isocrinids prefer hard substrates or cling to pieces of rubble and shell (see Chapter 29). After breakage of the stem, the animal could reanchor itself with the terminal cirri, a possibility that does not exist for crinoids fixed with an attachment disc. Disintegration after death occur more rapidly along cryptosymplexies than along symplexies, and this is the reason for the occurrence of pluricolumnals (several articulated columnals) in sediments. Because such stem segments are common in many sediments from the Palaeozoic onward, it may be assumed that the two types of ligaments were developed early in the history of crinoids (Baumiller & Ausich 1992).

A different type of articulation is characterized by two opposing bundles of long ligaments that are separated by a fulcral ridge (Fig. 13). Such articulations, called synarthries, first evolved during the Middle Ordovician. Synarthries were never a dominant column articulation type, but one or another crinoid group had synarthries from the Middle Ordovician until the pres-

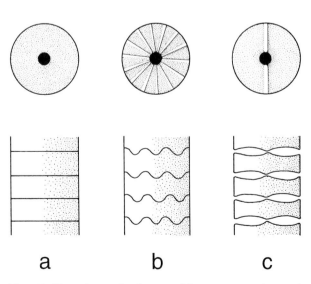

Fig. 13. Typical articular facets and long sections of crinoid stems. (a) Synostosis; (b) symplexy; (c) synarthry. (Redrawn from Donovan 1989a.)

ent. Furthermore, this columnal articulation style developed independently in four subclasses. Synarthrial articulations with fulcra aligned and unequal ligamentary areas on either side of each fulcrum produced a planispirally coiled column. Perhaps this served a protective function; and it evolved in the flexible *Ammonicrinus* (Fig. 14), and in the camerate *Camptocrinus* (Fig. 15). Possibly the most abundant crinoid with synarthrial articulations is *Platycrinites* (Mississippian to Permian) (Fig. 17). More or less circular articular facets with rather deep bifascial pits are a feature of the Bourgueticrinida, an order of articulates occurring from the Upper Cretaceous to Recent (Fig. 16), and in very young isocrinids and comatulids. Synarthrial-type articulations are also present in the cirri, as discussed later in the subsection on cirri.

Columnals of living crinoids have only a small central canal, but the lumen was very large in some fossil species, such as in the long stems of *Liliocrinus* (Figs. 8, 62) with their sometimes massive holdfasts. A wide canal does not lead to reduced strength.

Flexibility

Even when the stem serves solely for attachment, it must be either massively robust, as in Recent and fossil cyrtocrinids (Figs. 32–34), or else flexible enough to avoid fracturing due to stresses imposed by currents. The stereomic structure of crinoid ossicles enhances the resistance to fracturing of the calcite, but, nonetheless, the material of the skeleton is inherently inflexible. To overcome this constraint, the crinoid stem is divided into a series of rigid ossicles connected by flexible ligaments. It is interesting that stem flexibilities in Lower Mississippian crinoids are not correlated with hard-part characters such as stem diameter or columnal height (Baumiller & Ausich 1996), and ligament properties have been implied to be the most likely control of flexibility.

In some crinoids, such as the post-Palaeozoic encrinids and the isocrinids, the stem is most flexible a short distance below the crown and stiffer more distally, allowing for optimum positioning of the food-gathering arms in the current (Fig. 1). In other crinoids, such as the Jurassic millericrinids *Apiocrinites* and *Liliocrinus*, enlargement of the proximal columnals greatly reduces the flexibility near the cup. Instead, probably the whole stem, which reached a length of 2 m, was bent over by strong currents (Fig. 8). Seilacher *et al.* (1968) found that in the Lower Jurassic *Seirocrinus*, flexibility in-

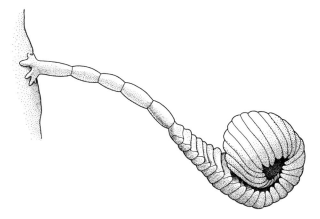

Fig. 14. Reconstruction of *Ammonicrinus doliiformis*, with crown hidden in enrolled stem. Stem is xenomorphic, with abrupt change between distal and middle part. Middle Devonian, Germany. (Redrawn from Ubaghs 1953.) ×1.5.

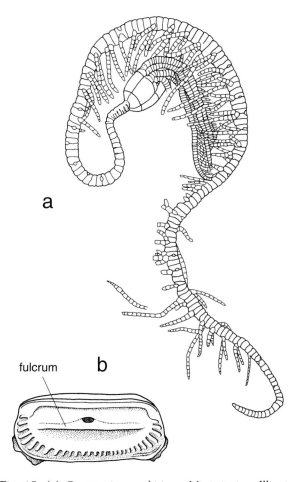

Fig. 15. (a) *Camptocrinus multicirrus*. Mississippian, Illinois. Complete specimen with coiled stem. (Redrawn from Ubaghs 1978.) ×1.5. (b) Articular, synarthrial facet of a columnal of *Camptocrinus compressus*. Lower Carboniferous, Scotland. (Redrawn from Ubaghs 1978.)×5.

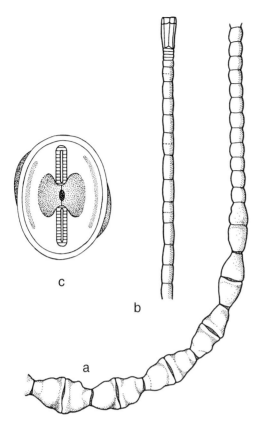

Fig. 16. (a, b) *Naumachocrinus hawaiiensis*, a Recent bourgueticrinid with synarthrial stem articulations. (Redrawn from Breimer 1978.) Approx. ×1. (a) Distal column with fulcral ridges on alternate pairs of apposed facets rotated by 180°; (b) proximal column with cup; (c) Articular facet of the bourgueticrinid *Democrinus rawsoni*. (Redrawn from Breimer 1978.) ×18.

creased toward the distal end of the stem. They interpreted this as an adaptation to a pseudoplanktonic mode of life (see Chapter 23).

Resistance to Tension, Torsion and Shearing

Among crinoids inhabiting environments where there is significant current activity, the stem may be subject to a range of stresses, which can broadly be grouped as tension (stretching), torsion (twisting) and shearing. In fossil stems, torsion is documented by twisted pluricolumnals of the Upper Ordovician *Plicodendrocrinus casei* (Donovan et al. 1995). Tensional stresses are resisted largely by the ligaments.

In crinoids with symplectial connections, shearing caused by lateral forces and torsion caused by twisting of the stem are resisted by the crenulae, a series of ridges

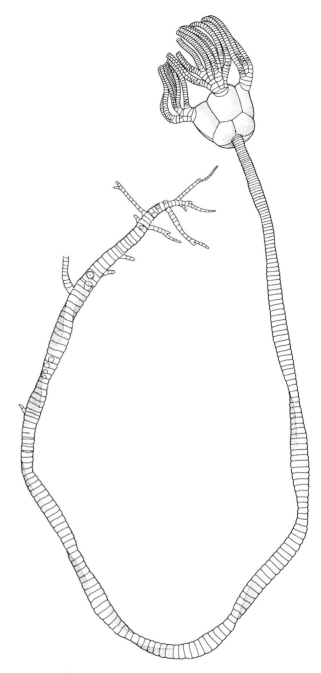

Fig. 17. *Platycrinites regalis*. Complete specimen with twisted stem carrying root-like radicular cirri (radices) distally. Mississippian (Burlington Limestone), Iowa. (Redrawn after Wachsmuth & Springer 1897.) ×0.7.

and grooves on the articulating face of one columnal that interlock with those on the opposing face of the next columnal (Figs. 10, 13). In circular columnals the crenulae are arranged around the margin of the articular facet; hence the number of crenulae is limited by the diameter of the columnal and the size of the crenulae.

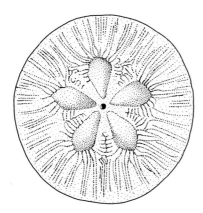

Fig. 18. Articular facet of internodal columnal of *Austinocrinus erckerti*. Senonian of Turkestan. (Redrawn from Rasmussen 1961.) ×2.

In pentaradiate columnals, such as those of the isocrinids, the crenularium has radial infoldings that greatly increase its effective length compared with that of circular columnals. This is particularly important in resisting shear stresses. In circular columnals there may be only a few crenulae actually oriented perpendicular to the direction of shear stress, whereas in pentaradiate columnals more than five times as many may have such an orientation. In some rounded columnals that evolved from pentaradiate ones, the crenulae may be greatly elongated in the radii (Fig. 18). This has been carried a stage further in the Early Jurassic *Seirocrinus*, where these radial spaces have developed significant rugosities that interlock with corresponding rugosities on the next columnal (Fig. 197). No such obvious skeletal adaptations exist in synostosial or synarthrial articulations to resist these forces.

ATTACHMENT AND SUPPORT

If the stem served only as an attachment structure for the animal, little would be gained by having a long stem. For crinoids, a longer stem confers feeding advantages. It elevates the food-gathering apparatus, the crown, above the sea floor into faster currents. Ausich (1980) and Ausich and Bottjer (1985) have documented tiering among crinoids, with different species having different stem lengths. High-diversity crinoid communities typically have species with various stem lengths, and this places different species at different feeding levels to minimize competition. The maximum number of tiering levels was reached when diversity, and presumably competition for resources, was also at a maximum (Simms 1990). At first sight it would appear that comatulids, the stemless form that dominates the modern crinoid fauna, might be adaptively disadvantaged. However, loss of the stem conferred much greater mobility on the comatulids. Many are able to clamber on to rocks, coral heads and other high points on the sea floor, using these objects as a surrogate stem. Many shallow-water species hide from predators during the day, and at night they crawl to favourable feeding perches (Magnus 1963). Some comatulids can even swim for short distances.

Attachment by distally tapered coiled stems was widespread during the Early Palaeozoic, particularly among the camerates and certain cladids. Examples are the Upper Ordovician *Pycnocrinus* (Fig. 90) and the Lower Devonian *Acanthocrinus* (Fig. 124). It appears that many of these crinoids either lacked a primary holdfast or possessed such a cemented attachment only during the juvenile stages and then broke free. There are many cases of these crinoids attaching to other objects by what appears to have been initially tapered stems (Fig. 103). Just how the juvenile crinoids were able to move around in order to locate a suitable host and then coil the distal stem around the object is quite enigmatic. One possibility is that crinoid larvae initially settled on upright stalks of bryozoans or other crinoids and then twisted the entire stem gradually around the upright posts, moving their crown in a spiralling motion coordinated with growth. After a point in time the attachment became permanent. This is inferred from the presence, in many of the distal coils, of wedge-shaped columnals that are thicker on the outside of the curving coiled stem.

Other crinoids do not appear to have coiled the distal end of the stem around any object, but rather to have laid out loose coils like a rattlesnake on the substrate to provide a base of stability (Fig. 124). Still more curious, but also very common, are many crinoids, including shoal-dwelling species during the Palaeozoic, that appear to have lacked any sort of holdfast whatsoever (Fig. 117). It is amazing that, without any appendages (radicular cirri) and with only a distally tapering stem, such crinoids were able to live in relatively high-energy environments. It seems plausible, but is purely speculative, that they simply dragged the column behind them, particularly if the crown achieved neutral buoyancy and was hence lower in density than the partially recumbent column.

Holdfasts

In any event, it is clear that certain crinoids had horizontally trailing stems because we find their holdfasts as creeping roots or runners along the substrate. Up to several centimetres of horizontal stem may be anchored to the substrate by small finger- or lobe-like protrusions of the stereom. Such specimens, which are common in many Silurian (Franzén 1977) and Devonian reef settings, typically attached to corals or stromatoporoids. The peculiar calceocrinids combined a stem that lay on the sea floor with an attachment disc (Fig. 29). This must have made them particularly vulnerable to burial and clogging of the ambulacra due to turbidity. However, the presence of a hinged crown, folding on the column, would have sealed the feeding surface; opening of the crown would have disengaged it from accumulated sediment (Brett 1984).

On muddy substrates, creeping stems are also common; these may be anchored to the substrate or attached to each other by strands of stereom. Only the distal parts of these stems were horizontal; the crown was borne by an upright stem. Examples abound in Middle Oxfordian calcareous mudstones of the Swiss and French Jura (Loriol 1877–1879, Pl. 12, Figs. 1–6).

Attachment of the stem may be either permanent or temporary. Attachment may be by a root or holdfast cemented to a hard substrate or by flexible outgrowths from the stem, known as cirri. The bewildering array of holdfasts, roots, accretion discs and so on shows the importance of fixation for benthic crinoids. Brett (1981) gave a comprehensive account of the variety of crinoid attachment structures.

The most primitive type of holdfast, a hollow tube made up of small irregular plates, is restricted to the archaic *Aethocrinus* (Fig. 9), but it was common in primitive blastozoans such as *Gogia* (Fig. 2). Such multi-plated tubes (*Hohlwurzeln*) appear to have grown downward into a muddy bottom to keep the animals upright and to counter-balance the crown. Smaller multi-plated holdfasts, such as *Lichenocrinus* (Figs. 87–89), were commonly cemented by a basal disc to solid substrates (hardgrounds or skeletal material). Many primitive disparids, some cladids and flexibles possessed a simple cone- or volcano-shaped attachment disc that was cemented to shells or hardgrounds. In a more advanced holdfast or radix, the distal end of the stem branches into root- or finger-like extensions. These are usually made up of segments and contain an axial canal; and again, they serve to anchor crinoids in unconsolidated sediments (Fig. 19). Roots of Jurassic millericrinids are commonly attached to dead corals, but they also occur on soft bottoms, where they reached a considerable length, growing stepwise in parallel with accumulating sediment (Figs. 8, 62).

Many crinoids living in agitated environments, such as reefs or flanks of reefs, were permanently attached. This is especially true of Palaeozoic crinoids that lacked efficient grappling devices for temporary attachment as present in comatulids and the isocrinids. The lack of attachment structures with contractile tissues prevented most Palaeozoic crinoids from actively moving around for better feeding positions or hiding from predators (Donovan 1993).

The cemented holdfast or radix structure of many Early Palaeozoic crinoids and other pelmatozoan echinoderms is by no means uncommon in later taxa. Recent comatulids pass through a sessile stage in early life. The larvae settle on some hard object, where they attach by a small disc and grow a stem during the so-called pentacrinoid stage before breaking away to assume a free-moving life. Among Mesozoic forms, the Triassic encrinids (Figs. 181, 183, 186) and Jurassic mil-

Fig. 19. Radicular cirri of *Rhizocrinus lofotensis*, a bourgueticrinid living on the muddy bottom. (Redrawn from Breimer 1978.) Approx. ×4

lericrinids (Fig. 8) had cemented holdfasts, as do fossil and extant cyrtocrinids, including *Holopus* (Fig. 31) and *Gymnocrinus* (Fig. 32). Additional weight may be added to roots by encrustment with secondary stereom. In *Liliocrinus*, such roots became very large blocks (Fig. 8) that held an animal with a total height reaching 2 m even in stronger currents.

The cemented form of attachment places much greater restriction on crinoids. Not only are the animals committed to that site of attachment during their life, but they are confined to hardground environments or sites where there are numerous hard objects to which they can attach and in which sedimentation rates are very low. The stability of the environment is of prime importance because crinoids would be unable to escape from any unfavourable change. From the palaeontologist's point of view, this has one considerable advantage: any sudden increase in sedimentation will entomb such faunas where they stand. There are many examples of hardground crinoid faunas preserved at the base of such sediment influxes. In some instances, crinoids with this mode of attachment appear to have broken free of the holdfast yet survived for some time after, as shown by the rounding of the distal end of the stem (Fig. 180). The length of stem remaining attached to the crown may vary from only a few to many columnals, suggesting that detachment occurred as a result of trauma.

Some additional holdfast types are worth mentioning

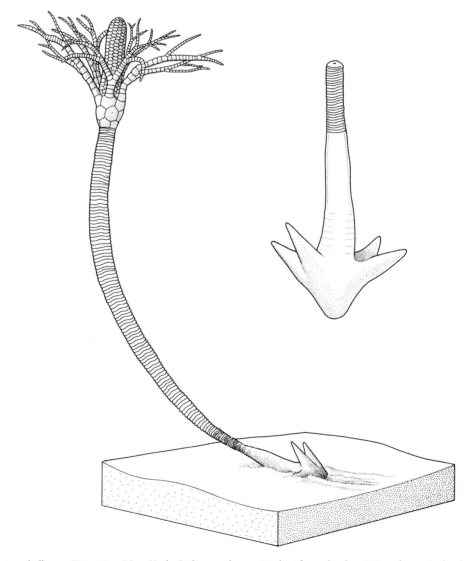

Fig. 20. *Ancyrocrinus bulbosus*. Devonian, New York. *Left*: complete animal; *right*, radicular cirri and terminal columnals ankylosed into the anchor-like holdfast. (Redrawn from Ubaghs 1953.) Complete animal ×2.

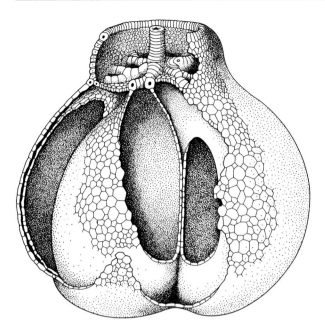

Fig. 21. Reconstruction of a lobolith ('*Camarocrinus*') of a scyphocrinitid. Lower Devonian, North America. The bulb is shown in presumably inverted life position, with the stump of the broken-off stem, contained in a collar, pointing upward (see Chapter 11 for further discussion). (Redrawn from Springer 1917.) ×1.

here. The grapnel- or anchor-like holdfasts of *Ancyrocrinus*, composed of fused columnals and their radicular cirri and crusted over with secondary stereom, prevented drag of the animal in moving waters (Fig. 20). Probably the most remarkable 'holdfasts' are the highly specialized chambered bodies or bulbs developed in scyphocrinitids (Fig. 21), large crinoids widely distributed in Silurian–Devonian boundary beds. Whereas some authors have argued that the bulbs (also called loboliths) may simply have provided anchorage on loose or soft substrates (Springer 1917; Brett 1984), others have assumed that these organs served as buoys to sustain a planktonic lifestyle, a theory that is adopted here (see Chapter 11 for further discussion).

Cirri

Cirri, flexible appendages arising at intervals from the stem of certain crinoids, are widespread among post-Palaeozoic crinoids (Figs. 11, 201–203, 210, 213, 214, 235, 236). True cirri, as they occur in isocrinids and comatulids, have synarthrial articulations, in contrast to the appendages of Palaeozoic crinoids with symplectial or synostial articulations (Donovan 1993); these are termed radicular cirri or radices. Cirri have some degree of variation and almost certainly evolved more than once among the different crinoid taxonomic groups (Brett 1981; Simms & Sevastopulo 1993). For attachment they have considerable advantages over the cemented holdfast. First, they are not permanently attached to the substrate. Crinoids could detach themselves at will and drift with the current in search of food. Second, crinoids with cirri are not necessarily confined to specific substrates. The cirri could be used to anchor in soft bottoms (see, e.g., *Paracomatula helvetica*, Chapter 25) or onto hard objects on the sea floor, such as sunken driftwood, rocks or other benthic organisms (Fig. 234). Finally, because the cirri in many crinoids arise at regular intervals along the stem, loss of its distal part did not entail the complete loss of the attachment structure, as was the case for crinoids with a holdfast or a root. The possibilities of this strategy were carried a step further by the post-Palaeozoic isocrinids, in which pre-formed rupture points (= cryptosymplexies) developed in the stem immediately beneath the nodals. This ensured the optimum positioning of the cirri in the event of the loss of the distal part of the stem. Efficient anchorage is provided by terminal cirri, as demonstrated by isocrinids in the Straits of Florida under quite high current regimes (Fig. 1). The strategy was perhaps carried to its ultimate conclusion in the comatulids, the most common extant group, in which the stem has been lost and cirri arise from the centrodorsal at the base of the cup (Fig. 22a). In comatulids, cirri show great diversity in form and size, usually correlated with the mode of life and habitat. Cirri are exceptionally closely spaced, long and sturdy in species of *Pentacrinites* (Fig. 201), presumably an adaptation to their lifestyles (see Chapters 22, 23 and 25 for further discussion).

The radicular cirri, or radices, in many Palaeozoic crinoids resembled miniature versions of the stem, with the same overall morphology of circular ossicles with symplectial articulations (Fig. 17). They may serve functions in addition to that of attachment, such as protection of the crown in myelodactylids (Figs. 15, 100) (Donovan & Sevastopulo 1989). By Late Palaeozoic times, the radices became more specialized structures (Simms & Sevastopulo 1993). Proximal ossicles developed a fulcral ridge type of articulation, imparting greater flexibility in a plane parallel to the long axis of the stem. In isocrinids and comatulids, the cirri developed a morphology quite distinct from that of the stem, reflecting the prevailing stresses imposed upon the different parts of each cirrus (Fig. 22b). The ossicles of

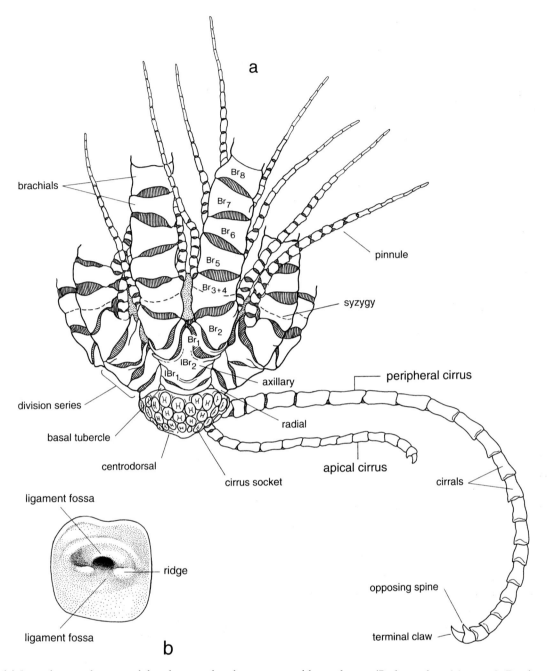

Fig. 22. (a) Lateral view of a comatulid with centrodorsal, two cirri and base of arms. (Redrawn from Messing & Dearborn 1990.) (b) Articular facet of second cirral of *Nemaster rubiginosus*. Recent. (After Donovan 1993.) Approx. ×30.

cirri, cirrals, are connected by ligaments composed of collagen fibrils. At each articulation there is an oral and an aboral ligament separated by a fulcral ridge. The cirrals are pierced by a central canal, a branch of the axial canal of the stem with coelomic and nervous extensions. In living isocrinids, distal cirri are used to anchor the stem (Figs. 1, 235, 236). In living comatulids, cirri may grasp the substrate with a terminal claw

or hook, aided by an opposing spine on the penultimate segment (Fig. 22a). Grasping is made possible by very fine contractile filaments within the coelomic epithelial cells of the axial canal (Grimmer et al. 1985). Because the axial canal of comatulids is below the synarthrial fulcra (i.e., it is situated aborally), contraction of the filaments curves cirri away from the crown towards the centrodorsal for clasping (Figs. 22, 234). In isocrinids, contraction of the filaments pulls the cirrus downward towards the stem. Relaxation of the filaments, combined with the elasticity of the upper, orally situated ligament, allows the cirrus to be detached and raised (Donovan 1989a). Cirri help isocrinids to crawl or climb to higher positions; they may also play a role in achieving and maintaining the vertical posture of the stem (Baumiller et al. 1991).

CUP OR CALYX

The aboral cup (or calyx) represents the link between the stem and the arms and is the site of the main organs of the digestive, haemal and nervous systems. It must provide both a rigid base from which the arms can operate efficiently and a protective housing for the vital organs.

In the great majority of cases, the morphology of the crinoid cup represents variations on a common theme – a series of two or three interlocking and offset circlets of five plates. The shape of the calyx may be altered by the addition or elimination of plates or by the modification of the size or shape of existing plates. Ausich (1988) recognized 11 basic calyx designs, some of which are shown in Figs. 23–27. These designs embrace constructional possibilities available to crinoids and have developed in parallel in different taxonomic groups. For example, the multi-plated bowl design, common in camerates (Fig. 38), also evolved in flexibles (*Sagenocrinites* and *Forbesiocrinus*, Fig. 27) and in articulates (*Uintacrinus*, Fig. 24, and *Apiocrinites*, Fig. 204).

The number of plates in each circlet is a consequence of the phylogenetic history of the crinoids. A single circlet of five plates would lack rigidity. The addition of a second circlet interlocking with the first circlet, and with the sutures between the plates offset by 36°, so that the suture in one circlet coincides with the centre of the plate in the other circlet, produces a much more rigid structure. The addition of a third circlet of plates would confer little extra advantage in terms of increased strength. This is perhaps borne out by the evolutionary history of the Crinoidea, in which there has been an increasing prevalence of two-circlet forms in parallel with the decline in diversity of three-circlet, and even four-circlet, forms of Early Palaeozoic faunas (Simms 1994a).

In addition to these basic plate circlets, various other plates may form an integral part of the cup in some crinoids. In most Palaeozoic taxa, the pentaradiate sym-

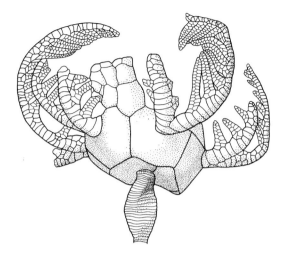

Fig. 23. Hand-shaped cup of the monobathrid *Eucladocrinus pleuroviminus* with elliptical, twisted stem. Early Mississippian (Burlington Limestone), Iowa. (Redrawn from Ausich 1988.) ×0.6.

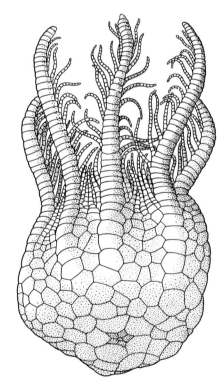

Fig. 24. Calyx of the articulate *Uintacrinus socialis*, a multi-plated bowl design. Upper Cretaceous, North America. (Redrawn from Ausich 1988.) ×1.

metry of the cup circlets is disrupted by the addition of one or more so-called anal plates (Fig. 27). The function of these plates is unclear, although they are lost in all post-Palaeozoic crinoids (Simms & Sevastopulo 1993).

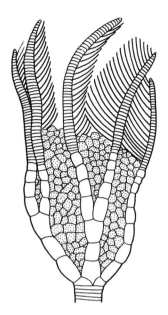

Fig. 25. Conical mosaic calyx of the monobathrid *Xenocrinus penicillus*. Upper Ordovician, Ohio. (Redrawn from Ausich 1988.) ×2.

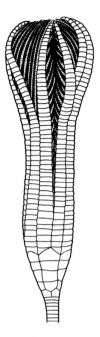

Fig. 26. *Ectenocrinus grandis*, a disparid with cylindric cup. Upper Ordovician, Ohio. (Redrawn from Ausich 1988.) ×1.

Technically, the aboral cup encompasses the radial plates and all other plates beneath the radials and above the stem, whereas the calyx includes all of the plates above the stem and beneath the point where the arms become free. In its simplest form, the cup is composed of two circlets of five plates each. The uppermost plates of the cup are radial plates. Radials typically define the five-part symmetry of crinoids and give rise to the arms. Interradially below the radials are the basal plates, which may form the base of the cup and articulate with the column (Fig. 11). In comatulids, basals may be only partly visible (basal tubercle, Fig. 22a) or hidden. In other crinoids, an additional circlet of plates, called infrabasal plates, is present between the basals and the stem. An aboral cup composed of radials and basals is termed 'monocyclic' (Figs. 11, 26), and a dicyclic cup (Figs. 20, 28, 35) is one that also has infrabasals ('mono'- and 'dicyclic' refer to the one or two circlets of plates, respectively, beneath the radials). Cladids, disparids, articulates and some flexibles typically have this construction. All of these plates are immovably joined to each other, with only rare exceptions, such as the calceocrinids (Fig. 29).

If the arms become free above the radials, the cup is equivalent to the calyx (Figs. 22, 23, 28, 30). Alternatively, proximal arm plates (brachials) may be sutured directly into the body wall of the crinoid rather than being part of the free, feeding arms. These brachials are

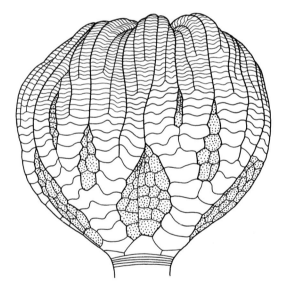

Fig. 27. *Forbesiocrinus wortheni*, a flexible with undifferentiated anal and interbrachial plates (stippled). Mississippian, Indiana. (Redrawn from Ubaghs 1978.) ×1.

called fixed brachials; where fixed brachials are present, the cup is only the lower part of the calyx. The fixed brachials greatly expanded the size and volume of the calyx. Crinoids with this construction may also be either monocyclic or dicyclic. This type of calyx is characteristic of most camerates (Figs. 38, 112, 161), many flexi-

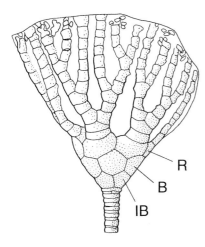

Fig. 28. *Eoparisocrinus siluricus*, a cladid with dicyclic cup. Upper Silurian, Indiana. Key: IB, infrabasal; B, basal; R, radial. (Redrawn from Ubaghs 1978.) ×1.5.

bles (Fig. 27) and also some articulates (Fig. 24). Calyxes with fixed brachials typically require extra plates (interradials or interbrachials) to fill in the area between adjacent rays (Figs. 27, 38). Extra plates are also typically present in the posterior interradius[2] and the reader is referred to the *Treatise* (Ubaghs 1978) for further details about these plates or about modifications from the standard plating described here.

Calyx shapes such as bowls, urns, cones and hands appear to be sensible constructions for food processing, but the very specialized bilateral recumbent constructions and the fists merit special attention and a short discussion. Bilateral recumbent designs evolved in a single, highly specialized family, the disparid Calceocrinidae, which lasted from the Middle Ordovician to Late Permian. These bent-down crinoids with well-developed hinges within the cup have engendered much interest in the function of this unique morphology. Most workers now agree that these crinoids lived with the stem along the bottom. In a resting posture, the arms folded

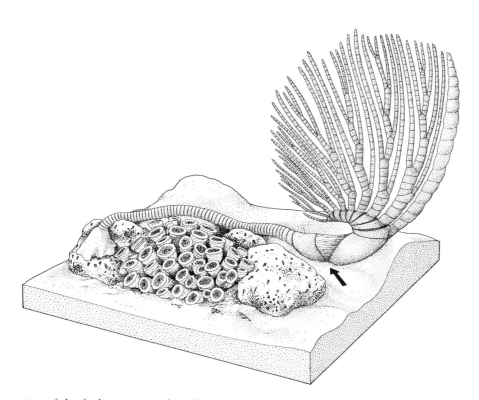

Fig. 29. Reconstruction of the feeding posture of a calceocrinid with opened crown on a coral reef (current from right). Radial circlet and arms are elevated above the substrate by opening above a hinge (arrow) between the basal and radial circlet. One lateral arm of this bilaterally symmetrical crinoid is shown; the second lateral arm behind is not shown. (Combined after Jaekel 1918; Harvey & Ausich 1997.)

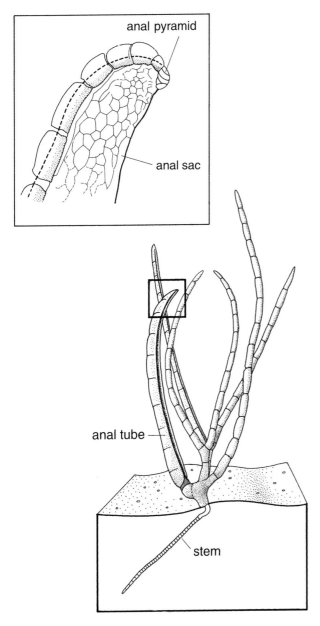

Fig. 30. Reconstruction of the calceocrinid *Senariocrinus maucheri*. Crown with strong bilateral symmetry, single dichotomous division of two arms and long unbranched arm; short, vermiform stem attached at extremity of single triangular basal plate. ×3. *Inset*: Enlarged distal part of anal sac attached to anal 'tube', with small anal pyramid at tip. Lower Devonian Hunsrück Slate, Bundenbach. (After Schmidt 1934; Moore 1962.)

back onto the stem, but the crown bent upward for feeding (Fig. 29). This interpretation was rejected by Schmidt (1934), who favoured an upright stem with a bent-down crown during times of rest. The crown would have been raised at a right angle into the current for feeding. Schmidt even thought that *Senariocrinus maucheri* (Fig. 30), a calceocrinid from the Lower Devonian Hunsrück Slate with a rudimentary, tapering stem and a long anal sac, was free-living; and he proposed that this peculiar crinoid used its short stem not only for temporary attachment but also for swimming, in conjunction with the arms and hinge. Muscle or other contractile fibres may have moved the hinges,[3] but we think it more likely that *Senariocrinus maucheri* lived on the muddy bottom, used the stem for attachment, folded the crown to protect the ambulacral furrows against predators and clogging by sediment and raised the arms for feeding, just as other calceocrinids did.

Fists include cyrtocrinids with a few commonly short arms. The cup may be asymmetric (hence the name for these crooked crinoids), and it is cemented to a substrate or borne by a short stem. Such a structure has been explained as an adaptation to wave action (reef forms), but these crinoids occur mainly in deeper waters (see Chapters 3 and 29). A recently discovered species from the Pacific is *Holopus alidis* (Fig. 31). The discovery of a living species of *Gymnocrinus* (Fig. 32) at depths of 300–500 m on seamounts off New Caledonia confirms that most cyrtocrinids were restricted to hard substrates

Fig. 31. *Holopus alidis*. Side view of complete individual with closed arms, dredged from a depth of 460–470 m off the Loyauté Islands. (Courtesy J.-P. Bourseau; from Bourseau et al. 1991.) ×3.

Fig. 32. *Gymnocrinus richeri*, dredged from 470 m off New Caledonia. Side view of individual with enrolled arms and pinnules. (Courtesy J.-P. Bourseau; from Bourseau *et al.* 1991.) ×1.

in rather sheltered environments. These living fossils have been observed coiling their asymmetric arms quite slowly, which may be an adaptation for protection against predators. For additional protection of the food grooves, the pinnules can be enrolled between projections of the brachials. Tegmen and soft parts within the cup are completely covered by lid-like oral extensions of the first primibrachials (Fig. 238). The food grooves are thus completely hidden in a tunnel if the arms are enrolled and the soft parts are inaccessible to predators. Similar cyrtocrinids with a short stem and a crown with arms that could be enrolled were widely distributed in the Jurassic and Lower Cretaceous of Europe. They have furnished a number of extraordinary forms. In some of these, the small arms could be hidden in a cavity formed by large median prolongation of the second primibrachials (*Eugeniacrinites, Lonchocrinus*) or by interradial processes of the radials (Fig. 33). Others (*Hemicrinus*, Fig. 34) are extremely asymmetric with a spoon-like cup and fused proximal stem, presumably adapted to constant unidirectional current. Possibly due to local conditions, individual species (e.g., *Cyrtocrinus nutans*; see Chapter 3) may vary considerably in degree of asymmetry. It is interesting that many individuals of the living *Gymnocrinus richeri* were in the course of regeneration when collected. Deformations, which may have been caused by a parasite, are common in some fossil cyrtocrinids

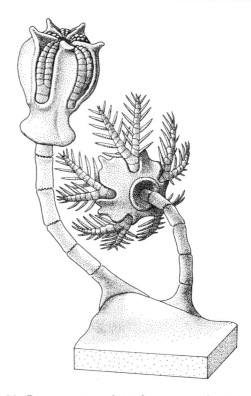

Fig. 33. Reconstruction of *Apsidocrinus moeschi*. Tithonian, Rogoznik (Poland) and Switzerland. (Redrawn from Pisera & Dzik 1979.) ×4.

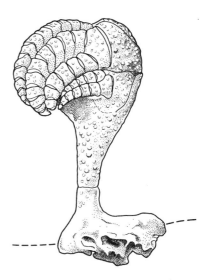

Fig. 34. Reconstruction of *Hemicrinus astierianus*. Lower Cretaceous, Var, France. (Redrawn from Jaekel 1918.) ×3.

(see Chapter 3). The advantage of protective morphology appears to have outweighed the reduced surface area for catching food, perhaps because of the small size of these crinoids, which occupied and still occupy a low tier, feeding on larger-sized particles. In Cretaceous and extant *Cyathidium*, the small arms may be completely sheltered underneath the closely fitting proximal arm pieces, so that these forms resemble a barnacle (Fig. 237; see also Chapter 29).

The cups of the stemless, pelagic Roveacrinida are, as a rule, composed of only thin radials that commonly have spines, ridges or wings to facilitate floating. In the somphocrinids, the cups may be prolonged into aboral spines, and long or short spine-like projections may have served to stabilize the animals in an upright position. The best-known representative of this group is *Saccocoma*, which is described in Chapter 26.

The oral surface of the cup may be covered by a variety of types of plates, in each case presumably serving to house and protect the vital organs. Five large oral plates are present in larval crinoids and persist into the adult stage of neotenous microcrinoids,[4] disparids and cladids (Fig. 35). In cladids and the post-Palaeozoic articulates, there is a tegmen (also called disc) composed of small tessellate plates or a membrane studded with calcareous granules (Fig. 6). In living crinoids, the tegmen is typically divided into five interambulacral areas by narrow ambulacral grooves passing into the arms (Fig.

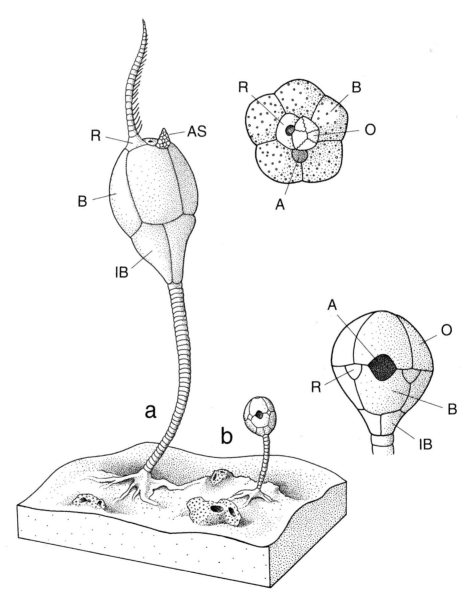

Fig. 35. Reconstruction of (a) *Monobrachiocrinus ficiformis granulatus* with single arm, with enlarged oral view of cup; and (b) armless *Embryocrinus hanieli*, with enlarged anal side view. Permian Basleo (also called Besleo) Beds, Timor. Key: A, anal opening; AS, anal sac; B, basal; IB, infrabasal; R, radial; O, oral plate. (Redrawn from Wanner 1920.) ×1.

5). The ambulacral grooves are commonly protected by covering plates. In the Upper Carboniferous camerate genus *Pterotocrinus*, the ambulacrals located at the points of bifurcation of the ambulacra are developed into conspicuous appendages, reaching a size similar to that of the whole crown. These wing plates may assume an extraordinary variety of shapes, from thin, knife-like wings to massive clubs and spines. Their function has been a matter of much speculation. According to a recent hydrodynamic analysis, they may have served as stabilizing fins or rudders, allowing the passive maintenance of an efficient feeding position in moving water (Baumiller & Plotnick 1989).

Many crinoids have an elongate anal sac or tube with the anal opening at the summit. The function of this sometimes highly ornate structure may have been to avoid faecal contamination by ingested food, because the anus and the mouth otherwise lie close together on the tegmen. Although both camerates and inadunates (cladids and disparids) have the anus on tubular extensions above the tegmen, different names are applied, because the anal tube and anal sac had independent evolutionary origins. The most extensive modifications of anal structure occur among inadunates (disparids and cladids), in which the structure is termed an anal sac. An example of a crinoid with an anal sac is the bilaterally symmetric calceocrinid *Senariocrinus*, which has an arm-like series of U-shaped anal plates, supporting a fine-plated sac with the anal pyramid at its tip (Fig. 30). The hinged anal sac may have been moved to direct the anal opening away from the arms and mouth of this crinoid, a helpful function in the more quiet waters of the Devonian Hunsrück facies. Late Palaeozoic inadunates developed elaborate, sometimes spinose anal sacs (Figs. 36, 141). Some authors have thought that spines were an anti-predatory adaptation, whereas Lane (1984) suggested that the entire anal sac, rising from the crown, may have been an offering to predators so that predatory attacks would be non-lethal.

A tegmen of a very different nature is present on the Palaeozoic camerates. Here, the multi-plated tegmen and the calyx plates are united into a rigid, commonly globose structure that rises above the arms (Figs. 37, 38). It resists disarticulation following death and, presumably, also resisted predator attacks just as effectively. The anal tube of camerates is typically a distal extension of the central tegmen. It has been suggested that such anal tubes, directed downcurrent, not only served to dispose of waste, but also helped to stabilize crowns deployed as a filtration fan (Baumiller 1990a). A long

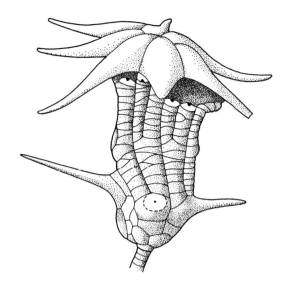

Fig. 36. *Bicidiocrinus wetherbyi*. The mushroom-shaped anal sac of this cladid is concealed by arms, except the distal spiniferous canopy of the sac. Note the large spiniferous axillary primibrachials. Mississippian, Kentucky. (Redrawn from Ubaghs 1978.) ×1.5.

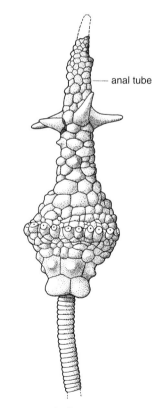

Fig. 37. *Uperocrinus nashvillae*, a camerate with large anal tube (distal end of anal tube and arms lacking); interbrachials are stippled. Mississippian, North America. (Redrawn from Wachsmuth & Springer 1897.) ×0.5.

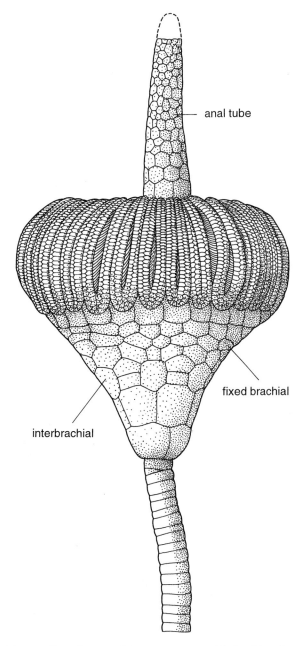

Fig. 38. *Eutrochocrinus christyi*, a camerate with fixed interbrachials and brachials; distal end of anal tube lacking. Mississippian (Burlington Limestone), Iowa. (Redrawn from Wachsmuth & Springer 1897.) ×1.

anal tube would have been ineffective for crinoids living in still water or where currents were very weak.

The size of the cup in many articulates is greatly reduced in comparison to typical Palaeozoic forms, thereby reducing the level of protection afforded to the vital organs. This may be compensated for in part by the relocation of the gonads onto the proximal pinnules, but also by the remarkable regenerative powers that have been documented in extant crinoids. D. L. Meyer (1988) found that comatulids can regenerate the entire visceral mass in only a few weeks, whereas Amemiya and Oji (1992) have documented even more remarkable instances where isocrinids have regenerated the entire crown from just the basal circlet of the cup. According to Donovan and Pawson (1998) the stem of Recent bourgueticrinids continued to survive following the loss of the crown and proximal column, presumably by absorbing dissolved nutrients through the ectoderm.

THE ARMS

The arms represent the food-gathering parts of the crinoid. The ultimate food-collecting structures are the tube feet; hence some crinoids can function quite effectively despite the presence of only one arm or in the absence of any arms (Fig. 35).

Although food gathering is the primary function of crinoid arms, it is by no means their only function. Arms are important for respiration and locomotion in some taxa, either swimming as in some comatulids or crawling as in comatulids and even in some stalked isocrinids (Messing 1985; Baumiller *et al.* 1991). Attachment is also an important function, particularly in multi-armed taxa such as comasterid comatulids.

Brachials and Their Articulations

The crinoid arm consists of a series of ossicles, called brachials, united by ligaments and/or muscles. A groove along the oral side of the arm is lined with tube feet (Fig. 7), which pass food particles down to the mouth. In most crinoids, the arm must retain both flexibility and mobility so that it can be deployed to maximum advantage in filtering food particles from the water. Flexibility is inherent in the multi-element construction, and mobility is achieved by antagonistic series of ligaments and muscles. Arm flexibility is to an extent controlled by brachial width. Hence, to maintain relative flexibility during growth, brachial height increases much more slowly than brachial width.

In Palaeozoic forms, ligamentary articulations are the most common mode of union between radials and arms and between arm plates. Articulations may be movable to some extent; they may also be immovable, permitting at the most slight movements. Movable articulations typically have three bundles of ligament separated by more or less well-defined elevations. In contrast to the

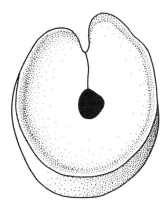

Fig. 39. Synostosial articulation on distal facet of third primibrachial of *Barycrinus rhombiferus*. Mississippian, Indiana. (Redrawn from Ubaghs 1978.) ×3.

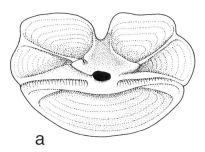
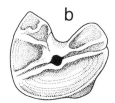

Fig. 40. Muscular articulations of *Marsupites testudinarius*. Upper Cretaceous, Europe. (a) Radial facet with straight articulation; (b) oblique muscular articulation of distal secundibrachial with pinnule facet. (After Ubaghs 1978.) ×6.

still further orally are two flanges for the attachment of the adoral muscle. The aboral ligaments serve to hold the arms flexed aborally, as is common in parabolic filtration fans, whereas the muscles are antagonistic to these ligaments and are used to draw the arms back in towards the oral surface of the disc. In crinoids such as comatulids that occupy changeable environments or that are highly mobile, the muscular fields on the brachials are large and well developed. In contrast, in taxa such as the pseudoplanktonic *Pentacrinites* and *Seirocrinus* that occupied very stable environments and did not need a great degree of arm mobility, the muscular fields are quite small, and the aboral ligament pit is greatly enlarged, occupying more than half of the total area of the brachial articulum (Fig. 41). Whether muscles existed in the arms of many Palaeozoic crinoids remains a topic of debate (Simms & Sevastopulo 1993). It appears that muscular arm articulations first evolved during the Middle Palaeozoic among cladid inadunates (Ausich & Baumiller 1993).

Immovable articulations, connected by short ligament fibres only, are synostoses with smooth surfaces

Fig. 41. Muscular articulation on brachials of *Pentacrinites fossilis*. Note comparatively small muscular fields (arrows); a rhomboid cirral is visible at lower left. Sinemurian, *obtusum* Subzone, Black Ven, Charmouth, Dorset. (Also figured by Simms 1989.) ×4.

bifascial synarthry discussed later in this section, such articulations did not act as a fulcrum. More or less immovable synostosial-like articulations are very common in Palaeozoic crinoids (Fig. 39). The reader is referred to the *Treatise* (Ubaghs 1978) for further details.

Post-Palaeozoic articulate crinoids have brachials with several types of articulations, with muscular articulations (Fig. 40) the most common. These articulations combine ligament fibres with muscles. They possess a transverse fulcral ridge, to the aboral side of which is a large ligament pit for the attachment of the aboral ligament. Immediately to the oral side of the fulcral ridge are two smaller pits for the interarticular ligaments, and

and syzygies with radiating ridges and furrows on otherwise flat surfaces (Figs. 4, 42). Instead of interlocking as in stem symplexies so that the suture is crenulate (see Fig. 13), the ridges of one face oppose the ridges of the other. Syzygies are irregularly spaced along the proximal part of the arms of articulates and are preferred breaking points (similar to the cryptosymplexy between the nodal and infranodal of the stem). In this way, the optimum position for rupture of the arm, following a predator attack or other trauma, can be determined in advance, thereby limiting the damage to the filtration fan as a whole.

Synarthries are superficially similar to those in the stem. They have a fulcral ridge running orally–aborally and two separate bundles of ligaments (Fig. 43). This structure allows limited movement. Synarthries are confined to the proximal parts of the arms in articulates and have been shown to represent modified syzygial articulations (Simms & Sevastopulo 1993).

Feeding Strategies and Arm Structure

Crinoids are passive suspension feeders, which means that they must rely exclusively on currents to bring suspended food particles to their food-gathering apparatus. In their pioneering work, Macurda and Meyer (1974) showed that living crinoids feed via a parabolic filtration fan with the oral surface downcurrent (Figs. 1, 236). This mode of feeding on fine particles was probably approximated in ancient crinoids, such as the strongly pinnulate camerates. However, because of lack of evidence for muscular articulations in most Palaeozoic crinoids, it is questionable whether the relatively flexible filtration fans of modern crinoids were developed in these ancient forms.

Dense filtration fans appear to be a highly specialized feature that became most successful in advanced cladids and modern articulates. Most of the early crinoids and nearly all of the disparids possess only atomous or slightly branched, stick-like arms that lack pinnules and have relatively broad food grooves. This indicates that the earliest crinoids may have possessed large, fleshy tube feet. These could not have trapped small particles in turbulent downcurrent eddies the way modern crinoids do with their filtration fans. Rather, they may have fed by means of very elongated tube feet. Other crinoids with permanently fused arm plates, such as *Petalocrinus*, could again not possibly have formed a filtration fan. Another example of crinoid that may have fed in an alternative way is the Silurian cladid *Crotalocrinites* with laterally united brachials. The resulting box-like structures may be interpreted as pumping chambers. It seems possible to surmise that these crinoids used a type of ciliary pumping mechanism to draw currents inside the enclosed chambers for feeding in much the way sponges and other chamber-feeding-filter feeders do. Likewise, certain other crinoids such as *Eucalyptocrinites*, with arms that could be only partially withdrawn from large tegminal plates, may also have been able to generate their own feeding currents and hence survived well in an environment of low turbulence (Brett 1984). The same is true of another camerate Silurian reef dweller, the highly specialized *Barrandeocrinus* (Fig. 99). Confirming the existence of these alternative feeding styles requires further research.

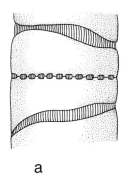

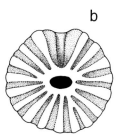

Fig. 42. (a) Schematic representation of two brachials united by syzygy; (b) syzygial articular facet. (Redrawn from Ubaghs 1978.)

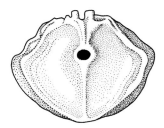

Fig. 43. Synarthrial articular facet of first primibrachial of *Himerometra bassleri*. Eocene, South Carolina. (Redrawn from Ubaghs 1978.) ×6.

Flexibles must have had still other feeding strategies. These crinoids never developed pinnules, but instead had arms that ranged from extremely broad, petal-like appendages to highly ramulate arms that branched out to tendril-like endings (Fig. 56). The arms are strongly incurved at the summit of the crown (Figs. 27, 158). Some of the forms have short, rather stubby arms with extremely broad food grooves, suggesting the presence of large, fleshy tube feet. These may have been able to capture relatively large prey, which could be enrolled within the arms to form a kind of digestive chamber overlying the mouth. These types of flexibles are reminiscent of cyrtocrinids such as *Holopus* (Fig. 31) and *Gymnocrinus* (Fig. 32). Living *Cyathidium* and *Holopus* may use their arms to pass food to the mouth (Grimmer & Holland 1990; Heinzeller & Fechter 1995), but *Gymnocrinus* appears to enroll its arms for protection only (Heinzeller *et al.* 1994); see Chapter 29. Reasoning by analogy, Meyer and Lane (1976) suggested that the feeding strategies of taxocrinid flexibles and cyathocrine cladids with highly branched, non-pinnulate arms were similar to that of basket stars (gorgonocephalid ophiuroids), but this appears unlikely in view of the absence of muscles in the arms of these Palaeozoic forms. Basket stars seize their prey by wrapping the tendril-like endings of the arms around passing macroscopic objects of 10–30 mm size; this catch is eventually moved towards the mouth by enrolling the arms.

Arm Branching

The arms are inserted on the radials, which form the base of a ray. Each ray consists of a series of brachials. In its simplest form, a single unbranched arm is inserted on each ray.

A primary function of the stem is to elevate the filtration fan above frictional effects at the sea floor and into zones of higher current velocity, where the fan can filter larger volumes of water per unit time. From the simple non-branching arms, finer-meshed fans are produced by arm branching. Different filters result from the vast array of brachial types and arm branching styles. If other features remain constant, the progression from non-pinnulate to uniserially pinnulate to biserially pinnulate (see next subsection) results in increasingly more dense filtration fans. Many different patterns of arm branching have developed among the Crinoidea, each of which may have its advantages and disadvantages. Increased filtration fan density is achieved through the following sequence of arm branching styles (Fig. 44):

unbranched or atomous (Fig. 45), simple isotomous/heterotomous branching (Figs. 25, 26), complex isotomous/heterotomous branching (Fig. 46), uniserial pinnulation (Figs. 22a, 33) and biserial pinnulation (Figs. 47, 48). In heterotomous arms the main branch, or ramus, carries smaller branches, called armlets or ramules (ramuli).

Increased density of branching does not increase the efficiency of crinoid feeding. Instead, filtration fans with different densities optimally capture different-sized food particles (Ausich 1980). Crinoids living together, but with contrasting filtration fans, capture different-sized populations of food particles and, therefore, do not com-

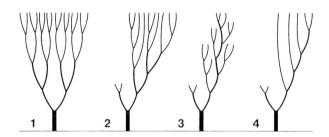

Fig. 44. Various types of arm branching: (1) isotomy; (2 and 3) heterotomy; (4) endotomy. (Redrawn from Ubaghs 1978.)

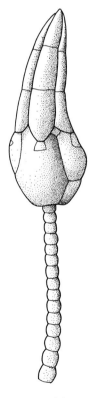

Fig. 45. *Parapisocrinus quinquelobus* with unbranched (atomous) arms. Silurian, Indiana. (After Springer 1926.) ×2.

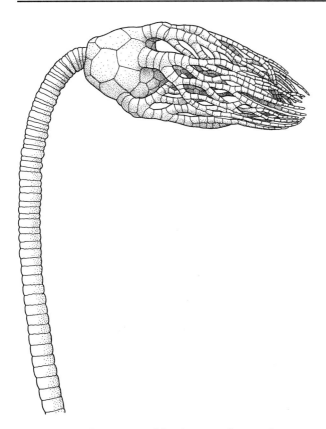

Fig. 46. *Cyathocrinites multibrachiatus* with complex arm branching. (Redrawn from Bather 1900.) ×1.2.

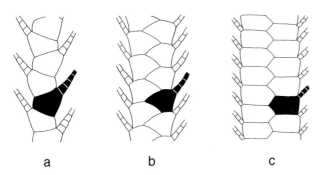

Fig. 47. Comparison of pinnule spacing on uniserial and biserial arms; one pinnule-bearing brachial is in black. (After Ubaghs 1953.)

pete for the same resources. In terms of filtration fan efficiency, if applicable, an endotomous pattern (Figs. 51, 189) has been considered to represent the optimum for transporting food to the mouth (Cowen 1981; Simms 1986), yet this configuration is encountered only under circumstances in which a parabolic or planar filtration fan may have been of limited use (see Chapters 22 and 25). The most common branching is simple isotomous branching, commonly repeated several times (Fig. 214), suggesting that this configuration, although far from the optimum for total filtration surface, is quite adequate for most crinoids.

In rays with arm branching, each series of brachials leading either from the radial to the first branch point or from a subsequent branch point to another is called a division series or brachitaxis. Such a division series terminates in an axillary with two distal articular facets, but it may also remain undivided to its distal extremity. In most extant and many fossil crinoids, the first brachitaxis (the primibrachials, or IBr) is composed of two brachials. Brachials that immediately precede an arm division are known as axillaries. The second primibrachial is usually an axillary (Fig. 22a, in Fig. 11 it is the fourth) and carries the secundibrachials (IIBr), followed by tertibrachials (IIIBr) after the next axillary, and so on.

Pinnules

One type of arm branching deserves further mention – that of pinnulation. Pinnules are small, usually unbranched offshoots of the arm that arise from each brachial plate (Fig. 49). Pinnulation produces the most dense filtration fans. The success of this strategy is evident from the fact that pinnulation arose independently at least three times – in the camerates, cladids and disparids. In Late Palaeozoic cladids and the post-Palaeozoic articulates, pinnules developed on alternative sides of successive brachials (Fig. 50a,b), except axillary brachials, or the distal brachial of a syzygial or synarthrial pair (Fig. 49). Hence, gaps exist in the rows of pinnules along each side of an arm – regular gaps on every alternate brachial and less regular gaps caused by the presence of syzygial articulations and axillary brachials. In a rather bizarre evolutionary twist, the Triassic encrinids *Traumatocrinus* and *Vostocovacrinus*, which have an endotomous pattern of arm branching, achieved almost complete pinnulation through the elimination of all muscular articulations and the development of exclusively syzygial articulations in the arms!

A second strategy for increasing the number of pinnules along a given stretch of crinoid arm is the reduction in height of the brachials. Such a strategy was adopted by the scyphocrinitids (Fig. 112), and it perhaps reached its acme in the bizarre Triassic genera *Vostocovacrinus* (Fig. 51) and *Traumatocrinus*. Brachials generally have a wedge-shaped section, with the pinnules arising from the high side of the brachial (Fig. 50a,b). Hence,

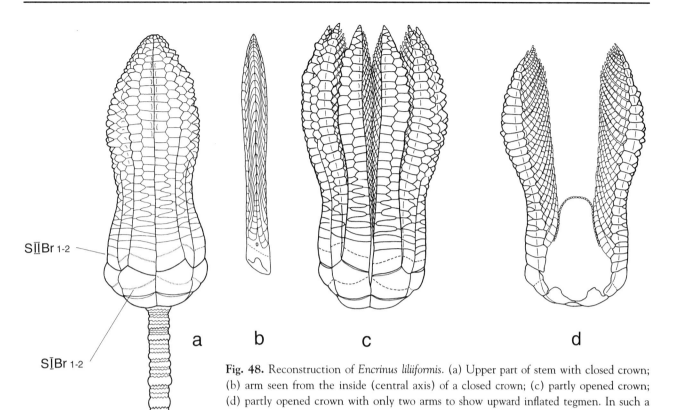

Fig. 48. Reconstruction of *Encrinus liliiformis*. (a) Upper part of stem with closed crown; (b) arm seen from the inside (central axis) of a closed crown; (c) partly opened crown; (d) partly opened crown with only two arms to show upward inflated tegmen. In such a position, the arms and pinnules of *Encrinus liliiformis* would have supported its role as an active filter feeder; see discussion in Chapter 21. Key: S, synostosis between first and second primi- and secundibrachials. (Redrawn from Jefferies 1989.) ×1.

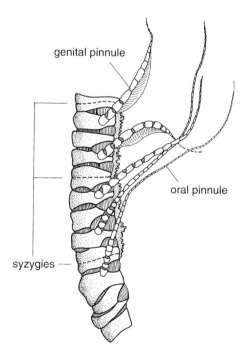

Fig. 49. Oral and genital pinnules on proximal arm of the Recent comatulid *Promachocrinus*. (Redrawn from Breimer 1978.) Approx. ×5.

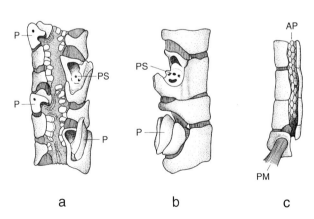

Fig. 50. Pinnular structures in the Recent *Neocrinus decorus*. (a) Oral side showing insertion of first pinnular ossicle on alternating arm ossicle; (b) side view of same; (c) four proximal pinnulars with ambulacral covering plates. Key: P, pinnular; PS, pinnule socket on brachial; PM, pinnular muscle; AP, ambulacral plates. (Redrawn from Breimer 1978.) Approx. ×5.

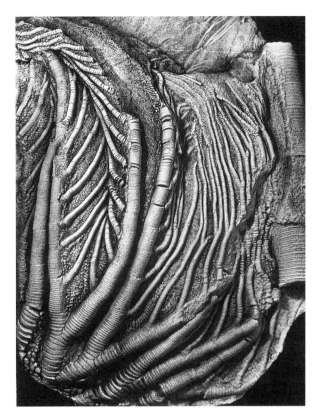

Fig. 51. Extremely low brachials with syzygial articula and complete pinnulation in the endotomously branched uniserial arms of the Triassic (Ladinian) encrinid *Vostocovacrinus boreus*. Kotel Island, Novosibirsk, Russia. (From Yeltisheva & Polyarnaya 1986.) ×0.65.

lowering the height of the opposite side of the brachial significantly reduces the spacing of the pinnules (Fig. 47). The inevitable development from this, present in many camerates and cladids as well as in the Triassic encrinids, is a biserial arrangement of brachials (Fig. 48). Biserial arms are formed as a result of the reduction in thickness, ultimately to zero, of the thin, non-pinnulate side of alternate brachials. Hence, the thicker, pinnulate sides of alternate brachials become juxtaposed in a single column. The overall effect of this is to form two parallel, offset columns of interdigitating, wedge-shaped brachials with a pinnule arising from each along either side of the arm (Fig. 47). Clearly this represents the most dense deployment of pinnules, but at the expense of arm mobility and flexibility as a consequence of the interlocking of the brachials between the adjacent offset columns. Again, the disadvantages associated with this strategy may have restricted it to taxa occupying specific niches (Simms 1990).

In articulate crinoids, one or more pair of pinnules near the arm base may be modified into oral pinnules (Figs. 6, 49). These lack an ambulacral groove and obviously serve to protect and sweep the oral surface. In a large number of Palaeozoic forms such as the scyphocrinitids (Fig. 11) as well as in some articulates (e.g., *Uintacrinus*, Fig. 24), the calyx includes pinnulate brachials that are firmly articulated with other plates of the calyx. Therefore, the calyx contains not only fixed brachials and interradials but also fixed pinnulars.

The arms form a framework for the support of the radiating extensions of the water-vascular and ambulacral systems used for feeding (Fig. 7). In addition, the arms of living crinoids carry the reproductive system. This is usually confined to the specialized genital pinnules in the lower part of the arms (Fig. 49). These pinnules contain the gonads, which are simply masses of sex cells filling the corresponding cavity. Crinoids are dioecious, but the sexes are usually indistinguishable.

The soft structures of the arms with the podia are lodged in a deep depression in the oral surface of brachials and pinnules; they may be protected by covering plates (Fig. 50c). Nichols (1994) suggested that parts of the arms with the gonads may have been sacrificed to predators to distract them from the more vulnerable organs within the cup. In support of this, he found that in *Antedon bifida* the gonads are maintained in ripe condition virtually all year round rather than just during the spawning season. This suggests that the gonads themselves have taken on this secondary function of predator distraction.

NOTES

1. Mutable collagenous tissues, unique to echinoderms, have variable tensile strength – i.e., the capacity to undergo rapid changes in mechanical properties. These tissues can, within less than a second to a few minutes, switch (e.g., at symplexies) between stiff and compliant conditions. In their compliant state, they allow movements without much force, for example, to attain a feeding posture; in their stiff state, they allow the animal to maintain the feeding posture even under strong currents. They can even disintegrate irreversibly to permit complete detachment of body parts during autotomy, e.g., at synostoses (Wilkie & Emson 1988).
2. 'Posterior' refers to interradius with anus or anal plates, generally wider than other interradii; see *Treatise* (Ubaghs 1978, p. T61) for a discussion of orientation.
3. To date, evidence for ligamentary tissue has been found only in the hinge structure of calceocrinids; and, at least in living echinoderms, ligaments lack active contractile ability.
4. The term 'microcrinoid' has also been used by several authors for post-Palaeozoic articulate crinoids (see Chapter 3).

2

Systematics, Phylogeny and Evolutionary History

MICHAEL J. SIMMS

The phylum Echinodermata has been divided into two distinct subphyla – the Pelmatozoa, characterized by the possession of a stem, and the Eleutherozoa, which lack any trace of a stem. During the Palaeozoic there were many different pelmatozoan groups, including, among others, crinoids, blastoids, rhombiferans and diploporans. The non-crinoid pelmatozoans, commonly called blastozoans (Fig. 2), were most diverse and abundant during the Middle Ordovician, but crinoids reached their acme somewhat later, during the Early Carboniferous. By the end of the Permian, both crinoids and non-crinoid pelmatozoans suffered a catastrophic decline (Fig. 3). Crinoids were reduced to a single lineage surviving into the Early Mesozoic, while all blastozoans became extinct by the close of the Palaeozoic. In today's oceans, crinoids are the only surviving pelmatozoans, but there are four extant eleutherozoan classes – the Asteroidea, Ophiuroidea, Echinoidea and Holothuroidea.

Crinoids were first recognized as a distinct group of echinoderms by J. S. Miller in 1821. Previously, they had been grouped together with the asteroids. Numerous classification schemes have been developed since to explain relationships among the thousands of fossil and living crinoid species. The goal of all biological classification in the latter half of the twentieth century has been to reflect natural evolutionary groupings deduced from morphological similarity. However, morphological similarities commonly have evolved by convergent evolution and, if these convergent characters go unrecognized, the classification may not reflect true evolutionary relationships.

The most comprehensive and widely accepted classification scheme currently in use is that of the *Treatise on Invertebrate Paleontology* (Moore & Teichert 1978) (Table 1). A major goal of current crinoid research is to reevaluate this in the quest for the ideal phylogenetic classification. An alternative high-level classification scheme has been proposed by Simms and Sevastopulo (1993) (Table 2), but at lower taxonomic levels much remains to be done.

For Miller (1821), the structure of the aboral cup had paramount importance for the classification of crinoids, a philosophy largely followed in all subsequent classifications (see N. G. Lane 1978). The fundamental division of the Crinoidea has been based upon a combination of characters such as the number of plate circlets in the cup, the rigidity with which they are sutured together, the presence and position of any additional plates within these circlets, the structure of the oral surface and the structure of the arms.

Five major morphological groups were recognized within the class Crinoidea in the *Treatise* (Ubaghs 1978). Four of these, the Camerata, Disparida, Cladida, and Flexibilia, occurred only during the Palaeozoic, and one, the Articulata, is exclusively post-Palaeozoic. Two minor groups, the Hybocrinida and Hemistreptocrinida, are confined to the Early Palaeozoic. Each of the major

Hans Hess, William I. Ausich, Carlton E. Brett, and Michael J. Simms, eds., *Fossil Crinoids*. © 1999 Cambridge University Press. All rights reserved. Printed in the United States of America.

Table 1. *Crinoid Classification by Moore and Teichert (1978)*

Class Crinoidea
  Subclass Echmatocrinea
    Order Echmatocrinida
  Subclass Camerata
    Order Diplobathrida
      Suborder Zygodiplobathrina
      Suborder Eudiplobathrina
        Superfamily Rhodocrinitacea
        Superfamily Dimerocrinitacea
        Superfamily Nyctocrinacea
    Order Monobathrida
      Suborder Compsocrinina
      Suborder Glyptocrinina
  Subclass Inadunata
    Order Disparida
    Order Hybocrinida
    Order Coronata (now recognized as a separate class)
    Order Cladida
      Suborder Cyathocrinina
      Suborder Dendrocrinina
      Suborder Poteriocrinina
  Subclass Flexibilia
    Order Taxocrinida
    Order Sagenocrinida
  Subclass Articulata
    Order Millericrinida
    Order Cyrtocrinida
    Order Bourgueticrinida
    Order Isocrinida
    Order Comatulida
    Order Uintacrinida
    Order Roveacrinida
  Subclass Hemistreptocrinoidea (proposed by Arendt 1976)
    Order Hemistreptocrinida

Note: Only categories mentioned in the text are listed.

Table 2. *Crinoid Classification by Simms and Sevastopulo (1993)*

Class Crinoidea (Miller 1821)
  Subclass Camerata (Wachsmuth & Springer 1885)
    Order Diplobathrida (Moore & Laudon 1943)
    Order Monobathrida (Moore & Laudon 1943)
  Subclass Disparida (Moore & Laudon 1943)
  Subclass Cladida (Moore & Laudon 1943)
    'Stem-group cladids'
    Infraclass Cyathocrinina (Bather 1899)
    Infraclass Flexibilia (Zittel 1895)
    Infraclass Articulata (Miller 1821)
  Incertae Sedis ('Subclass') Hybocrinida (Jaekel 1918)

groups has been further subdivided, with evolutionary relationships well resolved for some groups but still unclear for others.

Perhaps the most distinctive of the Palaeozoic crinoids have been grouped together in the subclass Camerata. Camerates are characterized by a rigidly sutured cup (of infrabasals, basals, radials and anal plates) and, in contrast to other crinoids, a heavily plated, rigid tegmen and a subtegminal mouth (i.e., not exposed on the surface). The name 'Camerata' comes from the chamber that this tegmen forms above the cup. Proximal parts of the arms typically are fixed, forming a calyx composed of the aboral cup, fixed brachials, interradials and, in some cases, interbrachials. One or more anal plates interrupt the pentaradiate symmetry of the cup. Camerates have been divided into two groups based upon the number of plate circlets in the cup. Those with two circlets in the cup (monocyclic) have been grouped as the order Monobathrida (Figs. 52, 53) and those with three circlets (dicyclic) as the order Diplobathrida (Figs. 54, 55). The monobathrids have been divided into two suborders – the Compsocrinina (Fig. 52), in which the lower circlet is hexagonal and the upper circlet is interrupted by the anal series, and the Glyptocrinina (Fig. 53), in which the lower circlet is pentagonal and the anal series lies above the upper circlet. Monobathrids also include the Xenocrinacea (Figs. 25, 92), presumably derived from ancestors of compsocrinine camerates. The diplobathrids have been divided into the Zygodiplobathrina (Fig. 54), in which basals alternate with radials in a circlet of 10 plates, all of them in contact with infrabasals, and the Eudiplobathrina (Fig. 55), with basals and radials alternating in two distinct circlets.

The subclass Flexibilia (Fig. 56) is characterized by a three-circlet (dicyclic) cup in which the infrabasals are reduced in number, usually to three plates, but sometimes to two or even one. Although this feature is seen in all flexibles, it is not unique to them, being found also in other Palaeozoic groups, notably some camerates and disparids. Anal plates are present in the cup, but these too may be reduced or absent in some taxa. The proximal parts of the arms are typically fixed in the calyx, a rigid tegmen is lacking and the mouth is exposed on the tegmen. Calyx plates in the Flexibilia are

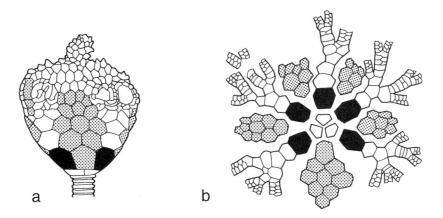

Fig. 52. Calyx structure of the camerate *Actinocrinites triacontadactylus* (order Monobathrida, suborder Compsocrinina). Lower Carboniferous, England. (a) Side view of calyx; (b) plate diagram. Radials black; anals, fixed brachials and interradials stippled. (Redrawn after Ubaghs 1978.) ×1.

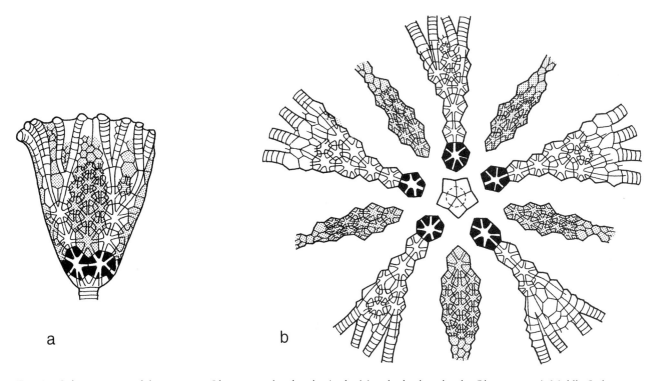

Fig. 53. Calyx structure of the camerate *Glyptocrinus decadactylus* (order Monobathrida, suborder Glyptocrinina). Middle Ordovician, USA. (a) Side view of calyx; (b) plate diagram. Radials black; anals, fixed brachials and interradials stippled. (Redrawn after Ubaghs 1978.) ×1.5.

only weakly united in comparison with the other major crinoid groups. Hence, fossil calyces are usually crushed, giving rise to the name 'Flexibilia'. The distal tips of the arms are commonly preserved coiled inward, further adding to the 'flexed' appearance of these crinoids. All flexible crinoids have uniserial arms without pinnules. Two orders of flexibles are recognized – the Taxocrinida, in which the anal plates are separated from the radials and proximal brachials by a series of small plates termed the perisome, and the Sagenocrinida, in which the anal plates are rigidly sutured to the radials and brachials.

The two remaining major Palaeozoic groups, the Disparida and the Cladida, were formerly united in a single subclass, the Inadunata (see Moore et al. 1978), so called because the arms were free above the cup. However, it is now generally accepted that this is insufficient

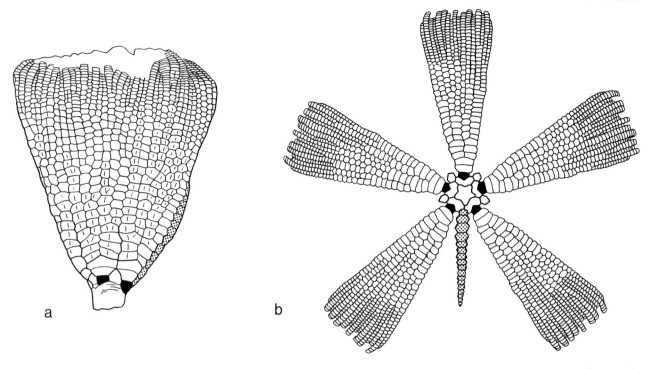

Fig. 54. Structure of the camerate *Cleiocrinus regius* (order Diplobathrida, suborder Zygodiplobathrina). Trentonian, Canada. (a) Side view of crown; (b) plate diagram. Radials black; anals stippled. (Redrawn after Ubaghs 1978.) ×1.3.

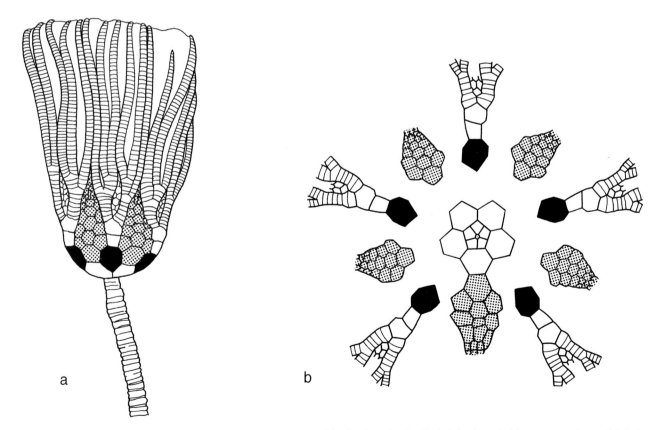

Fig. 55. Structure of the camerate *Rhodocrinites kirbyi* (order Diplobathrida, suborder Eudiplobathrina). Mississippian, Iowa. (a) Side view of crown; (b) plate diagram. Radials black; anals, fixed brachials and interradials stippled. (Redrawn after Ubaghs 1978.) ×2.

Fig. 56. Plate diagram of *Onychocrinus exsculptus* (infraclass Flexibilia). Lower Mississippian, Indiana. Radials black; anals stippled. (Redrawn after Moore 1978.) ×1.

reason for considering the two to represent a natural grouping, and both have been raised to subclass status (Donovan 1988; Simms & Sevastopulo 1993) (Table 2). Disparids have two-circlet (monocyclic) cups in which bilateral symmetry is developed to various degrees (Fig. 57). Most disparids have a cone- or barrel-shaped cup in which the anal plates lie above the upper circlet. In many disparids, one or more plates of the upper circlet are divided by a transverse suture; such plates have been termed compound radials. This unequal size of the cup plates is the origin of the group's name. The arms, which are usually erect, simple and non-pinnulate, typically do not have their proximal portion incorporated within the calyx. Further subdivision of the Disparida has been based mainly on the shape of the cup and the nature of its bilateral symmetry as deduced from the distribution of 'compound radials'.

Cladids are characterized by a cup with three circlets of plates (dicyclic) and by the presence, usually, of a radianal[1] within the upper circlet and a prominent anal sac (Figs. 58, 59). They lack a rigid tegmen, and the

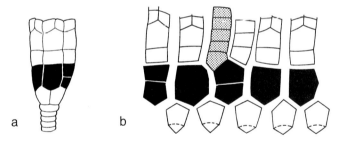

Fig. 57. Structure of *Columbicrinus crassus* (subclass Disparida). Upper Ordovician, Ohio. (a) Side view of calyx; (b) plate diagram. Note 'compound radials'. Radials black; anal series stippled. (Redrawn after Moore *et al.* 1978.) a, ×2.

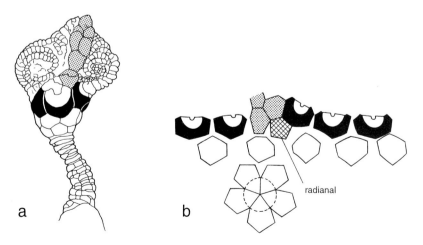

Fig. 58. Structure of *Euspirocrinus spiralis* (cyathocrinine cladid). Silurian, Gotland. (a) Side view of complete specimen; (b) plate diagram. Outer surface of radials black; anals stippled. (Redrawn after Moore *et al.* 1978). Ubaghs 1978. ×2.

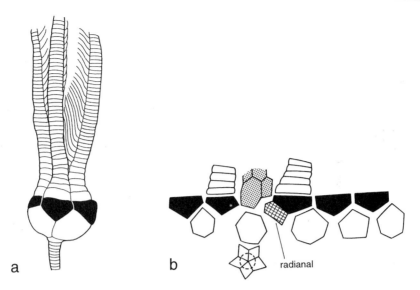

Fig. 59. Structure of *Cromyocrinus simplex* (advanced cladid, subclass Cladida, suborder Poteriocrinina). Upper Carboniferous, Russia. (a) Side view of crown; (b) plate diagram. Radials black; anals stippled. (Redrawn after Moore *et al.* 1978.) ×1.

mouth is exposed on the tegmen. Cladids have arms that are free above the radial plates. Primitive members have uniserial and non-pinnulate arms, but derived forms may be uniserial pinnulate and, in some instances, biserially pinnulate. It is the richly branched structure of the arms that gives the cladids their name (Greek *klados* = branch). In the *Treatise* the cladids were subdivided into the Cyathocrinina, Dendrocrinina and Poteriocrinina. The Cyathocrinina appear to be a distinctive group of cladids, typically having a bowl-to globe-shaped aboral cup with a convex base; a short, non-porous anal sac; narrow arm-base facets; and narrow, uniserial, non-pinnulate arms (Fig. 58). However, the dendrocrinines and poteriocrinines (Fig. 59) are merely parts of a morphological and evolutionary series. Moore and Teichert (1978) differentiated the dendrocrinines from the poteriocrinines principally on the basis of the presence of pinnules in the latter group, but it is probable that more than one dendrocrinine lineage developed pinnules. Hence, this subdivision, as configured in the *Treatise*, is inappropriate. Dendrocrinine and poteriocrinine cladids differ from cyathocrinine cladids in having a conical to bowl-shaped or even flat cup with a convex to concave base; commonly a tall, cylindrical or inflated anal sac that may have pores; arm-base facets commonly extending the full width of the radials; and rounded or flat, may be ramulate or uniserial pinnulate arms.

Two minor Palaeozoic groups were also listed among the inadunates in the *Treatise*, the orders Hybocrinida and Hemistreptocrinida. The hybocrinids (Fig. 60) have a cup of two circlets (monocyclic) with additional plates lying above the upper circlet. The arms are either simple, erect and free, or partly or wholly reduced to recumbent ambulacral grooves on the surface of the cup. The hybocrinid tegmen is composed of oral and ambulacral plates, with the mouth exposed on the surface. Arm number may be reduced to three.

The Hemistreptocrinida are known only from the Ordovician of Russia. The supposed cup is composed of three or more circlets in which each circlet is offset from those above and below by less than the 36° typical of other crinoids. Anal plates appear to be absent from within the cup. These forms are known only from the cups; arms and stem arising from the cup are unknown. It has been suggested that they may merely be meric pelmatozoan stems (Arendt & Rozhnov 1995).

Only a single group, the Articulata, were present after the close of the Palaeozoic. No single character uniquely defines this group, but all have pinnulate arms and a three-circlet or, more commonly, a secondarily two-circlet (dicyclic or cryptodicyclic) cup lacking any anal plates in the adult (Figs. 11, 22a). Articulates lack a rigid plated tegmen, and the mouth is exposed on the surface (Fig. 5). Several major groups are recognized, among them the isocrinids, the millericrinids, the comatulids and the exclusively Triassic encrinids. The encrinids are characterized most notably by the predominantly biserial arrangement of the arms, a character unique among post-Palaeozoic crinoids, although found to some extent in the Upper Jurassic to Lower Creta-

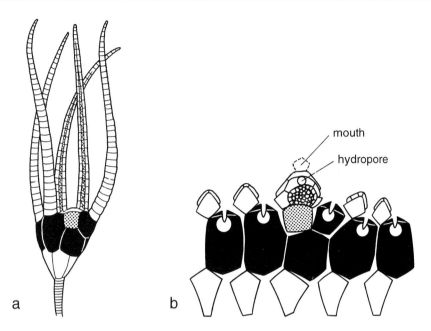

Fig. 60. Structure of *Hybocrinus conicus* (order Hybocrinida). Middle Ordovician, Ontario. (a) Reconstruction with appendages; (b) plate diagram. Radials black; note divided radial with anal (stippled). (Redrawn after Moore et al. 1978.) ×1.

ceous thiolliericrinids (see Chapter 3). Isocrinids retain a range of primitive characters, with whorls of cirri at regular intervals along the pentaradiate stem and long, slender, branching arms. Well-developed autotomy planes are present beneath each cirrinodal. Millericrinids are a comparatively derived group, lacking cirri and with predominantly multi-radiate stems. They are distinguished from other articulates by the presence of a syzygial articulation between the fourth and fifth secundibrachial (IIBr4–5) in the arms, whereas all other articulates have syzygies at IIBr3–4 (Taylor 1983). A number of small, specialized crinoids with protected arms, with a reduced number of plate circlets in the cup and reduced stems, grouped together as the order Cyrtocrinida, may have been derived from the millericrinids. The comatulids are perhaps the most successful of all articulate groups, comprising more than three-quarters of all extant crinoid species. They are characterized by loss of the stem in the adult and the development, beneath a reduced cup, of a small centrodorsal that usually bears cirri. The uintacrinids have a large, subspheroidal calyx with very long arms, and they lack a stem (Figs. 226, 227). It is assumed that they were adapted to life on the soft bottom of the Chalk seas (see Chapter 27). Bourgueticrinids are a relatively low-diversity group of stalked crinoids. They have a small cylindrical or conical aboral cup that lacks any trace of infrabasals and stems with an articular-ridge (synarthry) articulation and no cirri (Fig. 16). Roveacrinids are a diminutive

and much-simplified order of crinoids comprising three, possibly unrelated, groups – the Triassic somphocrinids, the Jurassic to Cretaceous saccocomids and the Cretaceous roveacrinids. They were mostly pelagic and displayed adaptations consistent with this mode of life, such as a very thin, delicate skeleton, greatly expanded flanges on basals or radials to increase frictional drag and long arms in some with flanges on the brachials (Figs. 219, 220).

RELATIONSHIPS AND PHYLOGENY OF THE MAJOR CRINOID GROUPS

Relationships among certain crinoid groups are now reasonably well understood, but most are not well known or remain unresolved. For example, it is now widely accepted that cladids gave rise to both the flexibles and the articulates (Lewis 1981; Kelly 1986; Simms & Sevastopulo 1993). Hagdorn (1983), Hagdorn and Campbell (1993) and Milsom et al. (1994) have made progress in understanding relationships within the articulates, but the evaluation of relationships within the flexibles or cladids remains incomplete. Nonetheless, it is clear that the division of the cladids into the Dendrocrinina and Poteriocrinina does not reflect any natural grouping. These two taxa are merely 'grade groups', with the poteriocrinines representing the descendants of various dendrocrinine lineages. The classification of Simms and Sevastopulo (1993) (Table 2), although not ideal, is closer to an evolutionary classification than that adopted by the *Treatise*.

Similarly, relationships within either the disparids or camerates have yet to be clarified. Sevastopulo (in Simms *et al.* 1993) considered it unlikely that the present disparid classification reflects their phylogeny, while the distinction between zygodiplobathrid camerates, in which the radial and basal plates alternate in a single circlet above the infrabasals (Fig. 54), and the more typical eudiplobathrid camerates no longer seems tenable (Brower 1975).

Attempts to undertake a comprehensive phylogenetic analysis of the major Palaeozoic crinoid groups have arisen through reappraisals of the homology of crinoid cup plates (Simms 1994a,b, 1995; Ausich 1996a). Since the time of Miller (1821), it has been assumed that two-circlet (monocyclic) crinoids differ from three-circlet (dicyclic) forms in lacking infrabasals, the lowest plate circlet in three-circlet cups. The arms were always considered to originate from the radial plates. However, new hypotheses suggest homologies that differ from this traditional view (represented in Figs. 52–60), proposing that any circlet may be absent. Simms (1994a,b, 1995) argued that many two-circlet crinoids lacked the radial circlet rather than the infrabasal circlet. Ausich (1996a) concluded that only the disparid and hybocrinid two-circlet crinoids lost radials. In fact, these are proposed to have lost all, some or no radials, to have lost the entire basal circlet, and to have retained a primitive circlet hitherto unrecognized. It remains to be seen whether either of these new hypotheses of crinoid cup plating will supersede the traditional views, but it is to be hoped that consideration of all of the options will lead to a more robust understanding of the phylogeny of the Crinoidea.

THE EVOLUTIONARY HISTORY OF THE CRINOIDEA

Crinoids have a history extending back more than 500 million years. It is obvious that the origin of the crinoids lies among the various other non-crinoid pelmatozoan groups with which they co-existed during the Early Palaeozoic, but there is no agreement on exactly which group is most closely related to the crinoids.

For more than a decade it was assumed that *Echmatocrinus brachiatus*, from the Middle Cambrian Burgess Shale, was the earliest known crinoid, even though it was more than 40 million years older than, and bore little resemblance to, any other known crinoid. However, it was no longer regarded as a crinoid, or in fact even an echinoderm, by Conway Morris (1993), Simms (in Simms *et al.* 1993) and Ausich and Babcock (1998), although Sprinkle and Collins (1995) still maintain their original view of the crinoid affinities of *Echmatocrinus*. The next oldest, and undisputed, crinoid is the (?) Tremadocian-Arenig genus *Aethocrinus* (Ubaghs 1969) (Fig. 9). A new subclass and a new order, Aethocrinea and Aethocrinida, respectively, were proposed by Ausich (1998) for *Aethocrinus* and related forms with four circlets of plates in the cup (see preceding section for a discussion of plate circlets). The Aethocrinea was a small, short-lived radiation, with a primitive cup design that is no longer found in the other crinoid subclasses: Cladida, Camerata, Disparida, Flexibilia and Articulata (Ausich 1998).

Diversification of the Crinoidea from this time appears to have been rapid, constituting the 'Lower Palaeozoic' crinoid fauna of Ausich *et al.* (1994). The ensuing Arenig Stage saw the establishment of several of the major crinoid groups, among them the disparids, cladids, hybocrinids and at least two groups of camerates. Most of the remaining Palaeozoic groups, other than the compsocrinine camerates, appeared by Middle Ordovician times. The morphological diversity of crinoids was high during this Early to Middle Ordovician interval, as taxa with primitive characters, such as *Aethocrinus* (Fig. 9) and *Colpodecrinus*, survived for a time alongside the more advanced forms, such as the disparids and diplobathrid camerates, which came to dominate the Palaeozoic (Fig. 61). In fact, the maximum level of morphological diversification reached in the Ordovician remained essentially unsurpassed (Foote 1995). Diversity continued to increase before suffering a substantial decline at the close of the Ordovician (Eckert 1988; Donovan 1989b). This end-Ordovician extinction has been attributed to effects associated with a major glacial episode (Brenchley 1989). Crinoids were poorly represented in Early Silurian (Llandovery) faunas, yet it is clear that there was a significant diversification during the Silurian, with a number of typically Silurian taxa having actually appeared before the end-Ordovician extinction. This radiation continued into the Devonian, and faunas were dominated by monobathrid camerates and, in some facies, cladids and flexibles. This 'Middle Palaeozoic' crinoid fauna (Ausich *et al.* 1994) appears to have entered into a decline during the Late Devonian, associated with the Frasnian–Fammenian extinction

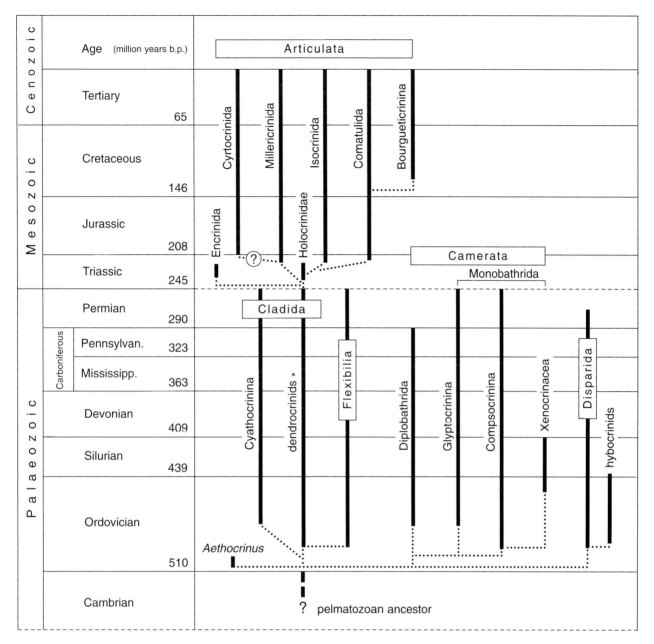

Fig. 61. Stratigraphic distribution and inferred phylogeny of crinoids. In contrast to Table 2, here Flexibilia and Articulata are regarded as subclasses. Dendrocrinids* refers to the dendrocrines and poteriocrines of the *Treatise*, but these clades are no longer considered to represent monophyletic groups.

event, but it had rebounded by the Early Carboniferous. A Late Devonian (Famennian) echinoderm fauna with 'Carboniferous' affinities recently discovered in northwestern China (Lane et al. 1997) indicates that echinoderm diversification and reradiation was well under way before the close of the Famennian. The acme of crinoid diversity was reached during Early Carboniferous times when crinoids substantially exceeded the total diversity of all other echinoderm taxa combined. In certain shallow-marine environments, they were the dominant element of the fauna, and their dissociated ossicles formed a major component of some sediment accumulations. The transition between the 'Middle' and 'Late Palaeozoic' crinoid faunas occurred during a period of rapid faunal turnover during the middle Early Carboniferous (Ausich et al. 1994). Late Early Carboniferous faunas were dominated by advanced cladids (Lane 1971; Waters & Maples 1991), and this dominance was maintained throughout the remainder of the Palaeozoic. There appears to have been a modest decline during the Late Carboniferous. Documented diversity declined still further during the Permian, although it was still relatively high. However, to what extent this apparent Late Carboniferous–Permian decline reflects the true evolutionary history of the Crinoidea remains unclear. Appropriate Upper Carboniferous and Permian facies are much less widespread than those of the Lower Carboniferous and have not been as thoroughly investigated. This is especially true of the Upper Permian, from which few crinoid taxa have been recorded or described. Nonetheless, it is clear that crinoids suffered a catastrophic decline before the start of the Mesozoic; only a single crinoid lineage appears to have survived into the Early Triassic (Schubert et al. 1992). Documentary evidence is insufficient to establish whether this extinction event was gradual or concentrated at the end of the Permian. Similarly, the cause of the mass extinction remains uncertain, although there have been many suggestions (Maxwell 1989).

This end-Permian extinction saw the disappearance of all of the Palaeozoic crinoid groups, the cladids, disparids, flexibles and remaining camerates. During the Early Triassic, there was only a single lineage representing the earliest of the subclass Articulata (Schubert et al. 1992) derived from a Late Palaeozoic cladid ancestor (Simms & Sevastopulo 1993). This was perhaps the most critical time in the history of the Crinoidea, representing a major evolutionary bottleneck. By the Middle Triassic, this lineage had begun to diversify and included the Holocrinidae with their rather isocrinid appearance. By early Late Triassic times, a range of distinctive taxa had appeared, among them the uniquely Triassic encrinids and somphocrinids, as well as the isocrinids and the millericrinids. However, this Late Triassic diversity peak was followed by another major extinction event, precipitated by climatic change, which caused the disappearance of the encrinids and most of the somphocrinids, leaving an isocrinid-dominated fauna not unlike that of the ensuing Jurassic (Simms & Ruffell 1990). During the Late Triassic and Early Jurassic, this fauna was supplemented by the earliest representatives of the comatulids, the stemless crinoids that dominate the modern crinoid fauna. Comatulids became increasingly dominant during the Late Jurassic and the Cretaceous, even though the stalked isocrinids appear not to have retreated to deeper water before the Late Palaeogene (see Chapter 28). Another group of stalked crinoids, the bourgueticrinids, became important beginning in the Late Cretaceous. They represent a neotenous offshoot of the comatulids, in which the stem is retained in the adult. Whereas the isocrinids are comparative generalists, the millericrinids include forms adapted to reef environments and others, such as those of certain Middle Oxfordian sediments, which inhabited soft bottoms in deeper water. Millericrinids were a particularly prominent element of some Late Jurassic reefs and still survive today at moderate diversities. Cyrtocrinids, a very diverse group of mostly small and compact crinoids, are common in some Jurassic and Lower Cretaceous sediments; they may have descended from Triassic millericrinids, and a few forms still live today.

The Late Triassic extinction event was the last major perturbation in crinoid history. Although many taxa have come and gone since then, the overall changes between the Late Triassic and the present day have been comparatively gradual. Even at the end of the Cretaceous, when so many other elements of the fauna experienced profound changes, crinoids appear to have suffered no great change in diversity, although it is perhaps significant that no pelagic microcrinoids survived the end-Cretaceous event.

NOTE

1. The radianal is at the base of one of the rays, adjoining the so-called anal X, which is situated between two radials and is the proximal plate of the anal sac.

3

Fossil Occurrence

WILLIAM I. AUSICH, STEPHEN K. DONOVAN, HANS HESS AND MICHAEL J. SIMMS

CRINOIDAL LIMESTONES

Disarticulated crinoid remains were important rock-building constituents during the Palaeozoic and Mesozoic, but especially during the Palaeozoic. Crinoidal limestones range in size and shape from thin, localized lenses to the remarkable regional encrinite deposits that are crinoidal limestones with minimum dimensions greater than a stratigraphic thickness of 5–10 m and an area extent of 500 km^2 (Ausich 1997). Regional encrinites are known from the Ordovician to the Jurassic, but they were most common during the Lower Mississippian. The Keokuk and Burlington (Chapter 17) Limestones from Iowa, Missouri and Illinois are the best-known examples of this facies. These two units have a combined thickness of up to 53 m and geographic extent of more than 74,000 km^2 – crinoids and crinoidal sediment dominated an entire shelf for a considerable length of time (H. R. Lane 1978; Thompson 1986; Ausich 1997). Other Palaeozoic examples of regional encrinites include the Middle Ordovician Holston Formation of Tennessee (Walker *et al.* 1989), the Devonian Sadler Ranch Formation of Nevada and California (Kendall *et al.* 1983) and the Lower Mississippian Kogruk Formation of Alaska (Sable & Dutro 1961).

In Mesozoic strata, crinoidal limestones decrease in importance, but commonly occur in the Triassic Muschelkalk (see Chapter 21) and in the Lower Jurassic. Typical examples are Liassic sediments of the southern Alps such as the Sinemurian Broccatello and the similar Hierlatz Limestone deposited on restricted shallow swells adjoining pelagic limestones with high proportions of sponge spiculites (Wiedenmayer 1979). The Broccatello (from brocade) was given its name by stone masons; this flaming, multicoloured 'marble' has been used in countless Italian and Swiss baroque and rococo churches. In the quarries of Arzo, southern Switzerland, crinoids account for up to half of the bulk of the Broccatello, which is usually a few metres thick. These crinoidal limestones interfinger with unbedded micrites to siltstones rich in brachiopods and calcareous sponges (Inozoa). The community in the Broccatello, with *Isocrinus* (*Chladocrinus*) *tuberculatus* and sparse columnals of millericrinids, seems to have grown on sediment composed largely of its own debris. Two later communities in the Arzo quarries are of Pliensbachian age; they were deposited in somewhat deeper water. One fauna is associated with a massive red limestone and bedded red nodular calcareous mudstones rich in cephalopods, and it contains numerous millericrinid roots and columnals. Another assemblage occurs in red calcareous mudstones (marls) of Late Pliensbachian (Domerian) age. Certain horizons are packed with mostly small remains of crinoids (millericrinids, isocrinids, cyrtocrinids, comatulids) that appear to have been washed down from seamounts or swells. Both of these facies types occur in fissure fillings in the Broccatello, caused by an Early Jurassic rifting phase. The crinoid faunas have not been

Hans Hess, William I. Ausich, Carlton E. Brett, and Michael J. Simms, eds., *Fossil Crinoids*. © 1999 Cambridge University Press.
All rights reserved. Printed in the United States of America.

described in detail. Massive Lower Jurassic crinoidal limestones also occur in other parts of the Tethys Ocean. Jenkyns (1971) assumed that such limestones were sand waves produced by tidal or other currents in water tens of metres deep on seamounts that formed after the break-up of pre-existing carbonate platforms and reefs. These migrating sand bodies, with material from temporary hardgrounds, commonly filled tectonic fissures or were displaced into basins, where they accumulated as turbidites. During early diagenesis, the more muddy sediments sometimes underwent segregation to produce a nodular limestone (Rosso Ammonitico Lombardo Formation, deposited in somewhat deeper water between the seamounts).

Rosso Ammonitico beds similar to those of the southern Alps have been reported from Bulgaria, the Crimea and the Caucasus (Klikushin 1987a), with remains of isocrinids, millericrinids (*Amaltheocrinus, Quenstedticrinus, Shroshaecrinus, Gutticrinus*, of uncertain status) and the cyrtocrinid *Cotyloderma*.

No regional crinoidal deposits are known from the Cenozoic.

Some crinoids have highly distinct columnals by which they can be identified to the level of family or, rarely, even genus or species. Examples include the Lower Palaeozoic disparids *Myelodactylus, Ristnacrinus* and *Iocrinus*; the Upper Palaeozoic cladid *Barycrinus* and the monobathrid platycrinitids; and the post-Palaeozoic articulate bourgueticrinids and isocrinines. However, because the biological names of crinoids are based largely on morphological features of the crown, dissociated columnals and brachials typically cannot be identified. In an attempt to classify the incredible wealth of disarticulated columnal material, alternative systems of naming and classification have been devised. These attach names to columnals on the basis of morphological features of the articular facet, side-view and, if known, patterns of columnal insertion within pluricolumnals. There are two principal schemes in existence, which we can call the 'Russian' and 'American' systems, devised by Yeltisheva (1955, 1956; comprehensively reviewed by Stukalina 1988) and by Moore et al. (1968a,b) and Moore and Jeffords (1968), respectively. The principal base of both systems is the morphology of the articular facet, particularly the outline of the lumen and facet. Only the 'Russian' system has found widespread use, mainly among workers in the former USSR. Such schemes are used as a convenience in settings where complete crinoids are, at best, rare, such as the Ordovician of the British Isles (Donovan 1995).

ASSEMBLAGES WITH COMPLETE SPECIMENS

A number of such assemblages are described in detail in Chapters 6–28. In the following account, important assemblages not discussed in these chapters are summarized. The choice is, by necessity, limited and subjective. For further details the reader is referred to the respective literature.

Palaeozoic

Diverse Middle Ordovician crinoid faunas are known from the North American carbonate platform. Exceptionally well-preserved crinoids occur in storm-deposited shaly interbeds of the Blackriveran (Mohawkian Series) Lebanon Limestone in Tennessee (Guensburg 1992). The crinoids are a hardground assemblage, and common crinoids include the disparids *Columbicrinus crassus* (Fig. 57) and *Tryssocrinus endotomitus* and two species of the camerate *Reteocrinus*. All have lichenocrinid-type attachment discs (Fig. 88). The small hybocrinid *Hybocrinus* has a large calyx and atomous arms (Fig. 60); it is attached by a single holdfast element and occupied a low tier. The diplobathrid camerates *Archaeocrinus* and *Gustabilicrinus* have distally coiled stems for attachment to other crinoid stems or to ramose bryozoans.

Another Middle Ordovician fauna with an abundance of well-preserved material occurs in the Trentonian (Mohawkian Series) Dunleith Formation of the Galena Group of northern Iowa and southern Minnesota (Brower 1992a,b, 1994, 1995a,b; Brower & Strimple 1983). The Dunleith fauna consists of a diverse crinoid assemblage containing, among others, the cladids *Cupulocrinus, Eoparisocrinus, Plicodendrocrinus* and *Praecupulocrinus*; the hybocrinid *Hybocrinus*; the disparids *Calceocrinus, Cremacrinus, Ectenocrinus* and *Ohiocrinus*; and the camerates *Cotylacrinna, Euptychocrinus* and *Abludoglyptocrinus*. Although relatively rare, these crinoids occur commonly as complete specimens on limestone bedding surfaces.

In the British Ordovician succession, only the Lady Burn Starfish Beds (Harper 1982) of Threave Glen in the Girvan District, Strathclyde, southwest Scotland, have produced a large fauna of fossil crinoids. Indeed, the echinoderm fauna of these Upper Ordovician (Ashgill) beds is diverse, including crinoids, rhombiferans, paracrinoid ossicles, asteroids, ophiuroids, echinoids, bothriocidaroids, cyclocystoids and edrioasteroids (reviewed by Donovan et al. 1996). Fossil echinoderms are commonly preserved as moulds in sandstones. Crinoids typically preserve the arms and at least the proximal column. Specimens are interpreted as having undergone at least some transport, as suggested by the general absence of crinoids retaining terminal attachment structures, probably in a turbulent, downslope setting (Ingham 1978). The crinoid fauna consists of about 20 described species, including at least 6 disparids, 6 cladids, a taxocrinid, 4 monobathrids and 2 diplobathrids. The original monographic study of the Lady Burn crinoids was that of Ramsbottom (1961), although a number of new species have been described subsequently (Brower 1974; Donovan 1989b, 1992b,c). The fauna includes many genera typical of North America, such as *Porocrinus, Plicodendrocrinus, Xenocrinus* and *Macrostylocrinus*. This is because much of Scotland is, geologically, part of North America. It was brought into contact with Europe when the Iapetus Ocean closed during the Middle Palaeozoic. This 'fragment of North America' was subsequently 'left behind' when the Atlantic opened during the Mesozoic and Cenozoic. The largest collections of echinoderms from the Lady Burn Starfish Beds are housed in the Natural History Museum, London, and the Hunterian Museum, University of Glasgow.

During the Middle Silurian, the Birmingham to Shropshire region of west-central England and the area extending to the south formed a shelf sea in which the first extensive, fossiliferous, bedded limestones of the English succession were deposited (Bassett *et al.* 1992). The Much Wenlock Limestone Formation of the Wren's Nest and Dudley Castle areas, Dudley, Staffordshire, has yielded a fauna of completely preserved Silurian crinoids. These were mainly collected when the area was actively quarried, between about 1750 and the 1920s (Hardie 1971). These crinoids occur in crinoidal and bryozoan limestone beds, 3–20 cm thick, that alternate with calcareous mudstones (Watkins & Hurst 1977). Complete crinoids were collected and can still be found, though rarely, on the upper surfaces of limestone beds. Still common are disarticulated brachials, cup plates, columnals and pluricolumnals, including the distinctive pluricolumnals of the disparids *Myelodactylus ammonis* and *Myelodactylus fletcheri* (Donovan & Sevastopulo 1989). Preservation by catastrophic storm sedimentation seems likely for this assemblage, which includes 10 nominal species of disparids, at least 15 species of cladids, 11 species of flexibles, 19 species of monobathrids and 5 species of diplobathrids (S. K. Donovan & C. R. C. Paul, research in progress). The most common, complete specimens belong to the cladid genus *Gissocrinus*. Dudley crinoids are typically 10–30 cm in height and thus occupied a feeding level above that of the smaller bryozoans, corals and brachiopods that contributed to the skeletal substrates. The only monographic study of the Dudley crinoids can be found in Ramsbottom's (1954) Ph.D. thesis. Notable collections of crinoids from the Dudley area include those of the Natural History Museum, London, the Lapworth Museum, University of Birmingham, and the Dudley Museum and Art Gallery.

A diverse fauna of Lower Carboniferous crinoids occurs in the Hook Head Formation, Hook Head, Ireland (Ausich & Sevastopulo 1994). This Late Tournaisian fauna, which is representative of similar faunas in England, Wales and Belgium, includes 48 species and is dominated by monobathrid camerates and cladids. These occur on a mixed carbonate and siliciclastic shelf/ramp. In shallow-water sediments, sturdy calyces of camerates are present, whereas in deeper water facies crowns with stems are more commonly preserved. The significance of the Hook Head crinoids lies partly in the fact that numerous specimens documenting population variability are now available that allow systematic housekeeping of species established during the 19th century (W. I. Ausich & G. D. Sevastopulo, in prep.).

The Permian occurrences are summarized in Chapter 20.

Mesozoic

In southern China the Middle Triassic has recently furnished complete crinoids of the remarkable *Traumatocrinus caudex*, an occurrence first mentioned by Mu (1949). The sediments belong to the Falang Formation (Ladinian) and are exposed in the Xinpu area, Guanling County, southwest of Guizhou (formerly Kueichow) Province. The Falang Formation is about 160 m thick and consists mainly of thin-bedded dark grey limestones, marls and black shales that were part of the Tethys, which extended to central Europe. This formation is underlain by shallow-marine carbonates of the Anisian Guanling Formation and overlain by mixed carbonate and siliciclastic coastal to offshore sediments of the Carnian Banan Formation (G. Liu, pers. comm., 1997). *Traumatocrinus caudex* belongs to the Encrinida and is characterized by a large crown with a spread of 30 cm. Stems are composed of very low columnals with a diameter of 2 cm; they reached a length of 1 m or more and were attached by a terminal disc. Arms are biserial with non-muscular articulations and spines on the side arms; branching is largely endotomous and strikingly similar to that of Lower Jurassic pentacrinitids (see Chapters 22 and 23). Its worldwide occurrence in black shales and other facies (Kristan-Tollmann & Tollmann 1983) as well as morphology indicates that *Traumatocrinus* was pseudoplanktonic, attached to driftwood (H. Hagdorn, in prep.). It is noteworthy that the pelagic scyphocrinitids share with *Traumatocrinus* a worldwide distribution, large crown and very flexible stem (see Chapter 11).

Complete specimens occur sporadically throughout the Lower Jurassic of England, although they are more common at certain horizons than at others. The occurrence of *Pentacrinites fossilis* is described in Chapter 22, but well-preserved specimens of this crinoid have been found at a similar stratigraphic level at other localities inland in southern England. Related species, such as *Pentacrinites doreckae*, *Pentacrinites dichotomus* and *Seirocrinus subangularis*, can be found wherever organic-rich mudstones are developed in the Lower Jurassic (Simms 1989). Indeed, the intact preservation of these taxa is a reflection more of their unusual pseudoplanktonic lifestyle than of local rapid burial conditions, as is so necessary for most other intact crinoid occurrences. On the Yorkshire coast, extensive bedding planes in the Upper Pliensbachian are covered with intact, although otherwise poorly preserved, specimens of *Balanocrinus gracilis* (Simms 1989). This is roughly the same level at which other well-preserved material has been found on the Dorset coast (see Chapter 22) and at several localities inland, and it appears to reflect a time of shallow seas in which storms commonly resulted in the catastrophic burial of crinoids and other echinoderms.

A remarkable occurrence of the isocrinid *Isocrinus nicoleti* and the comatulid *Palaeocomaster schlumbergeri* has been described by Bigot (1938) from the Middle Jurassic of Normandy. The lens, with a maximum thickness of 8 cm, is filled with remains of the crinoids. As judged from the lower surface of a slab, the relatively large *I. nicoleti* reached the remarkable density of 375 individuals per square metre, with scattered individuals of the comatulid. This occurrence is reminiscent of the *Paracomatula helvetica* bed described in Chapter 25, but here the isocrinid is the dominant species. In part of the lens, small specimens of the echinoid *Hemicidaris langrunensis* are present with lantern and spines still attached. More or less complete specimens of *I. nicoleti* are also present on the lower surface of a 4-cm-thick bed in Middle Jurassic strata near Montmédy in northern France. They were described by Loriol (1882–1889, Pl. 154–157) and Leuthardt (1930); unfortunately, no detailed stratigraphic observations are available.

Lower Kimmeridgian mudstones near La Rochelle (France) have furnished remains of the isocrinid *Balanocrinus maritimus* (Bourseau et al. 1998). The authors attribute the good preservation to rapid burial of the animals by pulses of muddy sediment at a depth of several hundred metres.

An interesting occurrence of the bourgueticrinid *Dunnicrinus mississippiensis* has been described from the Upper Cretaceous (Maastrichtian) Prairie Bluff Chalk of Mississippi (Moore 1967). Nearly complete crowns, commonly with long portions of stem and several articulated radicular holdfasts, occur on bedding planes of thin calcareous shallow-water siltstones in one of the outcrops.

ASSEMBLAGES WITH PARTLY DISARTICULATED SPECIMENS

Palaeozoic

One of the richest pelmatozoan faunas known occurs in the Middle Ordovician (Mohawkian Series) Bromide Formation from the Arbuckle Mountains of Oklahoma. Echinoderms are present in two members – the Mountain Lake Member and the Pooleville Member. Sprinkle (1982) monographed this echinoderm fauna. In total it contains more than 86 species from at least 67 genera and distributed among 13 classes. The Crinoidea is the most diverse echinoderm class, with 45 species. The most common crinoids are the hybocrinid *Hybocrinus*, the cladids *Palaeocrinus* and *Carabocrinus*, the diplobathrid camerates *Diabolocrinus* and *Archaeocrinus* and the disparids *Paracremacrinus* and *Apodasmocrinus*.

Silurian strata in east-central North America have yielded several important crinoid localities, including the Waldron Shale, reef faunas from the Great Lakes region, the Brownsport Formation and the Rochester Shale (Chapter 10). Unfortunately, the former three faunas are largely known from only calyces and crowns. The first comprehensive treatment of the Waldron Shale crinoid fauna was that of James Hall (1882). The Waldron Shale is a Wenlockian, blue-grey shale that is present in Indiana and Tennessee. The most common Waldron crinoids are the camerates *Eucalyptocrinites*, *Dimerocrinites*, *Lyriocrinus* and *Periechocrinus*; the cladid *Botryocrinus*; and the flexible *Lecanocrinus*. Waldron crinoids, along with other invertebrates, occur clumped in small bioherms or incipient reefs (Archer & Feldman 1986).

Large barrier reefs rimmed the basin margins, and patch reefs dotted the platforms to form the great Silurian reef tract of the Great Lakes region of the United States and Canada (Shaver *et al.* 1978). Tabulate corals, stromatoporoids and algae were the primary reef builders for these Llandoverian through Pridolian reefs, but crinoids were dominant reef dwellers. Crinoids produced substantial amounts of carbonate sediment on the reef cores, reef flanks and some non-reef facies. Crinoid faunas are known from, among other formations, the Hopkinton Dolomite of Iowa, the Racine Dolomite of Wisconsin and Illinois, the Huntington Dolomite of Indiana and the Cedarville Dolomite of Ohio. These crinoids are preserved principally as calyces and aboral cups. Furthermore, reef and reef-associated strata have been dolomitized, so that nearly all fossils are presented as moulds and casts. Despite this preservational style, a richly diverse and abundant pelmatozoan echinoderm fauna is known from the various reef and non-reef facies. For example, quiet-water reef phases were dominated by the disparid *Pisocrinus*; a transitional semi-rough-water fauna was represented by *Pisocrinus*, the cladids *Gissocrinus* and *Crotalocrinites* and the camerates *Eucalyptocrinites* and *Calliocrinus*; and the rough-water reef stages were dominated by large camerates such as *Eucalyptocrinites*, *Calliocrinus*, *Lampterocrinus*, *Siphonocrinus*, *Stiptocrinus* and *Periechocrinus* (Lowenstam 1957; Witzke & Strimple 1981; Witzke 1983). Common interreef crinoids include the disparids *Myelodactylus*, *Pisocrinus* and *Zophocrinus*; the camerates *Calliocrinus*, *Dimerocrinites*, *Macrostylocrinus* and *Periechocrinus*; and the cladids *Gissocrinus* and *Petalocrinus* (Lowenstam 1948; Watkins 1991; Lane & Ausich 1995).

Upper Silurian (Ludlovian) crinoids were first described from western Tennessee by Springer (1926). These crinoids are from a mixed carbonate–shale sequence in the Brownsport Formation. Common crinoids include the camerates *Lampterocrinus*, *Marsupiocrinus* and *Eucalyptocrinites*; the disparids *Pisocrinus*, *Calceocrinus*, *Cremacrinus* and *Myelodactylus*; the flexible *Lecanocrinus*; and the cladid *Gissocrinus*. The rhombiferan *Caryocrinites* and the blastoid *Troosticrinus* are also relatively common in the Brownsport Formation.

The Lower Devonian Hunsrück Slate (Chapter 13) with their famous echinoderm faunas is followed, in the Middle Devonian, by the equally classic crinoid occurrences of the western Eifel around Gerolstein. During Eifelian and Givetian times, a shelf area with local depressions and islands, with stromatoporoid and coral reefs, offered a multitude of habitats for brachiopods, molluscs, trilobites and crinoids (Jungheim 1995). The crinoids are most common in the fore reef areas (W. Meyer 1988) and include crowns of the compact cladids *Cupressocrinites abbreviatus* and the slender *Cupressocrinites gracilis*, the monobathrid camerate *Hexacrinites elongatus* and the diplobathrid camerate *Rhipidocrinus crenatus*, to name some of the species best known. The peculiar flexible *Ammonicrinus* with its stem coiled around the crown (Fig. 14) is represented by two species. The Eifel crinoids were recently described in a richly illustrated monograph (Hauser 1997).

Numerous important Lower Mississippian crinoid localities are present throughout North America, but only three are discussed in the assemblage chapters (Chapters 16–18). Examples of additional faunas include the Lodgepole Formation of Montana (Laudon & Severson 1953), the Anchor Limestone of Nevada (Webster & Lane 1987) and the Fort Payne Formation of Kentucky and Tennessee. The Lodgepole fauna is dominated by camerates (*Dichocrinus*, *Platycrinites* and *Rhodocrinites*). The Anchor Limestone fauna is dominated by members of the camerate family Actinocrinitidae and by the disparid *Synbathocrinus*.

The Lower Mississippian (Late Osagean) Fort Payne Formation of south-central Kentucky, along the shoreline of Lake Cumberland, has furnished rich crinoid faunas. The most abundant of these is associated with carbonate build-ups constructed above fossiliferous green shale mounds. More than 65 species assigned to 40 genera are present, and many of these also occur in the Edwardsville Formation at Crawfordsville (see Chapter 18). Camerates are most abundant and diverse in the carbonate build-ups and sheet-like limestone facies; the green shale facies are dominated by disparids and non-pinnulate cladids. The juxtaposition of several coeval facies with both autochthonous and allochthonous elements is of interest for understanding the preservation potential of different crinoids and the role of physical environmental factors for crinoid preservation (Meyer *et al.* 1989). Not surprisingly, the taphonomic spectrum resembles that of the Hook Head crinoids.

The Lower Carboniferous (Chadian Stage, Visean Series) 'reef-knolls' of the Clitheroe district, Lancashire, England, have yielded a large and diverse fauna of fossil crinoids, which were described in a series of papers by Wright (see, in particular, Wright 1950–1960). Westhead (1979) wrote that 'nowhere else in England have Carboniferous crinoids been found in such large numbers and also in such variety of genera and species...' as around Clitheroe. Crinoids are best known from three quarries, listed in ascending stratigraphic order, at Coplow (45 nominal species), Bellman (15 nominal species) and Salthill (34 nominal species) (Donovan 1992a). (The crinoid-rich beds of Salthill Quarry were not exposed until after the death of James Wright and were, therefore, not considered in his monographs.) Other echinoderms include blastoids and echinoids. Common crinoids of the Clitheroe area include the cladids *Cyathocrinites* and *Poteriocrinites*, the disparid *Synbathocrinus*, flexibles such as *Taxocrinus*, monobathrids such as *Actinocrinites, Amphoracrinus, Platycrinites* and *Pleurocrinus* and the diplobathrid *Gilbertsocrinus*. Identifiable crinoids are almost invariably preserved as calyces, without attached arms and stem; disarticulated ossicles of the crown and stem are abundant. It is interesting that the cladid *Barycrinus*, locally common in coeval deposits in North America, is known in Britain only from pluricolumnals at Salthill Quarry (Donovan & Veltkamp 1990), indicating that the true diversity of the crinoid fauna is greater than that indicated by cups alone. The principal collections of crinoids from the Clitheroe succession are housed in the Natural History Museum, London, and the Clitheroe Castle Museum.

Extensive carbonate platforms occurred in several areas of eastern North America between the Acadian and Appalachian orogenic events (Meramecian and Early Chesterian), yielding regional encrinites and carbonate grainstone deposits. Two examples include the Salem Limestone of Indiana and the Monteagle Limestone of Alabama.

The Salem Limestone is the famous Indiana building stone that has been used for many monuments and buildings, such as the Empire State Building in New York City. The Salem Limestone was deposited as a series of nearshore carbonate sand bars (Thompson 1990). The most comprehensive treatment of this fauna was that of Beede (1906). The Salem echinoderm fauna is dominated by the blastoid *Pentremites*, the camerates *Batocrinus, Dichocrinus, Platycrinites* and *Dizygocrinus* and the disparid *Synbathocrinus*. Similarly, the Monteagle Limestone is dominated by *Pentremites, Dichocrinus* and *Platycrinites* (Burdick & Strimple 1982).

The cyclical sedimentation so typical of the Pennsylvanian (Chapter 19) began during the Chesterian in the east-central United States. Chesterian environmental cyclicity ranged from terrestrial to marine facies, and marine facies were both carbonates and siliciclastics. Pinnulate cladid crinoids, such as *Agassizocrinus, Aphelecrinus, Pentaramicrinus, Phanocrinus, Tholocrinus* and *Zeacrinites*, dominated most units along with the blastoid *Pentremites* and the camerate *Pterotocrinus*, with highly modified 'wing plates' attached on the tegmen. Numerous Chesterian localities are known, but some of the most thoroughly studied include the Bangor Limestone of Alabama (Burdick & Strimple 1982), the Glen Dean Limestone of Indiana and Kentucky (Horowitz 1965; Chesnut & Ettensohn 1988) and the Pennington Formation of Kentucky (Chesnut & Ettensohn 1988).

Mesozoic

The soft, fine-grained limestones of the British Chalk lack the spectacular specimens of *Uintacrinus* found in North America, but intact cups of the related *Marsupites testudinarius* are common and preserved upright in life position at certain levels in the Santonian Chalk of Bridlington on the Yorkshire coast (Milsom *et al.* 1994).

Microcrinoids

'Microcrinoids' is an ambiguous term because these animals may be small growth stages of larger adult crinoids, or they may be small adults resulting from arrested development (neoteny), never attaining a larger size. We focus in this section on such true microcrinoids.

Palaeozoic microcrinoids are small stalked crinoids with a cup generally less than 2 mm in size. They have been reported from Silurian to Permian strata and include the disparid allagecrinids with basals and radials only and the cladid codiacrinids with infrabasals, basals and radials that may be lacking. Numerous forms have been described (Peck 1936; Strimple & Koenig 1956). A detailed morphological study was performed by Lane and Sevastopulo (1981) on a species of *Kallimorphocrinus*, an allagecrinid from the Mississippian New Providence Shale of Tennessee. The material is dominated by small armless individuals (the smallest measuring only 175 µm); most larger specimens up to 1.5 mm in height were three-armed. The cup has closely fitting orals. The authors assumed that the animals fed with podia protruding between the orals.

Mesozoic microcrinoids include stemless, presumably pelagic Roveacrinida. The Middle and Upper Triassic forms are assigned to the Somphocrinidae; species classified in the family Roveacrinidae are exclusively Cretaceous in age. In addition, diminutive stalked forms of uncertain affinity occur in the Upper Triassic (Hallstätter Kalk, Norian) of the Tethys, a tropical–subtropical ocean extending from Spain in the west to New Caledonia in the east (Kristan-Tollmann 1986). Some forms range from the eastern Alps to the Taurus Mountains of Turkey, the Himalayas and as far as the island of Timor. Kristan-Tollmann (1980, 1990) established a number of genera, such as *Tulipacrinus* (possibly an isocrinid), *Lanternocrinus*, *Nasutocrinus*, *Leocrinus* and *Bihaticrinus* (restricted to Timor). These short-stemmed, minute crinoids with a compact cup and greatly reduced arms are thought to have formed dense stands on the slopes or swells of the Hallstatt limestone ridges. Their remains were collected by dissolving limestone and by washing interbedded shales (Kristan-Tollmann 1990).

Triassic Somphocrinidae were first described by Peck (1948) from Mexico. *Somphocrinus mexicanus* has also been recorded from the Alpine region and from Timor (Kristan-Tollmann 1988). These roveacrinids have spine-like aboral processes (considered to be fused basals) that support the radials and 10 slender arms. The crinoids occur most abundantly in limestones and calcareous mudstones (marls) of the Upper Triassic (*Osteocrinus* facies, Carnian), which were deposited in shallow basins and on swells, but they are never preserved in reef settings (Kristan-Tollmann 1970, 1977). *Osteocrinus rectus* (first described as holothurian sclerites) occurs from the Middle (Anisian) to Upper Triassic (Rhaetian) and is widespread in the Tethys realm from the northern and southern Alps to Greece, Turkey, Iran, the Himalayas, southern China and Timor. Because of their widespread occurrence and their value as guide fossils, Triassic somphocrinids have been compared to Upper Jurassic saccocomids (see Chapter 26).

The rich material of Lower Cretaceous Roveacrinidae from Texas formed the basis of Peck's (1943) careful description of these pelagic crinoids with highly ornamental cups. The cups may have vertical (*Plotocrinus*, *Roveacrinus*) or horizontal flanges (*Roveacrinus*, *Poecilocrinus*). Flanges also occur on brachials to facilitate floating (Scott et al. 1977).

The Upper Cretaceous of the Gulf Coast area has furnished globular stemless and armless crinoids described by Peck (1973) as *Applinocrinus texanus*. This genus also includes a form from the Upper Cretaceous (Campanian) of England described by Bather (1924) as *Saccocoma cretacea*. Whereas the cup of *Saccocoma* is open, that of *Applinocrinus* is sealed by covering plates with a peculiar clockwise twist.

Mesozoic Cyrtocrinid Assemblages

The order Cyrtocrinida includes a wealth of small, stalked, mostly sturdy crinoids. The arms are commonly short and may be tightly enrolled for protection. Many of these highly variable and at times bizarre crinoids (see Figs. 31–34, 237, 238) have a more or less asymmetric cup, and deformations as well as pathological changes are more common in these than in other crinoids. Cyrtocrinids appear to have lived mostly on hardgrounds or cemented to hard objects in relatively deep water. Association with siliceous sponges and brachiopods is common in these deposits.

One of the earliest assemblages with cyrtocrinids occurs in the Lower Jurassic of the southern Alps (see section titled 'Crinoidal Limestones'). Middle Jurassic Bathonian calcareous mudstones of the Ardèche (France) contain one of the richest cyrtocrinid faunas, with *Cyrtocrinus nutans* (Fig. 70), species of *Eugeniacrinites*, *Lonchocrinus*, *Phyllocrinus* and the peculiar *Dolichocrinus (Tetanocrinus)* of doubtful affinity.

The Middle Oxfordian sponge facies of the Swiss Jura (Argovian Birmenstorf Beds) and Swabian Jura (Lochen Beds) contains numerous remains of cyrtocrinids [mainly *Cyrtocrinus nutans*, *Eugeniacrinites cariophilites* (*caryophyllatus*), *Pilocrinus moussoni* and *Tetracrinus moniliformis*], which were described in the classic monographs of Loriol (1877–1879) and Quenstedt (1874–1876) (see also Hess 1975). The clove-like cups of *E. cariophilites* have been known for a long time, and they were first figured by Scheuchzer in 1702. The corresponding strata are composed of carbonate and argillaceous mud with sponge biostromes and were deposited in deeper water (100 m or more); ammonites abound. The occurrence of *Balanocrinus subteres* and *Isocrinus cingulatus* may indicate that these long-stemmed isocrinids occupied a higher tier for feeding and were thus less susceptible to benthic predators or parasites than the cyrtocrinids cemented to the bottom. Most cups of *C. nutans* are moderately asymmetric, but some are spoon-like, with the articular facets for the arms at a right angle to the stem. In such cases the radials may be fused with the proximal columnal. The presence, at the same locality, of species with rather symmetric cups (*Eugeniacrinites*, *Pilocrinus*) and spoon-like cups indicates different niches; asymmetry may have been an adaptation to constant unidirectional current, with the oral side of the crinoid directed downcurrent.

A remarkable crinoid fauna from the Middle Tithonian of Rogoznik (Pienid Klippen Belt, Poland) contains isocrinids (*Balanocrinus*), cyrtocrinids (mainly *Hemicrinus*, *Phyllocrinus*, *Lonchocrinus*, *Psalidocrinus*, as well as some *Apsidocrinus*; see Fig. 33) and roveacrinids (*Saccocoma*). Pisera and Dzik (1979) beautifully illustrated this fauna, which occurs in a micritic reddish shelly limestone facies dominated by ammonites, brachiopods (*Pygope*) and irregular echinoids (*Cardiolampas*). Many of the fossils were broken or worn. This was attributed by the authors to sediment reworking below wave base, probably at the top of a seamount.

Perhaps the most famous cyrtocrinid fauna is the one from the Lower Cretaceous (Valanginian) of Stramberg in Moravia (Czech Republic). It served as the basis for one of Jaekel's fundamental monographs (1891) and for a number of subsequent papers (e.g., Zitt 1983). The crinoids occur in reddish calcareous mudstones that fill pockets of the eroded surface of white Upper Tithonian limestones. Crinoids are associated with corals and numerous calcareous sponges. Jaekel thought that the corroded rock surfaces were caused by the surf and that the crinoids lived on reefs (hence his conclusion that cyrtocrinids were reef forms, followed subsequently in the literature). He explained the many deformations by wave action. However, these rocks are now considered to be a corroded surface on seamounts. Dominant forms are *Sclerocrinus strambergensis* and *Eugeniacrinites zitteli*, but species of *Cyrtocrinus*, *Phyllocrinus* and the spoonlike *Hemicrinus* (Fig. 34), as well as the closely related *Torynocrinus*, are also present (Zitt 1983), as is *Hemibrachiocrinus*, a *Holopus*-like form first described from the Crimea by Arendt (1968), who wrote a monograph on the cyrtocrinids in 1974 (Arendt 1974). The creviced bottom of the Stramberg Sea must have offered a range of habitats for cyrtocrinids similar to that of the Middle Oxfordian sponge facies mentioned earlier. A Recent analogue can be found in the Pacific seamounts with *Gymnocrinus*, mentioned in Chapter 29, the section titled 'Cyrtocrinids: Living Fossils'.

A Thiolliericrinid Assemblage from the Lower Cretaceous of the Crimea

Thiolliericrinids are a group of Upper Jurassic and Lower Cretaceous crinoids that combine characters of comatulids (a centrodorsal, in some forms bearing cirri) and bourgueticrinids (a stem with synarthrial articulations). Klikushin (1987b) described a rich fauna of these poorly known crinoids from Berriasian and Valanginian strata of the Crimea, where they occur among coral bioherms.

The stem of most forms appears to have been very short, with a small number of columnals, and it was attached by means of a terminal disc. The brachials were united by muscular articulations only. Klikushin reconstructed these crinoids with crowns that could be firmly closed, similar to those of the Triassic *Encrinus liliiformis*. About 10 thiolliericrinid species were described from the Upper Berriasian of the Belbek Valley. These include three species of *Thiolliericrinus* with well-developed cirri that allowed them to move around freely, once they became detached from the stem. The presence of stalked and mobile forms indicates different habitats in this reefal setting, comparable to the habitats of different comatulid species in extant reefs.

CRINOIDS AS INDEX FOSSILS

Crinoids are not generally recognized as index or guide fossils. This is due in part to their occurrence as mainly disarticulated material and the resulting difficulties in identification to species level. Nevertheless, biostratigraphic potential has been attributed to certain taxa. These include pelagic Triassic and Jurassic roveacrinids (see the earlier section on microcrinoids). The potential of Silurian crinoids has been discussed by Sevastopulo *et al.* (1989). In particular, scyphocrinitids and their loboliths that occur around the Silurian–Devonian boundary may be of considerable value, once the taxonomy of these crinoids has been sorted out (Chapter 11). Crinoids have biostratigraphic significance in Osagean strata (Lower Mississippian) of North America, because crinoids are abundant and their stratigraphic distributions are well known in the type section along the Mississippi River valley (Chapter 17). Plates of calceolispongiids from the Permian of western Australia are easy to identify, and different species follow each other in a stratigraphic sequence of 2,000 m (Chapter 20). Finally, a crinoid biostratigraphy has been proposed for the Triassic Muschelkalk (Chapter 21).

Because the preservation of complete specimens is rare, a biostratigraphy based on crinoids must rely mainly on disarticulated skeletal elements, the commonest and most promising of which are the columnals. The most notable columnal-based stratigraphy is based on the faunas of the Ordovician–Permian rocks of the former USSR. These have been studied in great detail for more than 40 years, and a comprehensive biostratigraphy based on these elements has been formulated (see review in Stukalina 1988). Other Palaeozoic successions have not been studied in such detail, although other columnal biostratigraphies have been devised – for example, for the Ordovician of the British Isles (Donovan 1995). Although they have been used mainly in Palaeozoic stratigraphy, Rasmussen (1961) formulated a biostratigraphic scheme, based on the distinctive columnals of isocrinid crinoids, for the Middle to Late Cretaceous of Britain.

Taphonomy

WILLIAM I. AUSICH, CARLTON E. BRETT AND HANS HESS

PRESERVATION

Crinoids are rarely preserved as whole fossils. Instead, they commonly disarticulated after death and are preserved as partial crowns, cups, arm ossicles, stem fragments and individual columnals. Decay, disarticulation and preservation represent the field of study that we call taphonomy. Those who collect crinoids have probably always had an intuitive feeling for crinoid decay and disarticulation, but the scientific study of crinoid taphonomy did not begin until the 1960s. Cain (1968), Meyer (1971) and Liddell (1975b) were the first to document the fact that crinoids disarticulate very quickly on the sea floor after death. Within a few days after death, the muscles and ligaments of these animals decay, which leads to the disarticulation of the skeleton into isolated skeletal elements. For a complete crinoid to be preserved, it must be buried rapidly and deeply enough to prevent re-excavation by currents or disruption by scavengers or other burrowers (Lane 1971; Donovan 1991). Tempestites (storm deposits) and turbidites (submarine slides) represent common mechanisms for rapid burial, and many of the assemblages discussed in this book resulted from such types of deposition. Examples of studies of the sedimentology of crinoid localities include those of Franzén (1982), Brett and Eckert (1982), Schumacher and Ausich (1985), Aigner (1985), Parsons *et al.* (1988), D. L. Meyer *et al.* (1989), Brett and Seilacher (1991) and Schubert *et al.* (1992).

Despite the fact that crinoids disarticulate quite rapidly so that their ossicles become sedimentary particles, it appears that disarticulated crinoidal remains commonly become buried near the living site of the crinoids, as demonstrated in deep-water habitats (Llewellyn & Messing 1991) and in reef settings (Meyer & Meyer 1986).

Disarticulation of crinoids follows decay of connective tissues between plates. Each of these tissues could have a different rate of decay. Assuming that this leads to different rates of disarticulation, Baumiller and Ausich (1992) and Ausich and Baumiller (1993) were able to explain stem breakage and to infer the nature of soft tissues in Palaeozoic crinoids. By examining the stem, Baumiller and Ausich (1992) demonstrated that the stem of Early Mississippian crinoids had the same basic structure of ligaments as living isocrinids – that is, all columnals connected by short intercolumnal ligaments but regular lengths of columnals connected by long through-going ligaments. This arrangement results in the stem first disarticulating into segments of nearly equal length before disarticulating completely into isolated columnals. Also, by consideration of the disarticulation of the arms of crinoids in relationship to the stem, Ausich and Baumiller (1993) demonstrated that only advanced cladids had muscle tissues in the arms, whereas all other Early Mississippian crinoids presumably had only ligaments binding arm plates.

All crinoids do not have the same preservational potential. Given similar preservation conditions, groups of different crinoids disarticulate at different rates (D. L.

Hans Hess, William I. Ausich, Carlton E. Brett, and Michael J. Simms, eds., *Fossil Crinoids*. © 1999 Cambridge University Press. All rights reserved. Printed in the United States of America.

Meyer et al. 1989). With exceptions, Lower Mississippian crinoid resistance to disarticulation decreased in the following order: monobathrid camerates, disparids, cladids, flexibles. The camerates were most resistant to disarticulation because their calyx plates are commonly cemented together. Disparids and cladids have progressively less well-sutured plates, and calyx plates of flexibles were apparently held together primarily with connective tissues. These preservational differences are not a factor when crinoids are buried suddenly and preserved. However, under normal conditions of disarticulation, calyces of camerates should be preferentially preserved. Recognition of these potential biases is very important for palaeoecological studies.

ORIENTATION PATTERNS

In the assemblages described in this book, the position and orientation of the fossils on the bedding planes give important clues to the life habits of the animals. Examples where orientation of the fossils has been decisive for the recognition of lifestyles are: scyphocrinitids and their loboliths from the Silurian–Devonian boundary of Morocco (Chapter 11), *Encrinus liliiformis* bioherms from the Triassic Muschelkalk of Germany (Chapter 21), *Pentacrinites* and *Seirocrinus* attached to driftwood from the Lower Jurassic of southern England and Germany (Chapters 22 and 23), *Saccocoma* and corresponding coprolites from the Upper Jurassic Plattenkalk of Bavaria and lenses with *Uintacrinus* from the Niobrara Chalk (Chapter 27). The reader is referred to the respective chapters for more information.

ORGANIC PIGMENTS

A number of fossil crinoids are beautifully coloured. Outstanding examples are the Upper Silurian scyphocrinitids of Morocco (Fig. 109), the Lower Mississippian LeGrand crinoids (Chapter 16), the Middle Triassic *Chelocrinus carnalli* of Germany (Chapter 21) and some of the Middle Jurassic *Pentacrinites dargniesi* beds (Chapter 25). Perhaps the most spectacular occurrence is the Upper Jurassic millericrinids of Switzerland. The deep violet to purple colours are best preserved in the massive roots of large *Liliocrinus munsterianus* (Fig. 62), the dominant crinoids of the famous Liesberg beds. Blumer called the *Liliocrinus* pigments Fringelites after one of

Fig. 62. Root of *Liliocrinus munsterianus*, coloured with fossil organic pigments called Fringelites. The root, which is made up of partly uncoloured calcite, has grown stepwise, by successive accretion, in the muddy sediment. Liesberg Beds (Middle Oxfordian), Liesberg. (Natural History Museum, Basel; photograph S. Dahint.) ×0.5. To view this figure in colour, see the colour plate section following page xv.

the localities (Fringeli); he determined their chemical structure as early as 1960 (Blumer 1960). Fringelites are phenanthroperylene quinones, compounds with complex, polycyclic ring systems that were assumed by Blumer to result from repeated reduction and condensation of the original echinochrome-like compounds; these are red hydroquinone pigments, found in today's echinoderms. Phenanthroperylene quinones also colour Middle Devonian crinoids, Triassic crinoids such as *Chelocrinus carnalli* (Falk et al. 1994; Falk & Mayr 1997) and the Jurassic *Pentacrinites dargniesi* from the Tournus area (H. Falk, pers. comm., 1997); they include the gymnochromes of the extant *Gymnocrinus richeri* (DeRiccardis et al. 1991) and occur in other living organisms, both animals and plants. This strongly suggests that Fringelites are genuine chemical fossils (H. Falk, pers. comm., 1997); their extraordinary chemical stability and resistance to leaching appear to be the result of the unique complex formation behaviour with transition metal ions (Falk & Mayr 1997).

MODES OF OCCURRENCE OF FOSSIL CRINOID ASSEMBLAGES

Crinoids are normally preserved as disarticulated ossicles. During the Palaeozoic and parts of the Mesozoic when crinoids flourished in shallow-water areas, they were among the most important producers of carbonate sediment (see Chapter 3). Indeed, crinoids have sometimes been referred to as the '*Halimeda* of the Palaeozoic', by analogy to their role as principal sediment-formers on shallow carbonate platforms, the analogue of modern calcareous green algae. As Ausich (1997) has pointed out, there are many examples of widespread sheet-like deposits of limestone composed almost exclusively of disarticulated crinoid ossicles. Some of these limestone bodies encompass thousands of square kilometres and may be tens of metres thick. Excellent examples of these regional encrinites are found in the Mississippian Banff and Rundle Formation of Alberta or in the Burlington Limestone of the Mississippi Valley. Considering Macurda and Meyer's (1983) estimate that an average cubic metre of crinoidal limestone contains remains of about 15,000 crinoids, the total number of individuals represented in such deposits is staggering: on the order of 10^{13}–10^{16} crinoids! And yet very few can typically be recognized to the species level. Even in argillaceous limestones, crinoidal skeletal debris may account for 20–30% of rock volume. Crinoid debris in varying density has been reported in most studies of marine fossil assemblages from the Palaeozoic Era (for examples, see McKerrow 1978), and yet it is relatively rare of palaeocommunity studies to report specific crinoid taxa. This attests to the rapid breakdown of crinoid skeletons and the inherent bias against recognizable crinoid remains in reported fossil assemblages, despite evidence that these animals were abundant in many settings.

Assemblages of articulated crinoids, such as those described and illustrated in this book, are exceptional, and they are typically confined to particular bedding planes or localized lenses. They represent events of very rapid burial or smothering by sediment. For such beds Seilacher (1970) coined the term 'obrution deposits'; he pointed out that in such deposits echinoderms are over-represented (Seilacher 1990a), presumably because their ambulacral system becomes easily clogged by fine sediment. Brett and Seilacher (1991) presented a preliminary genetic classification of obrution deposits that is readily adapted to discussion of the types of articulated crinoid occurrences. These types of occurrences can be summarized as follows (see also Allison & Briggs 1991).

The Burlington type is epitomized by the classic occurrences in the Mississippian Burlington crinoidal limestones (Chapter 17) and other regional encrinites (Chapter 3). These include the Silurian reefs of Gotland (Chapter 9) and the Lower Devonian Coeymans facies (Chapter 12). These relatively massive limestone units with coarse-grained textures accumulated under shallow-water, high-energy conditions. The overwhelming majority of echinoderms in these assemblages are disarticulated, and the occurrence of abraded grains and ooids reflects frequent stirring and reworking of the carbonate sediments. Occasionally, well-preserved crinoid remains occur as scattered individuals or local lenses and pockets within these units. Many of the articulated remains in Palaeozoic strata are rather rigidly sutured calyces (e.g., of camerate crinoids and blastoids) that may have remained intact longer than other articulated parts, such as arms. However, the occurrence of complete crowns and individuals with columns intact demonstrates that occasionally individuals were killed, rapidly buried and, fortuitously, did not fall victim to later reworking. Probably, they accumulated in low spots between shifting skeletal sand dunes, or sediment that was shifted over a particular area was thick enough to protect the buried remains from later exhumation.

The Hauptrogenstein type is somewhat similar to the Burlington type, but the crinoid lenses are restricted to two particular horizons, with hardly any disarticulated material in the following strata (see Chapter 25). Most of the beds were formed on the muddy bottom at the margin of sand wave complexes.

The so-called Gmünd type occurrences consist of well-preserved crinoids and other fossils, buried along the top contacts of skeletal limestone bodies, typically in thin blanketing shale layers. The type example is a crinoid–echinoid–starfish bed preserved in dark shales above a thin transgressive conglomeratic and skeletal limestone from the Lower Jurassic of Germany (Rosenkranz 1971). The upper contact of the limestone bed may represent an early lithified sea bottom or hardground to which numerous oysters and echinoderms were attached. In these cases, a prolonged interval of minimal sedimentation and/or erosion preceded a final, abrupt burial event. During this period, skeletal debris (commonly including abundant disarticulated crinoid remains) accumulated and, in some cases, was cemented together, forming a firm pavement or hard sea floor to which sessile animals such as crinoids attached. The sudden influx of a particularly thick sediment layer, commonly fine siliciclastic mud, terminated the period of sediment starvation. Such conditions developed frequently during transgressions with fairly rapid deepening of the water. The influx of mud flows (turbidites or storm layers) following very major storms probably accounts for the rapid burial. Once such sediment was imported, it was likely to be protected from further reworking because of the relatively deep (and deepening) undisturbed setting. Similar occurrences are the shoal marginal assemblages of the Middle Ordovician Trenton Group in Ontario (Chapter 6), the Lower Devonian Manlius Formation of New York (Chapter 12) and crinoids overlying lenses of skeletal debris in the Devonian Arkona Shales of Ontario (Chapter 15). The Bradford hardground *Apiocrinites* (Chapter 24) and most of the occurrences of well-preserved crinoids in the Triassic Muschelkalk of Europe (Chapter 21) are also due to smothering by the catastrophic deposition of mud (Hagdorn & Schulz 1996).

Crawfordsville-type occurrences are typified by the famous crinoid beds of the Mississippian Edwardsville Formation near Crawfordsville (see Chapter 18); this type of occurrence probably accounts for the majority of well-preserved Palaeozoic crinoid assemblages described in this book – for example, most of those in the Middle Ordovician Trenton Group (Chapter 6), the Silurian Rochester Shale (Chapter 10) and the Middle Devonian Windom Shale in New York (Chapter 14). The Late Cretaceous Niobrara Chalk with its unique occurrence of the stemless, specialized *Uintacrinus* (Chapter 27) may also be included here. These occurrences are bedding planes in relatively fine-grained siliciclastic successions, mainly mudstones and siltstones. In contrast to the Gmünd-type assemblages, the lithologies of the under- and overlying strata are very similar; however, typically the well-preserved assemblage may lie upon a very thin (millimetres) and discontinuous layer of skeletal debris. This suggests that a minor discontinuity exists. During a brief interval of non-sedimentation, crinoids colonized the sea floor, using accumulated skeletal fragments as attachment substrates. After the passage of one or more generations of crinoids, a pulse of muddy sediment, probably associated with a storm, killed the benthic community and rapidly buried its remains. Thus, these types of assemblages are related to and may grade into the Gmünd-type assemblages.

In some cases, later diagenetic cementation of the muds, possibly related to anaerobic decay of the contained organisms, resulted in the formation of concretions that jacketed the remains and prevented compactional deformation. Such assemblages tended to develop on areas of the sea floor in the deeper parts of storm wave base where muds accumulated and sea floors remained undisturbed for many years but where it was still shallow enough to be affected by intermittent pulses of storm-resuspended sediment (Taylor & Brett 1996). Under these conditions, smothering evidently occurred repeatedly. Ironically, such low-energy, muddy environments may have been near the lower fringes of many crinoids' habitat ranges.

Hunsrück-type assemblages are more unusual and are exemplified by the famous crinoid assemblages of the Lower Devonian Hunsrück Slate of Germany (Chapter 13). Here, mostly delicate crinoids may be preserved locally as completely articulated and typically pyrite-replaced specimens. These assemblages appear to be related to the Crawfordsville-type assemblages, but they probably developed in somewhat deeper-water environments, below storm wave base; the iron-rich sediment rapidly became anoxic beneath the surface. Evidently, conditions on the sea floor were adequate to sustain a moderate abundance and diversity of crinoids. In this type, crinoids appear to be buried at or close to their life sites by muddy turbidites or tempestites. Anaerobic decay of the crinoids within the mud layers favoured early diagenetic pyrite, which impregnates the stereom of the skeletons. Anoxic conditions may have inhibited scavenging of the buried remains. A few assemblages in the Silurian Rochester Shale (Chapter 10) and Devonian Hamilton Group (Chapter 14) display comparable conditions of rapid burial in fine muds and early pyritization.

Posidonia Shale–type assemblages are typical stagnation deposits (Seilacher 1990a) and occur in black shale environments where only pseudoplanktonic or pelagic crinoids could live in the upper water layers. Occurrences from the Lower Jurassic have furnished some of the best-preserved and most spectacular crinoids of all time (Chapters 22 and 23). Beds and lenses with the pelagic scyphocrinitids (Chapter 11) may owe their preservation to similar anoxic conditions at the bottom, but the corresponding strata have not been studied as extensively.

The Solnhofen-type occurrence, classified by Seilacher (1990a) as a stagnation deposit, is exemplified by the crinoid-bearing Upper Jurassic Plattenkalk facies of Bavaria (Chapter 26), the Cretaceous of Lebanon and a few other Mesozoic examples. These limestones accumulated in partly closed basins with restricted water flow and, perhaps, in hypersaline bottom waters. As such, they represent a unique set of environmental conditions that developed only locally, particularly during the Mesozoic. The burial of fragile organisms in these sediments led to the preservation of completely articulated fossils, sometimes even with impressions of soft tissues. The absence of scavengers and reduced influence of chitinoclastic bacteria, due to hypersalinity and anoxia, may have been critical to the unique mode of preservation. Possible analogues of the Solnhofen-type assemblages occur rarely in the Palaeozoic, but may include an assemblage of articulated (but very poorly preserved) crinoids described recently from the Silurian interreef dolostones in Indiana (Lane & Ausich 1995).

5

Ecology and Ecological Interactions

WILLIAM I. AUSICH AND MICHAEL J. SIMMS

While marvelling at the beauty of fossils, it is easy to forget that they were once living animals. Rather than inanimate objects, when alive fossils were part of an ancient ecosystem and enmeshed in myriad biotic interactions, such as competition, predation, parasitism and commensalism. The study of this aspect of organisms is ecology (also termed palaeoecology for fossils), and ecology can be divided broadly into autecology, relating to individual organisms, and synecology, interactions among organisms.

Because ancient crinoids are known only from fossilized remains, many aspects of their ecology must be inferred from the study of living crinoids and verified by demonstrating consistency with data preserved in the geological record. For example, crinoids are exclusively marine today and can tolerate only very minor variations of marine salinity. Because fossil crinoids are found in rocks of oceanic origin and are associated with other organisms that also have a tolerance for only marine conditions, it can readily be assumed that all crinoids, ancient and living, had similar requirements. Similarly, crinoids lack specialized respiratory organs, but, unlike some echinoids and ophiuroids, they have a high oxygen demand. Consequently, crinoids are excluded from environments with low levels of oxygen; this has not limited them from spreading into a huge variety of ecological niches.

Throughout their history, crinoids have been passive suspension (or filter) feeders, so they must rely on ambient water currents for food. Crinoids never developed into either burrowing or boring habitats. Feeding strategies are one of the predominant factors of crinoid autecology. Suspension-feeding niches are subdivided by tiering (height above the ocean floor) and size selection for food, as illustrated in many of the chapters on assemblages in this book. Modes of attachment are another predominant factor in crinoid autecology. Were crinoid holdfasts adapted for attachment to hard substrates or soft substrates; were the attachments permanent or movable?

Crinoid morphology is of key importance in interpreting crinoid autecology (see Chapter 1). Some biotic interactions, such as the formation of a gall by a parasite, are directly preserved on fossils; and such interactions can be interpreted either by direct comparison to living crinoids or by inference. Direct observations of how fossil crinoids lived and interacted with their environment are impossible. We must infer how they lived through comparisons with living crinoids and from an understanding of how crinoid remains became incorporated into the fossil record (Meyer & Ausich 1983). Numerous examples are mentioned in the chapters on assemblages in this book. The following discussion of synecology centres on a few common examples of commensalism and parasitism that affected crinoids.

Hans Hess, William I. Ausich, Carlton E. Brett, and Michael J. Simms, eds., *Fossil Crinoids*. © 1999 Cambridge University Press. All rights reserved. Printed in the United States of America.

CRINOIDS AND PLATYCERATID GASTROPODS

The relationship between crinoids and platyceratid gastropods is one of the most commonly preserved and interesting interactions. Members of the Platyceratidae interacted with crinoids, blastoids and cystoids throughout much of the history of this group of archaeogastropods, from the Ordovician to the Triassic. These gastropods are almost always closely associated with stalked echinoderms, and they are typically preserved attached to a crinoid tegmen (Fig. 63). The traditional interpretation is that the gastropods were positioned over the anus and that they fed on crinoid faecal material. Thus, platyceratids were considered to be coprophagous commensals (Bowsher 1955; Breimer & Lane 1978). Ordovician platyceratids, such as *Cyclonema*, moved freely over the tegmen of the camerate crinoids *Pycnocrinus* (Fig. 95) and *Glyptocrinus*. *Platyceras*, the common later platyceratid, was permanently attached to the crinoid tegmen (Fig. 63). Camerate crinoids with a solidly plated tegmen were the most common hosts of platyceratids. Examples of crinoid–platyceratid relationships from the crinoid assemblages discussed in this book are listed in Table 3.

The coprophagous commensal relationship has been questioned by Rollins and Brezinski (1988) and Baumiller (1990b, 1993). Rollins and Brezinski argued that platyceratids were probably suspension feeders. In contrast, Baumiller suggested that they were shell-boring parasites. Baumiller attributed boreholes on crinoids and blastoids to platyceratids. Evidence for this is a firmly attached *Platyceras* sitting directly over a borehole through the tegmen of a *Macrocrinus mundulus*.

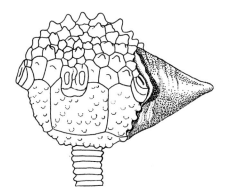

Fig. 63. The gastropod *Platyceras* over the anal opening of *Platycrinites hemisphaericus*. Edwardsville Formation, Crawfordsville, Indiana. (Redrawn from Wachsmuth & Springer 1897.) ×1.6.

Table 3. *Common Crinoid – Platyceratid Associations in Palaeozoic Assemblages*

ASSEMBLAGE	CRINOID	GASTROPOD
Indian Creek (Ch. 18)	*Macrocrinus mundulus*	*Platyceras*
Crawfordsville (Ch. 18)	*Platycrinites hemisphaericus*	*Platyceras*
	Actinocrinites gibsoni	*Platyceras*
	Gilbertsocrinus tuberosus	*Platyceras*
LeGrand (Ch. 16)	*Aorocrinus immaturus*	*Platyceras*
Silica Shale (Ch. 15)	*Arthroacantha carpenteri*	*Platyceras*
Cincinnati, Ohio (Ch. 8)	*Pycnocrinus dyeri*	*Cyclonema*

All crinoids listed are camerates. Platyceratids occur only rarely on non-camerate crinoids.

With the present data, it is difficult to convincingly reject any of the hypotheses for this biotic interaction, despite the fact that this crinoid–platyceratid association is known from very many fossil examples. The biology of this relationship is still under study, and different platyceratids may have had different feeding modes.

OTHER COMMENSAL RELATIONSHIPS

A variety of other relationships, generally regarded as commensal in nature, existed in ancient ecosystems. Commensals are organisms that gain an advantage from a host, without greatly affecting the host. Most commensals on crinoids were other suspension-feeding organisms that were epizoic in order to have a better suspension-feeding perch, to avoid settling on the bottom or perhaps both. These epizoans included encrusting organisms, such as bryozoans and corals, or mobile organisms, such as bivalves and ophiuroids. A variety of boreholes and galls have also generally been regarded as the result of commensal activities, although they may also have been caused by parasites.

Examples of suspension-feeding commensals on crinoids are tabulate corals and the ophiuroid *Onychaster*. *Onychaster* belongs to a group of suspension-feeding ophiuroids that first evolved during the Devonian. Perhaps it is not a coincidence that ophiuroid remains first become associated with fossil crinoids during the Devonian (Schmidt 1942). In Mississippian examples, such as at Indian Creek (Chapter 18), *Onychaster* is perfectly preserved entwined within the arms of *Actinocrinites multiramosus* (Fig. 64). At a slightly more recent locality

Fig. 64. The ophiuroid *Onychaster* wrapped around the anal tube of *Actinocrinites multiramosus*. Edwardsville Formation, Indian Creek, Indiana. (Redrawn from Wachsmuth & Springer 1897.) ×1.

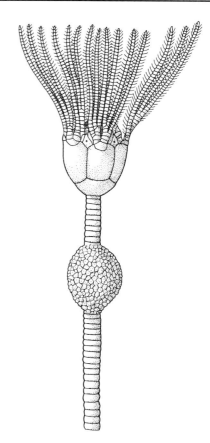

Fig. 65. Favositid tabulate coral encrusting around the stem of *Platycrinites*. Lower Carboniferous of Ireland. (Redrawn from Hudson et al. 1966.) ×1.

not discussed in this book (Canton, Indiana), *Onychaster* specimens are similarly preserved with *Actinocrinites multiramosus* and with *Scytalocrinus robustus* (Wachsmuth & Springer 1897). Ophiuroids are also associated with other Mississippian crinoids and with Jurassic crinoids (Austin & Austin 1843–1849). A presumably suspension-feeding asteroid preserved in a bed of *Chariocrinus andreae* (see Chapter 25) was described by Blake (1984). In today's oceans, suspension-feeding ophiuroids perch on living stalked crinoids (Macurda & Meyer 1974) to gain a feeding advantage, and it is presumed that these ancient associations were similar (Spencer & Wright 1966).

Tabulate corals also had a long association as epizoans on stalked crinoids. Tabulate corals were undoubtedly suspension feeders, so by being attached to an erect crinoid stem they gained an advantage by feeding in higher-velocity currents. In environments with a muddy substratum, a stem would also be a good hard substratum for larval attachment. In many cases, it is difficult to determine whether an encrusting epizoan attached to a living erect stem or to a dead stem lying on the ocean floor. However, in the tabulate corals discussed here and in some bryozoans, colonies grew symmetrically around the entire circumferences of the stem, which is most easily explained if the colony grew on an upright, living crinoid. Favositid tabulates grew in a spherical to ellipsoidal colony form with the crinoid stem encased along the vertical axis (Figs. 65, 106). Examples of this commensal association are known from the Ordovician, Silurian and Devonian, the time of maximum diversity of the tabulates, and also from the Lower Carboniferous.

During the Mississippian, auloporid corals were epizoic on crinoids. In contrast to the massive favositids, auloporids had a runner-type colony morphology where new coralites budded only from near the top of older coralites. With this growth mode, auloporids, such as *Cladochonus*, first grew to form a ring around the circumference of a stem and then grew outward into a three-dimensional, open bush (Fig. 66). In some cases, the

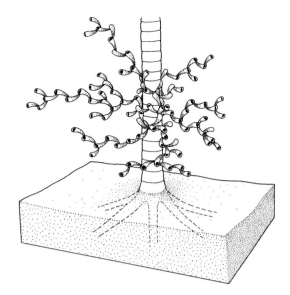

Fig. 66. The auloporid coral *Cladochonus* encrusting around a crinoid stem. Lower Mississippian Borden Formation, north-central Kentucky. (Redrawn from Kammer 1985.) ×0.5.

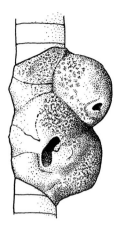

Fig. 67. A crinoid arm of Pennsylvanian age with probably a myzostome gall. Seminole Formation, Tulsa County, Oklahoma. (Redrawn from Welch 1976.) ×3.7.

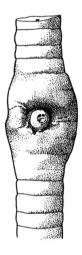

Fig. 68. A crinoid stem swollen from a *Phosphannulus*, with the broken upper part of the phosphatic tube in the gall enlarged. Indian Springs Shale, Upper Mississippian, Crawford County, Indiana. (Redrawn from Welch 1976.) ×3.

crinoid stem overgrew and completely encased the initial colony ring, which further demonstrates that the crinoid host was alive during encrustation.

PARASITISM

Many types of boreholes and galls occur on crinoid columns, calyx plates and arms from the Ordovician to the present. Some of these may simply represent a pathological response to an encrusting epizoan, but others are caused by specialized epizoans. Franzén (1974) and Brett (1978) summarized many types of galls on crinoids. On living crinoids, polychaete annelid worms called myzostomes are gall-forming parasites. Myzostomes live on the arms, pinnules and cups of living crinoids. They are most commonly associated with the arms and do not live on stems. On living crinoids, some myzostomes simply live in the ambulacral groove of the arms, whereas other myzostomes form galls on the aboral side of the arms. These galls are hollow, subellipsoidal swellings with a hole on the aboral side. Similar galls are preserved on arms of Palaeozoic crinoids (Fig. 67) and are presumed to be from myzostomid activity (Welch 1976).

Because myzostomes produce galls on living crinoids, myzostomes were commonly blamed for all galls on Palaeozoic crinoids, even the most common types, which are asymmetric swellings on stems. Welch (1976) recognized that such swellings were distinct from those of myzostomes and thought them to be caused by *Phosphannulus* (a genus of the order Hyolithelminthes). The centre of such galls either is hollow or contains *Phosphannulus*, a hollow phosphatic tube that penetrates into the crinoid stem, expanding at the centre (Fig. 68). Because it may penetrate into the lumen, it is possible that this epizoan was a parasite rather than a commensal. *Phosphannulus* is known throughout the Palaeozoic; a reported Jurassic occurrence is doubtful (Meyer & Ausich 1983).

Crinoid stems and cups have additional types of swellings with pits other than those containing the *Phosphannulus* tubes. They occur throughout the Palaeo-

Fig. 69. Stem of *Isocrinus basaltiformis* with galls caused by an unknown parasite (?). Lower Pliensbachian, Gloucestershire. (Also figured by Simms 1989.) ×3.5.

Fig. 70. Scanning electron photomicrograph of a cup of *Cyrtocrinus nutans*, two of the radials with pits. This specimen is one of 32 with pitted cups, out of 1,170 cups collected at this locality. Bathonian, La Pouza, Ardèche, France. (Hess Collection; micrograph M. Düggelin.) ×12.5.

zoic (see Franzén 1974; Brett 1978) and also during the Mesozoic. Examples among Jurassic crinoids include millericrinids from Middle Oxfordian mudstones (marls) of northern Switzerland and eastern France, figured in Loriol's classic monographs (1877–1879, 1882–1889) and isocrinid stems from the Lower Jurassic (Fig. 69). Similar malformations of millericrinid stems from the Lower Oxfordian of Crimea were figured by Klikushin (1996); this paper also contains figures of visible epizoans (serpulids, bryozoans, bivalves, sponges and corals) on crinoid stems. A number of cups of *Cyrtocrinus nutans* from the Bathonian of the Ardèche Department, France (see Chapter 3), have pits on more or less pronounced swellings (Fig. 70). These other swelling types were formed by epizoans of unknown identity.

Assemblages

6

Middle Ordovician Trenton Group of New York, USA

CARLTON E. BRETT

A QUARRY FOR COLLECTING TRILOBITES AND ECHINODERMS

Classic outcrops of the Middle Ordovician Trenton Beds are exposed along the Trenton Gorge of West Canada Creek and in Mill, Cincinnati, and other creeks tributary to the Mohawk River at the town of Trenton, Madison County, New York, north of the Mohawk River valley (Fig. 71). A small hand-operated quarry on the property of W. Rust, about 1 km east of Trenton Falls, was opened by the Rust family and Charles Walcott for the purpose of collecting spectacular trilobite and echinoderm fossils from the upper beds of the Trenton Group.

Limestones of the Trenton Group are of late Middle Ordovician age (Trentonian or Caradocian Series, Mohawkian Stage), about 460 million years before present. The productive strata for crinoids occur in the Rust Member of the Denley Formation (Figs. 72, 73).

SHALLOW PLATFORM, RAMP AND BASIN

The Trenton Group comprises some 100–130 m of highly fossiliferous, thin-bedded, grey limestones with thin interbeds of dark grey calcareous shale. Limestones include a variety of lithologies, such as pelmatozoan-rich skeletal and rubbly nodular limestones with remains of bryozoans and pelmatozoans and tabular, graded micritic limestone. The latter have sharp bases, internal planar to cross-lamination and, in some instances, perfectly preserved fossils, including crinoids.

The coarser skeletal limestone facies are considered to have been deposited in shallow shelf settings. These beds show various amounts of winnowing by storm-generated waves and currents. Nodular calcarenites have undergone thorough bioturbation and, in some cases, early diagenetic cementation. The fine-grained lime mudstone beds reflect rapid deposition from low-density turbidity or gradient currents.

The Trenton carbonates accumulated on a shallow, subtropical platform and east-dipping ramp that bordered a deeper-water peripheral foreland basin to the east (Fig. 72). The latter basin was created by overthrusting of an accretionary prism (Taconic allochthon) onto the eastern (then southeastern) margin of Laurentia (ancestral North America). The basin was the site of relatively deep dysaerobic to anoxic water where a thick succession of black shales accumulated during the late Middle Ordovician. Progressive westward subsidence of this basin produced instability that triggered episodes of shelf collapse and synsedimentary slumping and sliding.

Some of the most famous Trenton crinoid–trilobite occurrences are in thin lime mudstones that accumulated through turbidity currents on the unstable ramp of the Trenton shelf sea bordering the black shale basin. Eventually, the Trenton carbonate shelf subsided in

Hans Hess, William I. Ausich, Carlton E. Brett, and Michael J. Simms, eds., *Fossil Crinoids*. © 1999 Cambridge University Press. All rights reserved. Printed in the United States of America.

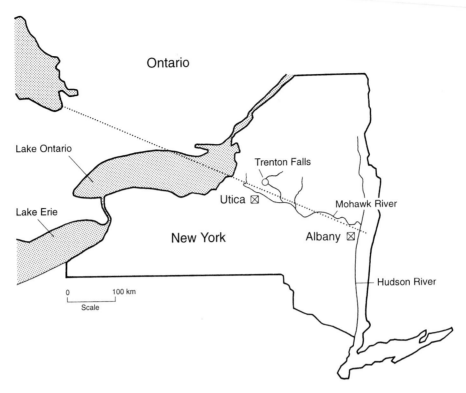

Fig. 71. Location map for Trenton Falls in the Mohawk River valley. Dotted line indicates profile of Fig. 72. (Modified from Mitchell *et al.* 1994.)

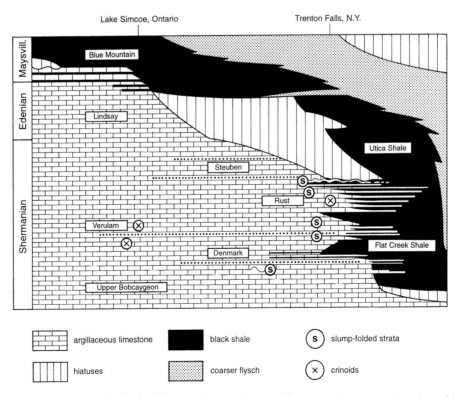

Fig. 72. Generalized stratigraphic profile for the Trenton–Simcoe Group and associated strata in Ontario and central New York State. Map in Fig. 71 shows the location of Trenton Falls. (Modified from Lehmann *et al.* 1995.)

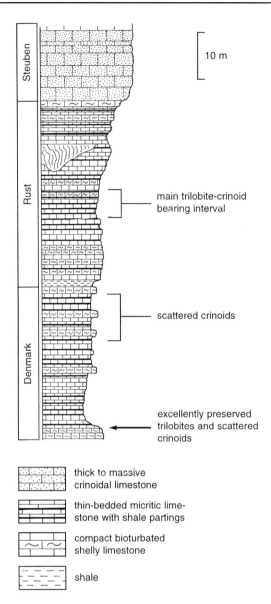

Fig. 73. Detailed stratigraphic column for Trenton Group at Trenton Falls. (Modified from Brett & Baird in press.)

New York. Beds rich in pelmatozoans tend to have faunas dominated by small ramose and fenestrate bryozoans and thin discoidal *Prasopora*. Brachiopods may be common and include the small strophomenids *Rafinesquina deltoidia*, *Sowerbyella* and the orthid *Paucicrura*. Abundant trilobites include exceptionally well-preserved, as well as disarticulated, material of *Isotelus*, *Flexicalymene* and, in some beds, *Ceraurus*.

Seventeen species of crinoids have been identified from the Rust Quarry beds. Crinoids are dominated by the disparids *Cincinnaticrinus* (formerly *Heterocrinus*) (Fig. 75) and *Ectenocrinus* (Fig. 76) and, rarely, the calceocrinid *Cremacrinus*. Camerates such as *Glyptocrinus* (see Fig. 53) are less abundant in most beds. One bed, consisting of a thin, turbidite-deposited lime mudstone, displays abundant, spread-out crowns of *Glyptocrinus* with the aboral side of the cup uppermost and the arms spread out with their ambulacral surfaces downward.

The pelmatozoans and trilobites of the Upper Trenton beds, specifically the Rust Quarry interval, are well articulated and composed of slightly recrystallized calcite. In the Rust Quarry beds, the disparids *Iocrinus* (Fig. 74) and *Ectenocrinus* may occur on the undersides of micritic beds, where they were apparently covered rapidly by the carbonate mud and silt. Other long-stemmed crinoids may occur on the tops of micritic beds, and a few are found obliquely through the thickness of micrite (up to 3 cm thick). Specimens of the rhombiferan cystoid *Cheirocrinus* are found in one layer with the cups oriented upright and stems downward, where they were apparently buried alive and in life position by the carbonate sediments. In some cases, crinoids with stems and crowns still intact are on the tops of nodular skeletal limestones buried by siliciclastic muds, resulting in calcareous shales. Associated complete trilobites (*Flexicalymene*, *Isotelus*, *Ceraurus*) occur on the bases of many micritic turbidites, where they are generally inverted (dorsal shield down). Other specimens occur on limestone bed tops, again buried by mud. Extraordinary specimens of enrolled *Ceraurus* in one bed at the Rust Quarry display calcified appendages.

These observations indicate that some of the Trenton communities were buried very rapidly when organisms were still alive or shortly thereafter. Such deposits were of two types – carbonate mud flows, probably the result of storm wave winnowing of shallow shelf areas, and siliciclastic mud turbidites (possibly also tempestites) that originated in the southeastern source area. Muds were probably flocculated and were deposited as silt-sized particles at rapid rates.

central New York such that the limestones are succeeded upward by black shales and then a thick molasse sequence.

A VARIED FAUNA OF BRYOZOANS, BRACHIOPODS, TRILOBITES AND PELMATOZOANS

Approximately 75 species of fossils, including bryozoans, brachiopods, trilobites and varied echinoderms are known from the Upper Trenton Limestone in central

Fig. 74. *Iocrinus trentonesis* on upper surface of slab; note plicate anal sac of this cladid. Middle Ordovician Trenton Group, Rust Limestone. Rust Quarry, Trenton Falls, N.Y. (Thomas Whiteley Collection.) ×3.

AN UNSTABLE ENVIRONMENT BETWEEN SHALLOW SHELF AND DEEP BASIN

The Rust Member appears to have been deposited mostly in a middle shelf environment at water depths of a few tens of metres. The occurrence of dasycladacian algae (cyclocrinitids) in some beds suggests shallow photic zone conditions, probably less than 50 m deep (Brett et al. 1993). Many layers exhibit evidence of intermittent winnowing, probably by storm waves. Furthermore, the typical Rust lithology is interbedded with thicker crinoidal limestones that also pass vertically into a massive, wave-rippled skeletal limestone (Steuben Member, Fig. 73). This facies was clearly deposited in a high-energy, shallow shelf environment approaching normal wave base.

On the other hand, the best-preserved crinoids, both in the Rust and underlying Denley limestones, occur in thin intervals of calcareous shales that reflect somewhat deeper shelf conditions below normal storm wave base and affected only by the deepest storm wave base (Fig. 73). As already noted, both the calcareous shales and muds seem to reflect events of fine-grained sediment accumulation in a generally low-energy, muddy bottom environment.

These beds appear to pass, within a few kilometres basinward, into barren calcareous mud and sparsely fossiliferous dark grey shales. The former have been interpreted as carbonate turbidites; the dark shales represent siliciclastic muds that accumulated in a dysaerobic environment. Crinoids, bryozoans and most brachiopods are absent from this facies.

Thus, the best-preserved pelmatozoan assemblages occur toward the upper end of a gently sloping ramp – that is, in transitional environments between a shallow, storm-dominated shelf and a dysaerobic, deeper slope to basin. Instability within this environment is evidenced by minor slump folding and convolute bedding. This suggests that submarine seismic-induced disturbance and slumping of sediment may have been a factor in the catastrophic burial of crinoid assemblages.

IMPORTANT COLLECTION IN THE UNITED STATES

Yale Peabody Museum, New Haven, Connecticut

Fig. 75. *Cincinnaticrinus* (formerly *Heterocrinus*) *heterodactylus*. Small delicate disparid; very common on upper surfaces of limestone beds in the Rust Limestone, Trenton Group. Rust Quarry, Trenton Falls, N.Y. (Thomas Whiteley Collection.) ×2.5.

Fig. 76. *Ectenocrinus simplex*, disparid with ramulate arms. Rust Limestone, Trenton Group. Rust Quarry, Trenton Falls, N.Y. (Thomas Whiteley Collection.) ×3.

7

Middle Ordovician of the Lake Simcoe Area of Ontario, Canada

CARLTON E. BRETT AND WENDY L. TAYLOR

Extraordinary occurrences of well-preserved crinoids and other echinoderms have long been known from the Middle Ordovician Trenton (or Simcoe) Group in the vicinity of Lake Simcoe in southern Ontario, Canada (Fig. 77). These rocks are exposed in several inactive and currently operating quarries; the most famous section can be found in the disused quarry just off the Trenton Canal locks west of Kirkfield, Ontario. Other excellent sections occur in the Carden Quarry near the town of Brechin, Ontario.

Pelmatozoan assemblages occur at several levels in the Middle Ordovician Trenton carbonates of the Lake Simcoe area, but two horizons may be singled out as particularly outstanding examples of exceptional preservation (Fig. 72):

Assemblages in the upper 3–4 m of the Bobcaygeon Formation (= Kirkfield Limestone of earlier workers) are associated with hardgrounds, resulting from interrupted sedimentation, on the tops of skeletal limestone beds.

One or more beds of *Cupulocrinus* are associated with crinoidal limestone in the lower 2–3 m of the Verulam Formation.

THE BOBCAYGEON HARDGROUNDS

The Upper Bobcaygeon (Kirkfield) Limestone consists of thin-to medium-bedded, commonly graded crinoidal

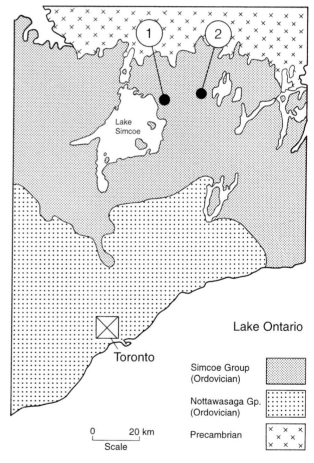

Fig. 77. Location map for crinoid localities near Lake Simcoe, Ontario. (1) Brechin; (2) Kirkfield. (Modified from Melchin et al. 1994.)

Hans Hess, William I. Ausich, Carlton E. Brett, and Michael J. Simms, eds., *Fossil Crinoids*. © 1999 Cambridge University Press. All rights reserved. Printed in the United States of America.

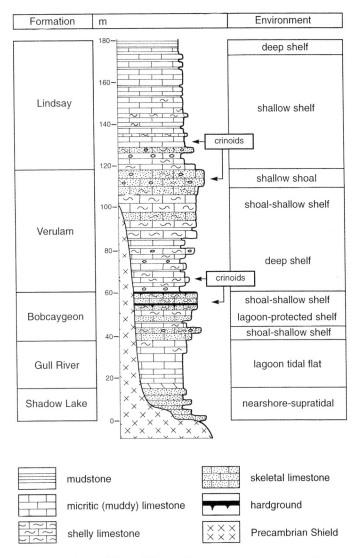

Fig. 78. Generalized stratigraphic section for Middle and Upper Ordovician rocks in the Lake Simcoe area of Ontario. (Modified from Brookfield & Brett 1988.)

limestones interbedded with medium grey calcareous shales and laminated calcareous silt beds. Horizons of intense burrowing are observed in some silty beds. Many beds display some evidence of normal grading from skeletal limestone upward into laminated calcareous siltstone. The tops of the skeletal-rich beds display irregular topographies representing burrowed hardground surfaces, especially noticeable on the tops of calcareous siltstones (Fig. 78).

Some bed tops are clearly hardgrounds. They display very sharply defined relief up to 5 cm with minor dark staining (phosphatic?) and abundant *Trypanites* borings. Some of these surfaces also show encrusting organisms, including patches of bryozoans, discoidal pelmatozoan holdfasts and, rarely, edrioasteroids (Fig. 83).

The echinoderm fauna associated with hardgrounds in the Middle Ordovician Bobcaygeon Formation is quite varied. The small hybocrinid *Hybocystites* is the most abundant crinoid (Fig. 79). The large camerate *Archaeocrinus* is present (Fig. 80), and various inadunates, including the newly described cladid *Illemocrinus amphiatus* (Fig. 81) and the disparid *Isotomocrinus typus* (Fig. 82). Much less common are the primitive calceocrinid *Cremacrinus* and the paracrinoid *Amygdalocystites*. Also found are cystoids, such as the rhombiferans *Cheirocrinus* and *Pleurocystites*, and edrioasteroids (*Edriophus levis*). These echinoderms are associated with abundant encrusting bryozoans (*Heterotrypa*, *Prasopora*) and the holdfasts of ptilodictyids. Brachiopods are uncommon in these assemblages but, locally, small *Platystrophia*, *Zyg-*

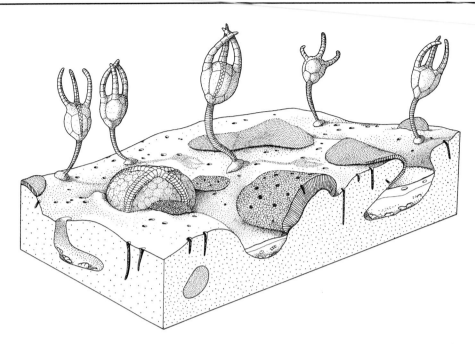

Fig. 79. Reconstruction of Middle Ordovician hardground community. Note irregular topography of hardground, abundant *Trypanites* borings and encrusting bryozoans. Echinoderms include the small, short-stemmed hybocrinid *Hybocystites eldonensis* and the edrioasteroid *Edrioaster bigsbyi* (left foreground). (After Brett & Liddell 1978.)

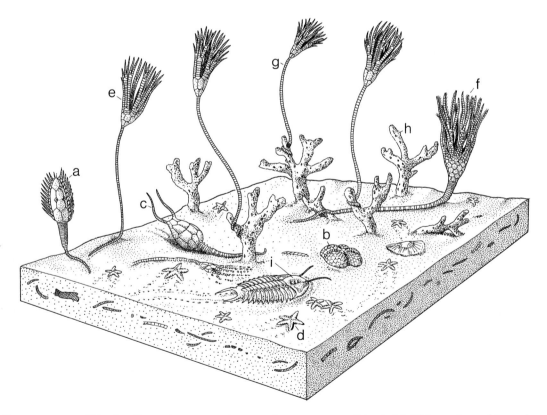

Fig. 80. Reconstruction of Verulam Formation community; area shown represents 0.1 m². (a) *Glyptocystites multiporus* (cystoid); (b) *Edrioaster bigsbyi*; (c) pleurocystid cystoid; (d) *Stenaster salteri* (asterozoan); (e) *Cupulocrinus jewetti*; (f) *Archaeocrinus pyriformis*; (g) *Ottawacrinus typus*; (h) *Batostoma* sp. (bryozoan); (i) *Ceraurus* sp. (trilobite). (After Liddell 1975a.)

locally attached clusters of the small crinoid *Hybocystites*. The echinoderms are tethered by short stems, typically only 1.5 cm in length, and they tend to be crowded into small, shale-filled pockets or irregularities on the surfaces. They are buried by grey shales. Less commonly, specimens of *Cremacrinus* are also found attached with their holdfasts, as is the paracrinoid *Amygdalocystites*. The bottom-dwelling and possibly vagrant *Pleurocystites* is present on several hardground bedding planes. Thus, despite long intervals of exposure of the hardground surfaces with little or no sediment cover, as evidenced by their strong encrustation and boring by *Trypanites* (Fig. 79), at times these surfaces were buried rapidly by silty clays. These sediments were probably transported by storms as distal gradient flow deposits from the weathered Precambrian bedrock surfaces to the north and northwest.

The details of depositional environments of the crinoid-encrusted hardgrounds from Kirkfield, Ontario,

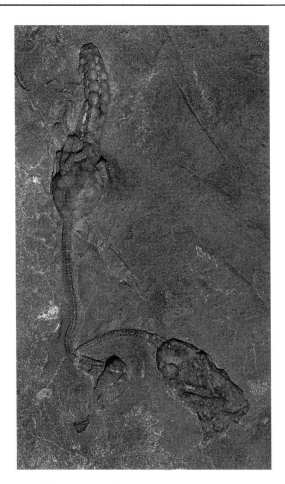

Fig. 81. *Illemocrinus amphiatus*, a small, simple cladid from the Bobcaygeon Limestone near Brechin, Ontario. Note complete short, pentameric stem and small, cemented discoidal holdfast, also stout anal sac. (Royal Ontario Museum.) ×1.5. To view this figure in colour, see the colour plate section following page xv.

ospira and inarticulate brachiopods are found attached to or encrusting upon the hardgrounds. Borings assignable to *Trypanites* typically riddle the irregular knobby surfaces of the Bobcaygeon hardgrounds.

On certain hardgrounds, echinoderms are represented only by holdfasts or attachment structures of varied cementing forms. Particularly large volcano-like structures, up to 2 cm in diameter, merit attention. These apparently belong to *Cleiocrinus* (Fig. 54), a long-stemmed camerate otherwise preserved as disarticulated crowns in shale-filled pockets on the surface. Edrioasteroids may also be preserved on many of the surfaces that are covered by grey mudstone. However, a small number of hardground surfaces are overlain by smothered bottom assemblages of the longer-stemmed crinoids and cystoids. Among the more spectacular surfaces is a persistent hardground in the Kirkfield, Brechin, region with

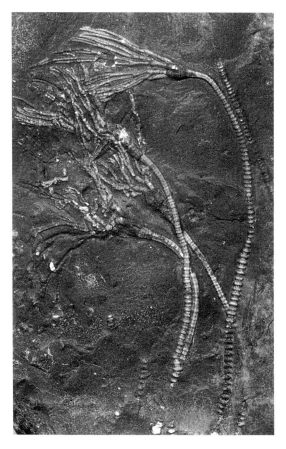

Fig. 82. *Isotomocrinus typus*, a small, slender disparid from the Bobcaygeon Limestone of the Carden Quarry, Brechin, Ontario. (Kevin Brett Collection.) ×1.3. To view this figure in colour, see the colour plate section following page xv.

Fig. 83. Hardground assemblage from the upper part of the Bobcaygeon (Kirkfield) Limestone showing the large edrioasteroid *Edrioaster bigsbyi*, the camerate crinoid *Archaeocrinus* sp. and the pinnulate arms of another, unidentified camerate. Quarry south of Brechin, Ontario. (Collection of Paleontological Research Institution, Ithaca, NY; photograph W. L. Taylor.) ×1.5.

have been discussed at some length by Brett and Liddell (1978). The carbonate sediments that constitute the hardgrounds were typically deposited very rapidly as graded storm layers of skeletal limestone and laminated to cross-laminated lime mudstone. After episodes of rapid reworking and deposition of the carbonate sediments, long intervals of non-deposition led to incipient cementation of the carbonate storm layers. Typically, the hardgrounds display relict burrow structures that were evidently produced by organisms in relatively firm, semi-consolidated carbonate silts and muds. Erosive scour of superficial layers exposed the burrow-galleried limestone on the sea bottom, where they formed firm to hard pavements suitable for the colonization of pelmatozoans attached by discs. In some cases, there is evidence for multiple generations of encrusting and boring organisms of skeletal remains attached to the hardgrounds.

The general depositional environments of the Bobcaygeon Limestone ranged from somewhat protected inner lagoon settings to relatively higher-energy shoal margin facies, characterized by closely stacked or amalgamated and graded crinoidal limestones. The hardground facies discussed here must have developed under rather quiet conditions, but occurred in areas that were subject to occasional storms that cleared the nearby skeletal beds from mud. The siliciclastic muds that buried the organisms appear to have been derived from Precambrian (Grenville) basement rocks. These were exposed with no vegetation cover, and were therefore deeply weathered, on nearby islands of the Precambrian shield. Some of the muds may also have been transported into the shallow shelf area from the active Taconic Orogenic belt that lay some 300 km to the southeast. The mechanism of transport of these muds remains enigmatic. Clearly, they represent event deposits of a different kind than those produced by a typical storm wave winnowing and wash-over from skeletal shoals. Possibly they reflect flood wash-off events in which the suspended mud was deposited relatively rapidly because of flocculation upon mixing fresh and salt water. Although Brookfield (1988) has argued that the Ordovician carbonates in the Lake Simcoe area were deposited under cool temperate climates, most authors agree that this area lay in a shallow subtropical belt perhaps 15–20° south of the palaeoequator.

Crinoidal Limestones of the Verulam Formation

A second type of echinoderm assemblage occurs more commonly on certain of the skeletal limestones of the Upper Bobcaygeon and Verulam Formations. These echinoderms are typically associated with skeletal debris layers that are composed largely of remains of the crinoids themselves. The most common of the crinoids is the cladid *Cupulocrinus jewetti*, which occurs locally in clusters of many hundreds of individuals near the base of the Verulam Formation (Fig. 84). Such beds were exceptionally well exposed in the higher layers of the now-abandoned Kirkfield Quarry. A large number of nearly complete *Cupulocrinus* specimens were obtained from thin shale partings within the skeletal limestones at one particular level. These crinoids were closely associated with a large number of the asteroid *Stenaster* and a much smaller number of other echinoderms, particularly the cladids *Cincinnaticrinus*, *Plicodendrocrinus* and *Carabocrinus*, and camerates such as *Archaeocrinus*. Ramose bryozoans are relatively common, although not as abundant as the crinoids on certain of these layers. Brachiopods are relatively scarce and are represented primarily by *Platystrophia*. Trilobites (*Ceraurus*) are moderately common.

Fig. 84. Three crowns of *Cupulocrinus jewetti*. This cladid with arm morphology similar to that of early flexible crinoids is the dominant echinoderm from the Verulam Formation. Kirkfield Quarry, Kirkfield, Ontario. (Thomas Whiteley Collection.) ×1.7.

These crinoid assemblages occur as high-density clusters on bedding planes typically associated with broad, shale-filled low spots or pockets. The crinoids appear to have been rapidly uprooted, toppled over and buried by muds. The fact that partial crowns and other remains of *Cupulocrinus* occur in the underlying skeletal limestones suggests that they were indeed a major component of the background fauna of these areas. Periodically, they were destroyed in large numbers and occasionally, when accompanied by pulses of mud, were buried almost intact. However, a remaining problem with these assemblages is the nearly complete lack of holdfasts on any of the crinoids. Moreover, none of the crinoids possess cirri that might have aided in anchoring the stems to the substrate. Rather, they all appear to have distally tapering stems. Conceivably, the crinoids may have had some flexibility in the distal portions of the stems that allowed them to coil around other objects or around each other; or the distal part of the stems formed semi-recumbent runners that resisted uprooting. However, they were also clearly vulnerable to dislodgement. This fact makes it all the more enigmatic that these crinoids seem to be closely associated with the well-washed crinoidal limestones that we normally associate with near-wave-base, high-energy conditions. Nonetheless, similar associations, also typically dominated by *Cupulocrinus*, have been documented in a number of other high-energy crinoidal limestones. Obviously, the crinoids had the ability to cope with such environments much of the time. Nonetheless, the intermittent strong storms dislodged and killed a large number of individuals in short periods of time. The crinoid remains, once toppled, were rapidly covered either with crinoid fragments, as in the case of certain specimens found intact within crinoidal limestones, or by siliciclastic muds. The latter typically served as substrates for the colonization of later generations of horizontal, sediment-feeding burrowers. These burrows commonly occur in masses within the burial muds overlying the crinoids. The burrow fillings are typically accentuated by early diagenetic cementation.

IMPORTANT COLLECTIONS IN NORTH AMERICA

Museum of Paleontology, University of Michigan, Ann Arbor, Michigan, USA

Royal Ontario Museum, Toronto, Canada

8

Upper Ordovician of the Cincinnati, Ohio, Area, USA

WILLIAM I. AUSICH

THE 'PUBLISHING AMATEURS'

The Upper Ordovician of North America is the Cincinnatian Series, named for the richly fossiliferous beds in the Ohio–Indiana–Kentucky tri-state area surrounding Cincinnati, Ohio. This area is near or beyond the southern extent of Pleistocene glaciation, so bedrock is exposed or covered only thinly with glacial sediments. Hillsides, stream beds and road cuts expose the interbedded shale and limestone Cincinnatian strata, and fossils virtually tumble from rocks into collecting bags. This physical backdrop of abundant fossils and layered strata has been an intellectual seed for more than a century and a half. Many now-famous palaeontologists grew up in Cincinnati, including R. S. Bassler, J. M. Nickles, C. Schuchert and E. O. Ulrich. Caster (1981) referred to early Cincinnati geologists as the 'publishing amateurs' or in the case of those just listed as the 'amateurs-turned-professionals'.

Bryozoans and brachiopods are the most prolific Cincinnatian fossils, but trilobites and echinoderms are the prizes of these strata. Crinoids are the most abundant echinoderms of the Cincinnatian, but asteroids, cyclocystoids, edrioasteroids, rhombiferans and stylophorans are also known from these beds.

STRATIGRAPHIC CYCLES

At an estimated 20°S palaeolatitude, Cincinnatian sediments were deposited in a tropical, shallow-water epicontinental setting, approximately 440 million years before present. The composite outcrop section in Ohio consists of more than 300 m of interbedded fossiliferous limestones and shales. Eustatic sea level fluctuations and storms were apparently the dominant physical factors controlling sedimentation, and this produced a hierarchy of cyclical strata.

Upper Ordovician strata (Cincinnatian) are dominated by four shallowing-upward sequences (Fox 1962; Holland 1993; Davis & Cuffey in press). The base of the Cincinnatian begins with the Kope Formation (Edenian Stage) and facies shallow upward to the Bellevue Formation in the Middle of the Maysvillian Stage (Fig. 85). The deepest Kope facies still appears to have been within storm wave base. Deposits of a second short cycle comprise the remainder of the Maysvillian, and a third cycle results in the lowest part of the Richmondian Stage. The sequence is capped by the remainder of the Richmondian, the fourth cycle. It begins with the deepwater Waynesville Formation, which gradually shallows to the unfossiliferous, mud-cracked Elkhorn Formation.

Hans Hess, William I. Ausich, Carlton E. Brett, and Michael J. Simms, eds., *Fossil Crinoids*. © 1999 Cambridge University Press. All rights reserved. Printed in the United States of America.

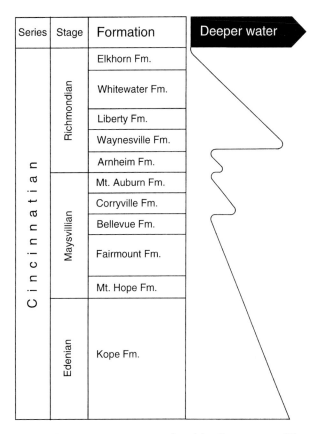

Fig. 85. Representative stratigraphy of the Cincinnatian (Upper Ordovician) outcrop belt in southwestern Ohio, with interpreted relative changes in sea level. (Modified from Davis & Cuffey in press.)

The termination of Cincinnatian deposition resulted from a global regression of sea level due to the latest Ordovician continental glaciations.

The Cincinnatian sequence is dominated by tempestites, fewer in the deep-water settings and many more in the shallow-water environments. Furthermore, tempestites are bundled into repeated packages, producing cyclic stratigraphic patterns (Tobin 1986; Jeanette & Pryor 1993).

CINCINNATIAN CRINOIDS

In comparison with many Middle Ordovician echinoderm assemblages, the Upper Ordovician Cincinnatian is a rather odd fauna. In the Cincinnatian, blastozoans are very rare. Furthermore, some of the characteristic Middle Ordovician crinoids, such as calceocrinids and hybocrinids, are absent and the generic diversity of diplobathrid camerates is much reduced. In total, approximately 37 species assigned to 20 genera are currently recognized from Cincinnatian strata. The most common crinoids are the diplobathrid *Gaurocrinus*; the monobathrids *Glyptocrinus*, *Pycnocrinus* (Figs. 90, 94, 95) and *Xenocrinus* (Fig. 25, 92); the disparids *Cincinnaticrinus* (Figs. 86, 89), *Ectenocrinus* (Fig. 26) and *Iocrinus* (Fig. 74); and the cladids *Cupulocrinus* (Fig. 93) and *Plicodendrocrinus*. The multi-plated disparid lichenocrinid-type holdfast, which cements to shells and hardgrounds, is also quite common (Figs. 87–89).

Fig. 86. *Cincinnaticrinus pentagonus*. Richmondian from southwestern Ohio. (From Ausich 1996b.) ×4.

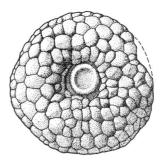

Fig. 88. Lichenocrinid-type holdfast with stem insertion, based on a specimen from the Cincinnatian, southwestern Ohio. ×4.

Fig. 87. Lichenocrinid-type holdfasts cemented on a brachiopod shell. Cincinnatian, southwestern Ohio. (From Ausich 1996b.) ×2.

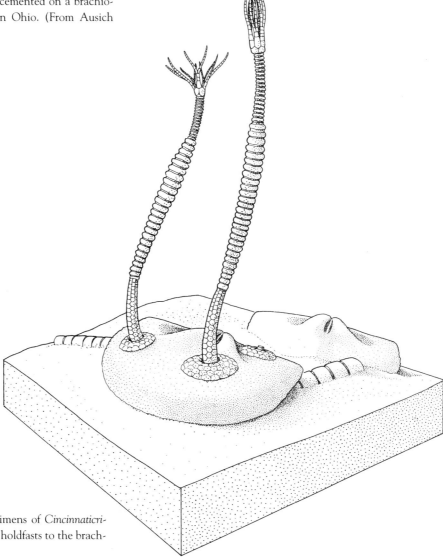

Fig. 89. Reconstructions of two specimens of *Cincinnaticrinus*, attached with their lichenocrinid holdfasts to the brachiopod *Rafinesquina*. ×1.

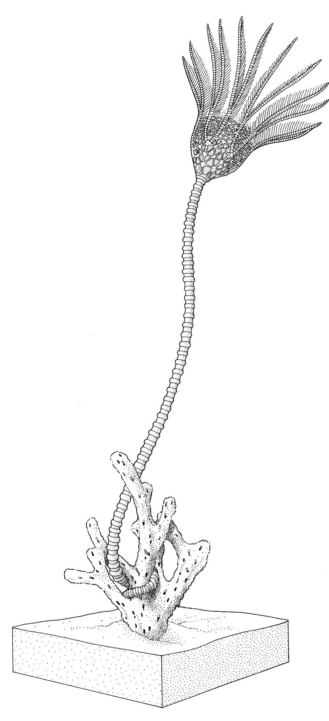

Fig. 90. Reconstruction of *Pycnocrinus* attached to a bryozoan. ×0.75.

Stalked echinoderms tend to be more common in the intermediate-depth facies, such as the shallower parts of the Kope, Fairmount, Corryville, Waynesville and Liberty Formations. Crinoid crowns were typically preserved on smothered bottoms, as transported elements of turbidites (Meyer 1981; Schumacher & Ausich 1985; Schumacher & Meyer 1986) or as large log jams of crinoid stems and crowns that were presumably swept into channels. These preservational modes were probably the result of episodic, catastrophic storm deposition (Meyer 1981). Cincinnatian crinoid assemblages are typically rather low in diversity. It is not unusual for monospecific stands to occur, and assemblages with more than five species are rare. Crinoids may occur in densities as high as 400 per square metre.

INTERESTING DISPARIDS

Two Cincinnatian disparids, *Cincinnaticrinus* and *Ohiocrinus*, are morphologically fascinating. *Cincinnaticrinus* (formerly *Heterocrinus*; see Warn & Strimple 1977) is a very common Cincinnatian crinoid, so it is very well known. It has an exceedingly small crown. In fact, the crown seems to be an afterthought on the stem rather than the focal point of the animal. Cincinnatian species of *Cincinnaticrinus* are *C. varibrachialus* (Edenian to Maysvillian) and *C. pentagonus* (Maysvillian to Richmondian) (Fig. 86).

Cincinnaticrinus has a lichenocrinid-type holdfast (Fig. 89). This holdfast is a low, convex, multi-plated structure with a central depression for stem attachment (Fig. 88). It commonly cemented to brachiopod shells, but it also attached to other skeletal debris, including crinoid columns, hardgrounds and lithified nodules. A progression of morphologically dissimilar columnals comprise the *Cincinnaticrinus* stem. The oldest (attached to the holdfast) columnals are pentameric with vertical ranges of pentameres offset laterally and joined along a zig-zag suture (Fig. 89). This arrangement gradually changes to one with offset pentameres and straight sutures, and then to one with pentameres aligned laterally with straight sutures. After the crown was mature, growth of the column was complex and is only partially understood. Apparently, pentagonal, pentameric columnals were added. Each of these enlarged in height and width and became circular. Subsequently pentagonal, pentameric columnals may have been added between these larger circular columnals; and these intercalated columnals also grew, becoming indistinguishable from the other, larger columnals.

With the arms closed, the crown, including the aboral cup with its high radial circlet, may be narrower than

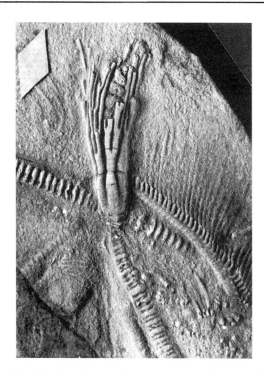

Fig. 91. *Ohiocrinus brauni*. Cincinnatian, southwestern Ohio. Note spiral anal sac. (Reprinted by permission of the National Museum of Natural History, Washington, D.C.) ×3.

the stem (Fig. 86). The anal sac in *Cincinnaticrinus* is short, narrow and arm-like in appearance. The arms divide isotomously on the fourth or fifth primibrachial and then heterotomously several times. Adult *Cincinnaticrinus* crowns attached to the holdfast are unknown. This has suggested to some authors that the adult *Cincinnaticrinus* broke free from the holdfast (Warn & Strimple 1977), but it is difficult to envisage how this crinoid survived without attachment, because it lacked cirri and shows no evidence of being distally coiled as in *Pycnocrinus dyeri*, discussed in the next section.

In total, the crown seems disproportionately small for the stem, the filtration fan was quite small and the body cavity in the cup was minute, yet this was a highly successful Cincinnatian crinoid. Perhaps its small size and encrusting habit arose paedomorphically, enabling it to adapt to an opportunistic mode of life.

Ohiocrinus, another Cincinnatian disparid, is very rare. The aboral cup and crown are more normally proportioned than in *Cincinnaticrinus*, but *Ohiocrinus* has a unique anal sac. This structure is composed of small plates arranged into a spiral structure (Fig. 91). Presumably this housed a greatly enlarged digestive system, but the reason for such an adaptation remains a mystery.

CAMERATES AND PLATYCERATID GASTROPODS

Pycnocrinus and *Glyptocrinus*, especially *Pycnocrinus dyeri*, are the most common Cincinnatian camerates. With a high calyx, many fixed brachials, many interradial plates and several biserially pinnulate arms, these are typical Ordovician camerates.

Pycnocrinus dyeri has a distally coiled stem that was used as an attachment around bryozoans or other erect crinoid stems (Fig. 90). The column is composed of circular columnals with nodals separated by varying numbers of internodals. The calyx has a high bowl shape and is distinguished by a median ridge along each ray and a star-shaped ridge pattern on interradial plates (Figs. 94, 95). The second primibrachial is axillary, but the fixed brachials do not divide again. The arms become free after six or seven secundibrachials, and after another three or four free secundibrachials the arms divide again. Fixed pinnules are incorporated into the interradial areas, and the free arms are biserial with long, delicate pinnules.

These crinoids provide one of the best early glimpses at the special biotic interaction between crinoids and platyceratid gastropods, a relationship that lasted throughout the Palaeozoic. Platyceratids are a morphologically diverse group of archaeogastropods that are most commonly preserved attached to a crinoid tegmen. Because these gastropods attach directly over the anus, platyceratids have traditionally been considered to be

Fig. 92. *Xenocrinus baeri*. Richmondian, southwestern Ohio. (From Ausich 1996b.) ×1.5.

coprophagous commensals (Bowsher 1955; Breimer & Lane 1978). In some Mississippian examples, it is clear that the gastropods were permanently attached over the anus. Alternative explanations of this association, such as active drilling by the gastropods, are now under consideration (Baumiller 1990b; Morris & Felton 1993). Perhaps different platyceratids fed in different ways.

Cyclonema and *Naticonema* are the two Cincinnatian platyceratids, with *Cyclonema* (Fig. 95) the most common (Morris & Felton 1993). Unlike younger platyceratids, these gastropods were not permanently affixed to the tegmen, but apparently moved about (Bowsher 1955), so the crinoid–platyceratid relationship was less specialized during the Ordovician than later.

IMPORTANT COLLECTIONS IN THE UNITED STATES

Cincinnati Museum of Natural History, Cincinnati, Ohio
Field Museum of Natural History, Chicago, Illinois
Miami University, Oxford, Ohio
National Museum of Natural History, Smithsonian Institution, Washington, D.C.

Fig. 93. *Cupulocrinus polydactylus*. Richmondian, southwestern Ohio. (From Ausich 1996b.) ×1.

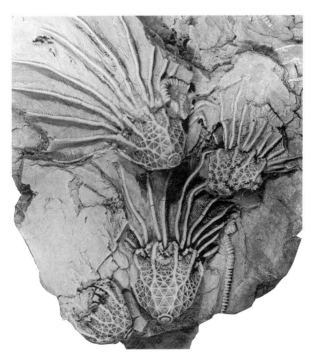

Fig. 94. *Pycnocrinus dyeri*. Maysvillian, southwestern Ohio. (From Ausich 1996b.) ×0.85.

Fig. 95. *Pycnocrinus dyeri* with the platyceratid gastropod *Cyclonema* sp. positioned on the tegmen. Cincinnatian, southwestern Ohio. (From Bowsher 1955; reprinted by permission.) ×1.5.

Silurian of Gotland, Sweden

HANS HESS

STROMATOPOROID AND CORAL REEFS IN THE BALTIC SEA

The Swedish Island of Gotland is well known to vacationers, many of whom are also fossil collectors. They cannot miss the numerous fragments of crinoid stems spread over much of this picturesque island. The Silurian beds of Gotland, a succession of 13 stratigraphic units, range in age from Late Llandovery to Late Ludlow (around 420 million years before present) and reach a thickness of about 500 m (Fig. 96). The oldest rocks, the Lower Visby beds, are along the northwestern coast. They are composed of soft, bluish grey mudstones and contain nodules, lenses and layers of marly limestones. To the south and southeast, successively younger strata, including three elongated reef belts separated by flat mudstone areas, were laid down in a shallowing sea. The limestone reefs, predominantly composed of stromatoporoids, are surrounded by bedded, bioclastic sediments consisting largely of crinoid remains. Reefs started to grow as bioherms during the Early Wenlockian, the time of deposition of the Högklint beds. Shallowing seas and a reduced supply of terrigenous material favoured the growth of extensive reefs, which are exposed in cliffs on the northwestern coast of the island. The lower part of a typical Högklint reef was initiated by tabulate corals growing as a patch reef in deeper water on cross-bedded lenses of crinoid remains. Further reef growth was dominated by laminar stromatoporoids and topped by dome-shaped stromatoporoids, leading to a pronounced vertical profile of the structure (Kershaw 1993). These bioherms are comparable to modern reefs, their build-up being the result of the interplay between sedimentation and the growth of organisms. Reef growth continued into later Wenlock and Ludlow times. In contrast, the Hemse reefs in southeastern Gotland have low profiles and are composed entirely of stromatoporoids that grew in shallow water. Their substrate was again formed by remains of crinoids, deposited presumably by storms. The Sundre Beds, the youngest unit on Gotland, comprise mostly limestones, including stromatoporoid bioherms. Collecting is unrewarding in these beds; only two species of crinoid have been identified. Gotland contains the richest crinoid assemblage from the Silurian, with 193 species assigned to 55 genera. Most of the well-preserved fossils were found during earlier quarrying. In the few quarries still open today, manual work has been replaced by mechanical operations, making fossil collection unproductive. Ubaghs (1956a,b, 1958) devoted three papers to camerates. The study of all Gotland crinoids has been brought up to modern standards by Christina Franzén (1983).

REEF FLANKS COLONIZED BY CRINOIDS

Crinoid populations vary to some extent in the stratigraphic sequence of Gotland. The Visby Beds (Late

Hans Hess, William I. Ausich, Carlton E. Brett, and Michael J. Simms, eds., *Fossil Crinoids*. © 1999 Cambridge University Press. All rights reserved. Printed in the United States of America.

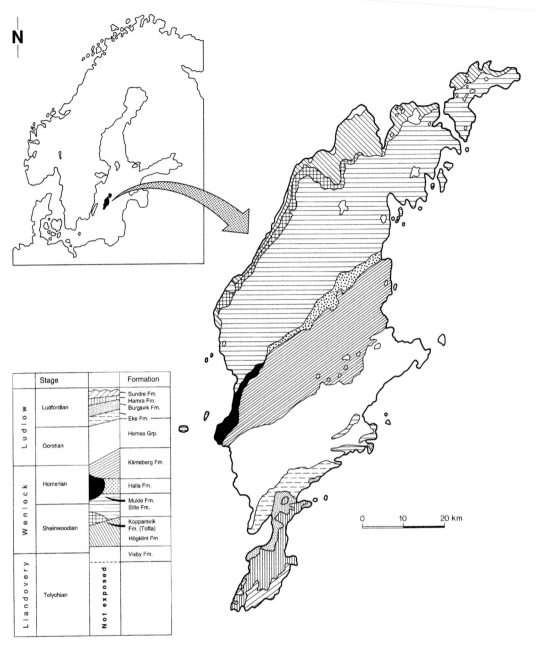

Fig. 96. Geological map and stratigraphy of Gotland. (Redrawn from Kershaw 1993.)

Llandoverian and Early Wenlockian), deposited in deeper waters at the beginning of the sequence, have yielded mostly disparids and cladids. Camerates are absent from these marlstone deposits. The Visby Bed crinoids, like those of other soft-bottom communities in deeper water, are rather small and delicate. They were anchored with branched, root-like cirri (radices) or attached with discs to hard objects on the bottom. Most of these crinoids are well preserved, but disarticulated, indicating slow sedimentation rates. The Visby marlstones were followed by the Högklint and Slite Beds. The flanks of Högklint reefs contain an abundant and diverse crinoid fauna. On the neighbouring sea floor, conditions for crinoids were less favourable. In stratified limestones and marly limestones, which were deposited at greater distances from the reefs, remains of crinoids are scattered through the sediment together with other marine invertebrates, such as brachiopods, bryozoans, corals and rare molluscs (Manten 1971). The Högklint reefs are highly fossiliferous, and well-preserved cups and even crowns occur in marly pockets within the reef structure. However, holdfasts, many of them in growth position, are considerably more common than crowns or cups. The crinoid fauna is dominated by the camerates *Calliocrinus*, *Dimerocrinites* and *Eucalyptocrinites*. In the stratigraphically higher Slite Beds, the crinoid fauna is still more diverse. Intact crowns with part of the stem occur in coarse-grained bioclastic debris composed of worn fragments of bryozoans, corals and stromatoporoids (Fig. 97). In these sediments, as in the higher strata, *Eucalyptocrinites* and the cladids *Crotalocrinites* and *Enallocrinus* are the dominant genera (for a complete listing, see Franzén 1983). Crinoid diversity diminishes in the succeeding beds. However, the best-known crinoid assemblage from Gotland is the spectacular 'När slab' (Fig. 98), a bioclastic limestone lens from the Ludlovian Eke Beds.

The Silurian rocks of Gotland were deposited in a shallow epicontinental sea with a water depth never exceeding 200 m. Crinoidal limestones formed at depths of from 5 to 50 m. Dense stands of crinoids must have covered the flanks of the larger reefs where the crinoids grew on their own debris. These faunas are dominated by camerates, but the two cladid genera mentioned earlier are also commonly found. Turbulence was high in such surroundings, leading, as a rule, to rapid disarticulation. In rare cases, more or less complete specimens were also preserved.

On a surface of approximately 1 by 1.2 m, the När slab contains 260 crinoid specimens belonging to 4 species (Franzén 1982). With the exception of three specimens of the flexible *Haereticotaxocrinus asper*, they are all monobathrid camerates. The crinoids are current-oriented and lie on a coarse-grained bed of sorted skeletal gravel made up mostly of worn *Coenites* (tabulate coral) fragments, but also with crinoid ossicles. Other fossils are absent. The crowns and stems were originally covered by *Coenites* fragments and probably also by silt. This layer, which was removed during preparation, appears to have been deposited by a storm; it rapidly buried the crinoids and protected them from scavengers. The absence of holdfasts, the orientation of the crinoids and the occurrence of crowns torn from the stem indicate that the crinoids were transported for a short distance by the current before burial. The two *Carpocrinus* species include juvenile and adult individuals, but *Desmidocrinus* specimens are adults; the reason for this difference is unknown. Stem length varies in the crinoids found on the När slab; in the camerates *Carpocrinus* and *Desmidocrinus* the stem was about 20 cm, whereas the longest stem from the flexible *Haereticotaxocrinus* measures 65 cm. The different lengths of the stems indicate that the animals were tiered, collecting food at different levels above the sea floor.

A WEEPING WILLOW CRINOID

Among the camerates from Gotland, *Barrandeocrinus sceptrum* is the most unusual. This crinoid, which is restricted to the Slite Beds, was redescribed by Ubaghs

Fig. 97. Coarse bioclastic limestone of Slite Beds (Wenlock) from Follingbo with crown of the monobathrid camerate *Abacocrinus tessellatus* (left). Naturhistoriska Riksmuseet, Stockholm. (Courtesy C. Franzén.) ×1.

Fig. 98. *(On facing page)* Part of upper surface of the När slab with the conspicuous, multi-armed monobathrid camerate *Desmidocrinus pentadactylus* (bottom of photo) and the smaller 10-armed monobathrids *Carpocrinus angelini* (with bowl-shaped calyx, stout stem and arms, which are widest in the middle) and *Carpocrinus petilus* (with conical calyx, slender stem and arms). The two long stems belong to the flexible *Haereticotaxocrinus asper* (crowns outside picture). *Coenites* fragments that originally covered the slab were partly removed to expose the crinoids. Eke Beds (Ludlow). (Naturhistoriska Riksmuseet, Stockholm; courtesy C. Franzén.) ×0.9.

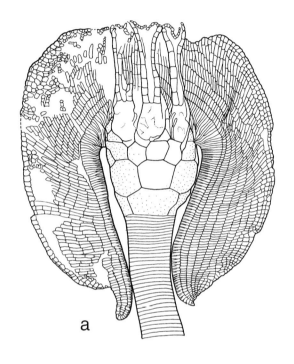

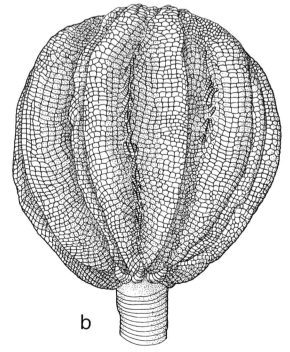

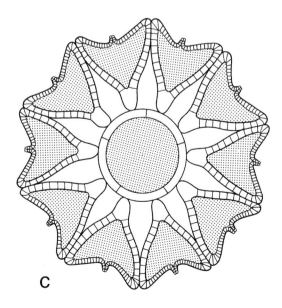

Fig. 99. *Barrandeocrinus sceptrum*. Slite Beds (Wenlock), Follingbo. (a) Partly dissected crown (original in the Naturhistoriska Riksmuseet, Stockholm) with three of the pendent arms removed to show the structure of arms, cup and uppermost stem; (b) reconstructed crown in lateral view, with recumbent arms, their pavement of pinnules completely covering the cup; (c) cross section of crown showing cup surrounded by the 10 brachial chambers, formed by rigid portions of the pinnules. (Redrawn from Ubaghs 1953.) ×2.

(1956a). Its 10 arms are permanently pendent and enfolded, giving the crown a hood-like appearance. The very short brachials carry close-set pinnules that are folded at midlength and attached together to form an external protective pavement enclosing 10 brachial chambers (Fig. 99). The most distal pinnules are free and would have allowed sea water to enter the chambers. It is thought that the water was expelled through a common aperture at the tip of the crown. This structure suggests that *Barrandeocrinus* may have created its own current for feeding. The longest preserved stem, measuring 14 cm, is composed of massive, circular columnals. Towards the distal end, radices were attached, as proved by the corresponding scars, and would have fixed the animal to the bottom or to some part of the reef structure. *Barrandeocrinus sceptrum* may have lived in sheltered pockets within the reef where currents were sluggish (Franzén 1983). It is not surprising that the reefs comprised a number of ecological niches, offering possibilities for crinoids with widely different morphologies and ways of feeding.

MYELODACTYLUS: A COIL ON THE BOTTOM

Of special interest are remains of the disparid *Myelodactylus* (formerly *Herpetocrinus*) from the Lower Visby Beds, discussed by Donovan and Franzén-Bengtson (1988). The stem of these crinoids is composed of bilateral columnals and has a double curvature, enabling the crown to fold back against the stem (Fig. 100). The small and slender crown is concealed by two rows of radices (radicular cirri) that are attached to the margins of the inner surface in the distal part of the coiled stem. Many theories have been put forward for the lifestyle of these peculiar crinoids. They have been thought to be free-moving, swimming, pelagic, similar to the ammonites with their coiled shells, or to have used their radicular cirri as legs for crawling or swimming. Based on the microstructure of the columnals, Donovan and Franzén-Bengtson (1988) could find no evidence in *Myelodactylus fletcheri* for the presence of contractile fibres, which would have allowed active movements of the stem. In addition, the shape of the columnals indicates coiling for life. Therefore, these authors suggested that *M. fletcheri* lived flat on its side on the sea floor.

Fig. 100. *Myelodactylus fletcheri*, lateral view of crown and column. Note long cirri in the middle (upper) part of the stem, partly covering the crown; the double row of radicular cirri is clearly visible on the distal part of the stem (right). Slite beds (Wenlock), Burgen in Endre. (Naturhistoriska Riksmuseet, Stockholm; from Bather 1893.) ×2.

The lower fringe of radicular cirri protected the crown from direct contact with the sediment, and the upper radices formed a covering roof. The curvature of the animal would have given it a fairly stable position on an unconsolidated substrate, preventing sinking into the mud. The crown could not have been raised, so that this crinoid would have had the unique properties necessary to feed with a closed crown enveloped by fringes of cirri. Feeding is thought to have been made possible by close contact with a surface layer of nutrients on the bottom, including the absorption of dissolved organic matter through the epidermis. In the absence of a modern analogue, this theory is purely conjectural.

IMPORTANT COLLECTION IN SWEDEN

Naturhistoriska Riksmuseet, Section for Paleozoology, Stockholm. The spectacular När slab is exhibited at the entrance.

10 Middle Silurian Rochester Shale of Western New York, USA, and Southern Ontario, Canada

WENDY L. TAYLOR AND CARLTON E. BRETT

A LONG HISTORY OF COLLECTING

Extraordinary assemblages of echinoderms and trilobites are preserved in the Rochester Shale of western New York and southern Ontario, Canada. Aside from being one of the first formally designated formations in North America, the Rochester Shale is also noted for its exceptional fossils. Outcrops around the city of Lockport, New York, were famous for their well-preserved crinoids and cystoids as early as the 1820s. Excavations were carried out by the Lockport physician Eugene Ringueberg (1890) and later continued by Frederick Braun (1911, 1914), who was employed by Frank Springer. The specimens are now housed in the Springer Collection of the National Museum of Natural History. These sites have produced an astonishing array of intact crinoids, cystoids, asterozoans and edrioasteroids. More recent study by the authors of this chapter at a commercial quarry run by the Caleb family in Middleport, New York (Fig. 101), and several other localities continues to produce rare echinoderms and spectacular trilobites.

The Rochester Shale, Upper Clinton Group (Wenlockian), crops out along the east–west-trending Niagara Escarpment in western New York and southern Ontario and is composed of Middle Silurian age (Late Sheinwoodian to Homerian) calcareous shales and limestones, deposited about 415 million years before present (Fig. 101). The major intervals of echinoderm *Lagerstätten* (*Homocrinus* Beds) occur in the Lower Rochester Shale or Lewiston Member, with only a few sporadic occurrences of well-preserved fossils noted in the upper, dominantly unfossiliferous Burleigh Hill Member (Brett & Eckert 1982; Brett 1983; Taylor & Brett 1996; Brett & Taylor 1997).

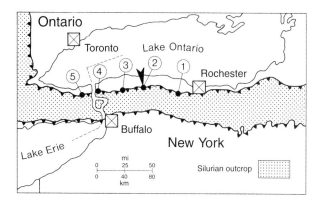

Fig. 101. Crinoid *Lagerstätten* horizons (*Homocrinus* Beds) from the Rochester Shale occur in western New York State and southern Ontario, Canada, along the Silurian outcrop belt (stippled) and bounded by the Niagara (north) and Onondaga (south) Escarpments, shown as toothed lines. Locations are as follows: (1) Wegman's Plaza, Brockport, N.Y.; (2) Jeddo Creek tributary, Middleport, N.Y. (arrow); (3) Lockport, N.Y.; (4) Niagara Gorge, Lewiston, N.Y.; (5) Welland Canal, Thorold, Ontario, Canada. (After Taylor & Brett 1996.)

Hans Hess, William I. Ausich, Carlton E. Brett, and Michael J. Simms, eds., *Fossil Crinoids*. © 1999 Cambridge University Press. All rights reserved. Printed in the United States of America.

BARREN SHALES WITH HORIZONS OF EXQUISITE FOSSILS

The Rochester Shale consists of up to 40–45 m of medium grey calcareous shale and mudstone with interbedded limestones. Fossil content is variable, ranging from nearly barren shales and limestones to highly fossiliferous skeletal limestones with pelmatozoans and bryozoan debris.

Sedimentological and taphonomic evidence indicates that benthic communities were rapidly buried by storm-generated siliciclastic muds and carbonate silts. The resulting rapid burial produced horizons with well-preserved invertebrate fossils, intact crinoids, cystoids and trilobites. In addition to the taphonomic evidence for rapid burial, scanning electron microscopy of the burial muds indicates that these sediments were deposited as aggregate particles of flocculated silt and clay (O'Brien et al. 1994). The mudstones generally lack internal lamination and good fissility and, in most cases, overlie shell accumulations of variable thickness. Carbonate silt deposits show a characteristic set of sedimentary structures. These indicate rapid deposition, including planar- to cross-lamination or small-scale hummocky cross-stratification, normal grading, basal scour marks and obliquely embedded pelmatozoans and trilobites.

Fossils present in the Rochester mudstones vary greatly in their preservation, ranging from totally disarticulated long-term background accumulations during periods of slow sedimentation, to well-preserved, smothered bottom assemblages that exhibit signs of slight disturbance and current alignment (Fig. 106). Calcitic skeletons are recrystallized with a variable degree of flattening. The *Lagerstätten*, several so-called *Homocrinus* Beds, are outstanding for their intact crinoids, cystoids and edrioasteroids, frequently preserved in life position (Fig. 102). Camerate crinoids (*Dimerocrinites*, *Macrostylocrinus*), disparids (*Homocrinus*, *Crinobrachiatus*), cladids (*Dendrocrinus*) and the rhombiferan cystoid *Caryocrinites* occur within the lower 1–2 cm of mudstone, commonly with stems attached to the underlying shell debris. In certain horizons, the flexible *Asaphocrinus* is abundant and anchored to the brachiopod *Striispirifer* or to a branching (ramose) bryozoan colony (Fig. 103). Embedded in thicker burial muds are the remains of free-living echinoderms such as asteroids and ophiuroids, commonly preserved in contorted, possibly escape positions. Complete trilobites (*Dalmanites*, *Bumastus*, *Arctinurus*) also occur within the mudstones in clusters, commonly inverted with the dorsal surface down. In certain horizons, the large lichiid *Arctinurus* may carry extraordinary microcommunities of encrusting brachiopods, bryozoans and worm tubes on the dorsal shield.

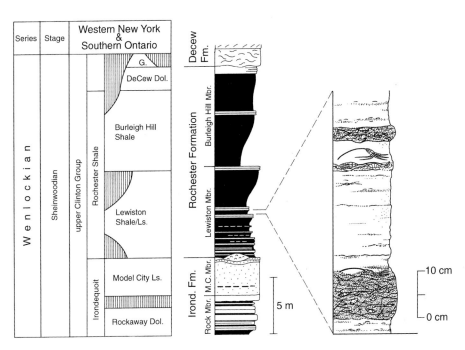

Fig. 102. Chronostratigraphic and generalized stratigraphic column of the Rochester Shale. Key: G, Glenmark Shale; Irond., Irondequoit; Rock, Rockway; M.C. Model City; Ls., limestone; Dol., dolomite. Note the smothered crinoid assemblages in the Lewiston Member. Black indicates dark grey shale; ledges indicate tabular thin carbonate layers; stipple indicates shell-rich units.

Fig. 103. Two intact individuals (adult and juvenile) of the flexible crinoid *Asaphocrinus ornatus* attached to a branching bryozoan. Note the brachiopod *Striispirifer* (left) and string-like stems of *Homocrinus*. Homocrinus Beds, Lewiston Member of the Rochester Shale, Lockport, N.Y. (National Museum of Natural History, Washington, D.C.) ×1.

The preservation of free-living ophiuroids and trilobites, as well as of extremely fragile, sedentary crinoids and cystoids, is dramatic evidence of rapid and deep burial of benthic communities. The burial layers themselves are composed of fine-grained, siliciclastic muds and carbonate silts that were transported into the basin by way of storm-generated sediment flows. Rare evidence of the activity of burrowing organisms can be seen in the occasional disruption of the delicate arms of crinoids, as noted in the arms of the camerate *Macrostylocrinus* (Fig. 104).

Fig. 104. Complete crown of the camerate *Macrostylocrinus ornatus*. Note the bioturbated arms on upper right (arrow). Homocrinus Beds, Lewiston Member of the Rochester Shale, Lockport, N.Y. (National Museum of Natural History, Washington, D.C.) ×1.6.

A MUD-BOTTOM FAUNA OF BRACHIOPODS, BRYOZOANS, TRILOBITES AND ECHINODERMS

More than 200 species of invertebrate fossils are known from the Rochester Shale in New York and southern Ontario. A significant proportion of species are represented by brachiopods, including the spiriferids *Striispirifer* and *Eospirifer*, the rhynchonellids *Stegerhynchus* and *Rhynchotetra*, the strophomenids *Leptaena*, *Coolinia* and *Amphistrophia* and the orthids *Dalejina* (now *Mendacella*) and *Resserella*. The ramose bryozoan *Cheilotrypa* and the fenestrate and encrusting forms *Fenestella* and *Lichenalia*, respectively, are also abundant in some horizons, and may form local thickets associated with the pelmatozoans *Caryocrinites*, *Crinobrachiatus* and *Stephanocrinus*. The trilobites *Dalmanites*, *Calymene*, *Bumastus* and *Arctinurus* commonly occur as disarticulated material; however, in rapid-burial beds they may be exceptionally well preserved.

A variety of echinoderm groups are present and include crinoids, cystoids, asterozoans and edrioasteroids. Crinoids are represented by a diversity of forms, including the camerates *Macrostylocrinus*, *Dimerocrinites*, *Saccocrinus* and *Eucalyptocrinites* (Figs. 105, 106). Cladids are represented by the long-stemmed *Dendrocrinus* with its tall anal sac and highly branched, non-pinnulate arms. Disparids occur as the tiny *Homocrinus*, recumbent calceocrinids (*Calceocrinus*) and the coiled, bilaterally symmetric *Crinobrachiatus*. Flexible crinoids are also present, with *Asaphocrinus*, *Icthyocrinus* and *Lecanocrinus*. In addition, the rhombiferan cystoid *Caryocrinites* is very common in most beds both as disarticulated material and as intact, even-rooted specimens, which in many cases formed clusters associated with bryozoan thickets. The edrioasteroid *Hemicystites* occurs in certain burial layers (*Homocrinus* Beds) attached to *Striispirifer* valves, and recent excavations have produced rare stalked edrioasteroids, part of a group of extinct multiplated echinoderms popularly known as seated stars.

The Rochester Shale communities were dominated by suspension-feeding organisms such as brachiopods, bryozoans, crinoids and cystoids. These tiered or vertically stratified communities were composed of organisms feeding at different levels. Low levels, several centimeters above the bottom, were occupied by brachiopods, calceocrinids, *Asaphocrinus* and encrusting bryozoans. Branching bryozoans, most crinoids and cystoids exploited higher levels, from 5 cm to more than 1 m above the sea floor; they include *Dendrocrinus*, *Dimerocrinites*, *Icthyocrinus* and *Saccocrinus*. Several species (*Caryocrinites*, *Crinobrachiatus*) favoured the bryozoan thickets, commonly perching on the bryozoan colonies (Fig. 107).

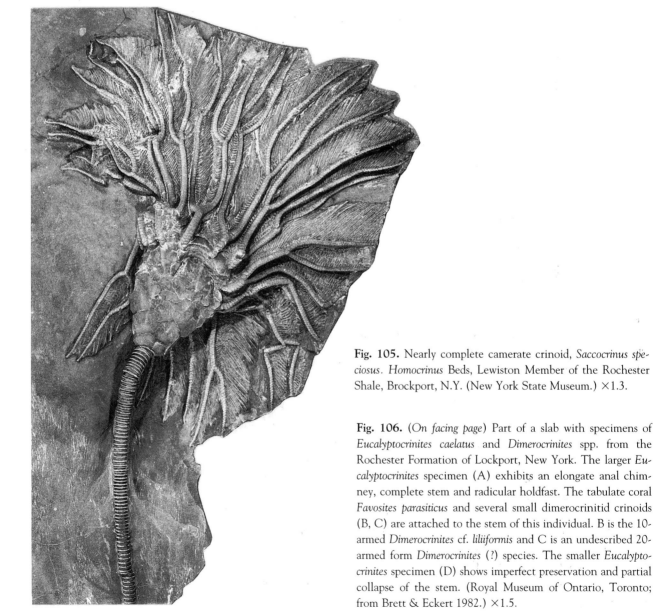

Fig. 105. Nearly complete camerate crinoid, *Saccocrinus speciosus*. *Homocrinus* Beds, Lewiston Member of the Rochester Shale, Brockport, N.Y. (New York State Museum.) ×1.3.

Fig. 106. (*On facing page*) Part of a slab with specimens of *Eucalyptocrinites caelatus* and *Dimerocrinites* spp. from the Rochester Formation of Lockport, New York. The larger *Eucalyptocrinites* specimen (A) exhibits an elongate anal chimney, complete stem and radicular holdfast. The tabulate coral *Favosites parasiticus* and several small dimerocrinitid crinoids (B, C) are attached to the stem of this individual. B is the 10-armed *Dimerocrinites* cf. *liliiformis* and C is an undescribed 20-armed form *Dimerocrinites* (?) species. The smaller *Eucalyptocrinites* specimen (D) shows imperfect preservation and partial collapse of the stem. (Royal Museum of Ontario, Toronto; from Brett & Eckert 1982.) ×1.5.

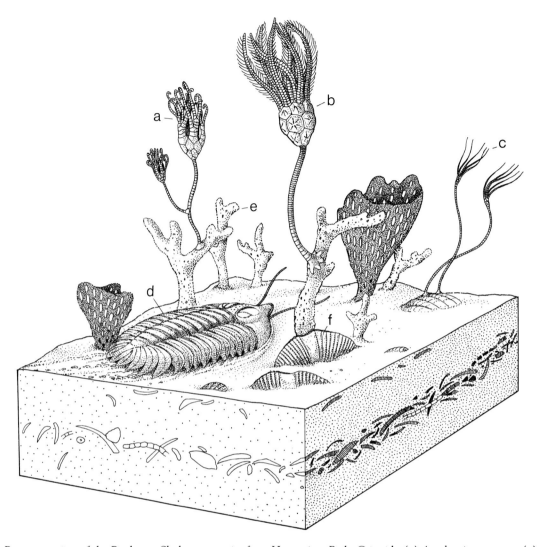

Fig. 107. Reconstruction of the Rochester Shale community from *Homocrinus* Beds. Crinoids: (a) *Asaphocrinus ornatus*; (c) *Homocrinus parvus*. Rhombiferan cystoid: (b) *Caryocrinites ornatus*. Also figured: (d) *Arctinurus* (trilobite); (e) *Cheilotrypa* (ramose bryozoan); (f) *Striispirifer* (brachiopod). A fenestrate bryozoan is between crinoids b and c.

Sediments of the Rochester Shale accumulated in a shallow, subtropical, muddy sea approximately 15–20° south of the palaeoequator. Siliciclastic muds were derived from the erosion of the Taconic highlands to the southeast of New York State and deposited into the Appalachian Foreland Basin. In addition, carbonate silt was transported into the basin from a carbonate platform – crinoidal shoal complex (Wiarton Shoal) to the northwest. Faunal and sedimentological evidence, as well as the presence of endolithic algal borings from Rochester Shale fossils, indicates a normal marine setting of these platform sediments below storm wave base. The finest preservation occurs within rapidly buried mudstone and calcisiltite beds of the Lower Rochester Shale, which record the episodic events of storms within a relatively shallow (50–100 m) subtropical continental sea.

IMPORTANT COLLECTIONS IN NORTH AMERICA

Buffalo Museum, Buffalo, N.Y.
New York State Museum, Albany
Rochester Museum and Science Center, Rochester, N.Y.
Paleontological Research Institution of the University of Rochester, Rochester, N.Y.
Royal Ontario Museum, Toronto, Canada
Springer Collection of the National Museum of Natural History, Smithsonian Institution, Washington, D.C.

Scyphocrinitids from the Silurian–Devonian Boundary of Morocco

HANS HESS

SEA LILIES IN THE DESERT

Slabs with beautifully preserved scyphocrinitids from the Sahara region regularly appear at mineral and fossil fairs; most of these fossils are labelled, 'Erfoud'. Very little work has been undertaken on the detailed stratigraphy of the scyphocrinitid beds of Morocco; hence, the information provided in this chapter is limited. One would hope that future work will increase our knowledge of these remarkable occurrences.

The western Sahara (Fig. 108) underwent several marine transgressions during the Palaeozoic. The extended Silurian transgression started during mid-Llandovery times with the deposition of graptolitic shales. Subsidence increased, so that the thickest part of the succession accumulated during Ludlow times. At the southern edge of the Anti-Atlas in the Dra Plain, in the Tafilalt area near Erfoud and in the Ougarta chains of western Algeria, the Silurian is composed largely of black, sometimes bituminous shales with graptolites. These shales weather on the surface to brighter colours. In the Pridolian part of the series (uppermost Silurian, about 410 million years before present) they contain bands of blue, fossiliferous limestones with *Scyphocrinites*, orthocone nautiloids ('*Orthoceras*'), bivalves (*Cardiola*) and gastropods. Upper Silurian to Lower Devonian argillaceous shales or laminated clays with limestone intercalations or lenses containing scyphocrinitids and orthocone nautiloids are in fact found in a wide area, from the coast near Casablanca and Rabat, across the High Atlas, to the Dra Plain, the Tafilalt, and the Ougarta chains of western Algeria. The limestones commonly occur as nodular concretions in large cushion-like masses that are interbedded in the shales. They appear to be the result of diagenetic accumulations of calcium carbonate by migration of material towards a suitable nucleus in a fairly homogeneous calcareous clay; the beds with calcareous nodules thus represent extremely slow deposition (Destombes et al. 1985). The thickness of the Silurian succession decreases from a maximum of 1,500 m in the west to 100–300 m in the Erfoud area (but reaches 500 m in the Ougarta, where strong subsidence occurred). The Silurian is followed without a break by Devonian sediments, mainly shales, which reach a thickness of 5,000 m in southern Morocco.

MINING FOR COLLECTIONS

The following information on the main crinoid site east of Haroum, about 15 km southeast of Erfoud, has been gathered from visitors to the extensive mining operations, performed by local workers who make their living selling fossils (Fig. 109). The horizons that have yielded most of the specimens are made up of a series of lenses of crinoidal limestone that strike northeast–southwest for several kilometres in a barren plane; this plane is

Hans Hess, William I. Ausich, Carlton E. Brett, and Michael J. Simms, eds., *Fossil Crinoids*. © 1999 Cambridge University Press. All rights reserved. Printed in the United States of America.

fringed by low hills and crossed by mostly dry river beds (called oueds). The sequence is as follows, from bottom to top (B. Imhof, pers. comm., 1993): (1) 'Orthoceras' limestones that crop out to the north, forming a low ridge; (2) 30 m of mudstone; (3) 10-cm-thick scyphocrinitid bed; (4) 1 m of mudstone with interbedded 'Orthoceras' limestone; (5) main scyphocrinitid bed; (6) 0.6-m-thick layer of mudstone; (7) 'Orthoceras' limestone. The scyphocrinitid lenses of the main bed can be located at the soil surface by the occurrence of disarticulated crinoids. In such places the crinoidal limestone, which has a dip of 30–50°S (as the southern part of an eroded anticline), is excavated to considerable depth. When excavations began in the early 1980s, the lenses were close to the surface; in order to reach them now, at 4–5 m below the surface, pits are dug manually. The lenses have a diameter of 0.5–3 m; most have a thickness of about 10 cm in the centre, but some may reach a thickness of 15 or even 20 cm. Intact crowns with attached stems occur on lower bedding planes, whereas the interior and the upper surface contain mainly stem fragments and ossicles. A layer of soft marl covers the fossils. Other fossils, including shells of orthocone nautiloids, small brachiopods, gastropods and bivalves, are also present in lesser abundance. Certain parts of the beds may contain only poorly preserved or recrystallized specimens. Intact plate loboliths (Fig. 21; see the section titled 'Remarkable New Discoveries' for an explanation of lobolith type) occur about 20 km to the east in lenses or fields with a diameter of up to 50 m. Newly discovered beds 80 km to the southwest have loboliths still attached to stems and accompanied by crowns (Fig. 110).

THE ERFOUD SLABS

The largest Erfoud slab or lens on exhibit in a museum, presumably from the horizon with the lenses mentioned earlier, measures about 2 m^2 and contains 75 crowns and stems of *Scyphocrinites* sp.; it can be admired in the museum of the University of Göttingen. This remarkable specimen has been described by Haude *et al.* (1994). As already mentioned, the crowns and attached stems are preserved on the lower surface of a crinoidal limestone. On the surface, a few lobolith fragments of the cirrus type are visible, but no distal part of any of the stems with an attached lobolith is preserved. The stems are mostly bundled together and oriented with the crowns pointing in the same direction. From this preservation, Haude *et al.* (1994) assumed that the crinoid stems, with their length of several metres, were entangled before being embedded. The crinoids are all of similar size. They came to rest on their sides, with the

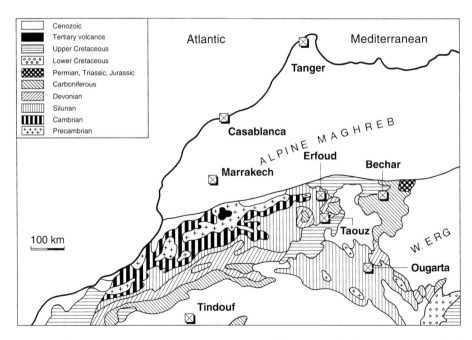

Fig. 108. Geological map of the Anti-Atlas area in southern Morocco and western Algeria. Strata north of the line separating the Anti-Atlas and Sahara regions from the Alpine Maghreb (High Atlas and Rif zones to the north) are omitted. (After Trümpy 1957.)

Fig. 109. Lower surface of a slab with *Scyphocrinites* sp. in natural colour. Upper Silurian, from an excavation east of Haroum near wadi of Oued Ziz. (Natural History Museum, Basel; photograph S. Dahint.) ×0.15. To view this figure in colour, see the colour plate section following page xv.

exception of three stemless crowns, which were embedded with the oral side downward and the arms splayed out, so that the aboral side of the calyx disappears in the bed. Similar preservation is shown by some of the 44 crowns on the large slab of the Naturmuseum Senckenberg (Plodowski 1996).

A slab with a surface of approximately 1 m² was donated to the Basel Natural History Museum (Fig. 109). It contains 21 complete crowns of different sizes. The largest individuals have a crown height of 25 cm and a proximal stem diameter of 0.7 cm. Five juvenile specimens have crown heights of between 1 and 4.5 cm and a stem diameter of 0.1 cm. Two of the juvenile stems reach a length of 1.2 cm, without being complete. All individuals lie on their side and are to some extent oriented as on the Göttingen slab. The stems are invariably broken. Broken and flattened remains of cirrus loboliths are also preserved on this slab.

REMARKABLE NEW DISCOVERIES: INTACT CROWNS AND STEMS WITH LOBOLITHS

Recently, new excavations have been made by local workers in the Jissoumour area between Tazoulet and Alnif, approximately 80 km southwest of Erfoud, at the

Fig. 110. Lower surface of a slab with mostly juvenile scyphocrinitids and loboliths of corresponding size. Several loboliths (L) (the largest one sideways) are on the bedding plane. The structure of the loboliths, including a wall composed of two layers (see Fig. 21), indicates that they belong to the plate type. Arrows point to the terminal distal stem with radicular cirri and stump. Primary roots disappear here and merge into bulbs (best visible at lower left and right, respectively); attached loboliths are crushed. A platyceratid snail is also exposed. The smallest loboliths (not figured) on the surface of the lens have a diameter of only 5 mm. On the basis of the type of lobolith and slightly elliptical cross section of stem in the median part, the specimens from this locality appear to belong to *Marhoumacrinus legrandi*. However, some characters of calyx and arms (a more or less pronounced biserial arrangement of tertibrachials is restricted to juvenile specimens at this location) make such an assignment doubtful. Silurian–Devonian boundary, Jissoumour area between Alnif and Tazoulait. (Natural History Museum, Basel; photograph S. Dahint.) ×0.85.

foot of hills rich in Devonian horned and giant trilobites. The beds are lenses with a maximum thickness of 15–20 cm. Some of them carry large crowns on the lower surface; locally they are packed with loboliths of the plate type reaching a diameter of 10 cm. Smaller, more or less intact loboliths occur on the lower surface of other lenses, where they are accompanied by complete crowns, many of them juvenile (Fig. 110). This is apparently the first report of loboliths still attached to the stem. A few of the smaller specimens have strongly sculptured calyx plates and a heteromorphic proximal stem, with nodals and internodals, a unique feature of scyphocrinitids (Fig. 111). The reason for this aberrant morphology is not known. The exact age of the bed has not been determined yet, but the trilobite faunas in the following strata are Early Devonian.

WIDE DISTRIBUTION OF SCYPHOCRINITIDS

Scyphocrinites elegans was first described by Zenker (1833) from Bohemia, where it occurs in black mudstones of Upper Silurian (Pridolian) and Lower Devonian age. Springer (1917) provided a detailed description of American material that he assigned to the Bohemian species. According to Prokop and Petr (1986), *S. elegans* is limited to Bohemia. Several species and genera of scyphocrinitids have been described from the Upper Silurian of England, France, Spain, Russia, Burma and China. In North America (Tennessee, Missouri, Oklahoma, West Virginia), they range from Upper Silurian to Lower Devonian (Bailey Formation). Prokop and Petr (1987) proposed for the Algerian scyphocrinitids (Upper Silurian to Lower Devonian) a new genus and species, *Marhoumacrinus legrandi*. The Erfoud specimens (Fig. 109) are assigned here to *Scyphocrinites* on the basis of the type of lobolith (see the next section).

The material described by Springer (1917) presumably from Lower Devonian strata, and now in the National Museum of Natural History, was excavated from the bluffs of the Mississippi River, near Cape Girardeau, Missouri. This occurrence is worth mentioning because it established for the first time the connection between well-preserved scyphocrinitids and their bulbous roots. At this locality, the crinoids occur in the lower third of a 15-cm-thick bed of argillaceous limestone overlying a seam of clay. The fossiliferous part of the layer was limited to a small area and considered by Springer to be 'the remnant of a thickly crowded colony, suddenly killed by some change in the water or movement of the sea bottom, and embedded in the soft mud without much disturbance by currents'. The main slab, 120 by 160 cm, contains 24 crowns, several with more or less broken stems, as well as a few flattened or fractured bulbs (loboliths, *Camarocrinus*; see the next section). The bulbs are free of stems, indicating 'that they were near the edge of the colony, and when the forest of crinoids went down . . . the stalks fell away from these roots, leaving them imbedded in the mud where they grew.' The bulbs were on the lower side of the layer, showing their rounded, non-stalked ends; 'hence they stood with the stalked end uppermost, as they naturally would if growing in or resting upon the soft bottom.' Springer did not find a single case in which the stem was still connected to the bulbous end.

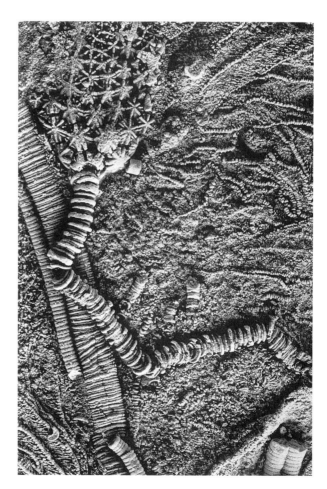

Fig. 111. Small scyphocrinitid with heteromorphic stem and strongly sculptured calyx. Lower bedding plane of a thin lens. Note part of the wall of a plate lobolith at lower right. Silurian–Devonian boundary, Jissoumour area between Alnif and Tazoulait. (Hess Collection; photograph S. Dahint.) ×2.

SOLUTION TO THE LOBOLITH ENIGMA: KEY TO THE LIFESTYLE OF SCYPHOCRINITIDS

In adult individuals, the column ends distally in a large, bulbous, chambered organ, which reaches a diameter of 20 cm. These bulbs were described, from U.S. material, by Hall in 1869 as *Camarocrinus*. Around the same time, Barrande (working on material from Bohemia) called them *Lobolithus*. Originally they were of unknown affinity. These curious, bladder-like structures were definitely associated with the camerate *Scyphocrinites* by both Jaekel and Bather around the turn of the century when the idea of their function as buoys was first proposed. According to Bather (1907), 'It is ... believed by many that this swelling was hollow and served as a float from which the crinoid hung, arms downward. The latter hypothesis explains why it is that in various parts of the world the loboliths occur unassociated with the crowns to which they are supposed to have belonged; following death, the gradual decay of the animal would cause the crown to drop off and sink to the bottom, while the lobolith floated on.' Schuchert (1904) derived his opinion that *Camarocrinus* served as the float of an 'unknown crinoid' from the fact that at that time loboliths in the United States had never been found associated with crinoids, and scyphocrinitids were unknown from U.S. strata. The discovery of well-preserved *Scyphocrinites* in the Bailey Formation at Cape Girardeau, Missouri, on which Springer's monograph (1917) was based, disproved Schuchert's notion that 'the great majority of the bulbs are found in strata with the stalked end downward.' As already discussed, the bulbs were embedded in the mud with the stalked end directed upward, into the bed.

Haude (1972, 1992) reanalyzed the loboliths. He recognized two types – orange-shaped cirrus loboliths with numerous chambers of unequal size, and lobed plate loboliths with a few large chambers, a curved, bilaterally symmetric root trunk, a simplified wall structure and a characteristic collar (Fig. 21). Both types are cosmopolitan. According to Haude, plate loboliths evolved from Silurian cirrus loboliths at the Devonian boundary, and he suggested that lobolith morphology might be useful in biostratigraphy. However, the occurrence of both cirrus and plate loboliths in the same horizon in Bohemia (R. J. Prokop, pers. comm., 1995) suggests that lobolith evolution may have been more complex. Mainly on morphological grounds, Haude followed and strengthened the conclusion of Schuchert and Bather that the loboliths must have served as buoys. In spite of the excellent preservation of the Erfoud scyphocrinitids, stems with attached loboliths have been found only at one locality so far.

The Bohemian *Scyphocrinites elegans* of Upper Pridolian and Lower Devonian age appears to be associated with cirrus loboliths only. Prokop and Petr (1987) reported plate loboliths from the Tafilalt (Erfoud) area and such loboliths are also on the Jissoumour slab (Fig. 110). However, according to Haude *et al.* (1994), the loboliths on the Göttingen slab from Erfoud are of the cirrus type, and this is also true of those on the Haroum slab (Fig. 109). A cirrus lobolith starts from primary roots at the end of the stem; these divide repeatedly into ever finer branches to form innumerable elements (spicules) that build up the walls of the inner chambers as well as the outer covering wall. The stronger outer and the thinner inner layers, which are composed of small spicules, are supported by a middle layer with larger elements. Crystallized calcite may fill the smaller, peripheral chambers (also in loboliths from Erfoud), suggesting that these chambers were sealed. Larger chambers around the base of the stem were probably open, as shown by infilled sediment (Haude 1992). Haude proposed a hypothetical model for lobolith evolution starting from a dense, tangled cirrus network of a crinoid rooted in muddy sediment. Small open spaces in the network then swelled into soap-bubble-like chambers (with the possibility of floating) and eventually into the large diving-bell-like chambers of the plated loboliths. As noted later, this sequence, starting from a rooted crinoid and ending as a planktonic one, is purely conjectural. The generation of gas to fill the bulb has so far not been explained in a satisfactory way and the issue of whether it actually occurred appears unresolvable at present. Haude (1972) assumed that gas accumulated from the surrounding water by diffusional process, as in floating algae, but failed to define the driving force. It may be speculated that the openings at the root bifurcations inside the projecting collar (see Fig. 21) served for gas exchange – for example, to relieve pressure in the case of overproduction of gas or its expansion due to temperature changes. How the smaller, presumably sealed chambers of the cirrus loboliths became filled with gas from the outside is an enigma, unless one assumes that gas formed from symbiotic cells growing in the chambers, one of many possibilities in the absence of evidence.

It appears that the stems were easily broken off near the bulbs, leaving plate loboliths with a typical stump that may be surrounded by a collar (Fig. 21). The occurrence of small encrusting roots, attached to mature bulbs (Springer 1917, Figs. 7–10; Haude 1992, Fig. 3), shows that larvae settled on mature bulbs so that juvenile individuals were anchored by branching roots, as in other crinoids. Aggregative larval attachment is very characteristic of abundant solitary animals (such as mussels), and loboliths must have presented a suitable environment for juveniles to grow. Larvae settling on loboliths may have started the aggregation of large numbers so typical of these crinoids, which include – on the Erfoud slabs – individuals of all age-classes. In Bohemian plate loboliths, roots of juvenile animals occur within the protected space of the collar (rarely also on the outer side of the collar), suggesting some kind of nursery (R. J. Prokop & V. Petr, in prep.; see also Haude 1992). Strimple (1963) suggested that larval scyphocrinitids started with roots and formed loboliths after breaking off the mature bulbs. However, no juvenile specimen with roots and a budding lobolith has ever been described. On the other hand, a small lobolith, with a diameter of only 5 mm, occurs on the slab of Fig. 110. For Haude (1992), the small rooted scyphocrinitids may represent sexual dimorphs, but this would be unique among crinoids. In any case, the formation and development of loboliths offer a fascinating area for future research.

LONG STEM AND LARGE CROWN WITH PAVED CALYX

Scyphocrinitids are large crinoids. Stems that may have reached 3 m in adult individuals supported a calyx 10 cm high and arms more than 30 cm long. The primibrachials and secundibrachials, as well as their pinnules, are all fixed in the calyx, forming a network of similar plates through surface sculpturing; the arms become free above some tertibrachials (Fig. 112). There are up to four more bifurcations above the tertibrachials; and because the free brachials are all very low, with a biserial tendency, the effective food-gathering surface of this magnificent crinoid was very large indeed. The multi-plated tegmen and the distinct anal tube are usually not visible in the Moroccan specimens, but occasionally platyceratid snails are attached to the tegmen (Fig. 110). The proximal stem is cylindrical, composed of extremely low columnals and thus very flexible. It tapers gradually from the cup until its diameter is reduced to about half at the distal end; through this length, the columnals gradually increase in height. The axial canal is obtusely pentagonal in the proximal and middle part of the stem, where it becomes very large. It diminishes to a sharply stellate opening distally. Although several scyphocrinitid species and genera have been distinguished in the literature, the clear morphological similarities between them indicate that these amazing crinoids had very similar lifestyles.

SCYPHOCRINITIDS: THE ONLY TRULY PELAGIC SEA LILIES?

The North African scyphocrinitid lenses occur in pelagic sediments that were deposited at moderate depth, possibly on a submarine ridge or platform in a basin with meagre sedimentation. This basin was open to the north and northeast and probably also to the southwest (Destombes et al. 1985). The wide distribution of scyphocrinitids in such sediments, as well as the peculiar transformation of the root cirri into chambered bulbs, has been taken by a number of authors as an indication of a planktonic lifestyle. The similarity between the long-stemmed scyphocrinitids with their well-developed arms (branching isotomously in scyphocrinitids) and the pseudoplanktonic Lower Jurassic *Seirocrinus* with similarly developed crowns and even longer stems is another point in favor of a floating lifestyle. Both crinoids appear to have grown rapidly to a large size, suggesting an ample supply of nutrients and a lack of other filter feeders in the oxygenated zone above the black shale environment. The presence of long stems indicates that the chances of food collection increased with depth, as in today's oceans, where plankton-rich zones are well below the surface. The fan or funnel formed by the crown possibly captured plankton moving to or from the surface. The huge number and wide occurrence of these crinoids suggest that efficient predators were lacking.

The planktonic lifestyle of scyphocrinitids was made possible by the development of the extraordinary gas-filled bulbs. The animals appear to have formed dense mats, presumably as a consequence of the preference of larvae for settling on adults. In the case of damage to some of the bulbs in such a colony or after water penetrated into the chambers in stormy weather, the gas would escape and the mat would sink to the bottom, perhaps like the great mats of sea grass found on the deep sea bottom in today's oceans. As a rule, the waterlogged bulbs became detached from the stem, breaking off at the weakest point close to the collar (where stem

Fig. 112 Scyphocrinitid, Upper Silurian, Erfoud (exact locality unknown). Specimen with calyx and arm structure well exposed by arid weathering. (Hess Collection; photograph S. Dahint.) ×1.

diameter is reduced). The bulbs eventually sank to the bottom with their collar upward, like short-necked, water-filled bottles. Haude et al. (1994) concluded that the isolated crowns observed on some of the Erfoud slabs reached the bottom first, oral side downward. They were followed by the group of more intact crinoids with entangled stems, which had sunk in a horizontal position, and then by more broken stems. According to this view, differences in transport by currents led to the deposition of crowns with attached stems in one place, whereas the loboliths usually accumulated elsewhere, sometimes in huge numbers. Preservation of the scyphocrinitids – articulated crowns with stems on the lower side, disarticulated ossicles and some stem fragments on the top surface – suggests that burial occurred some time after death.

The hypothesis of a planktonic lifestyle for scyphocrinitids (Fig. 113) is consistent with the morphological

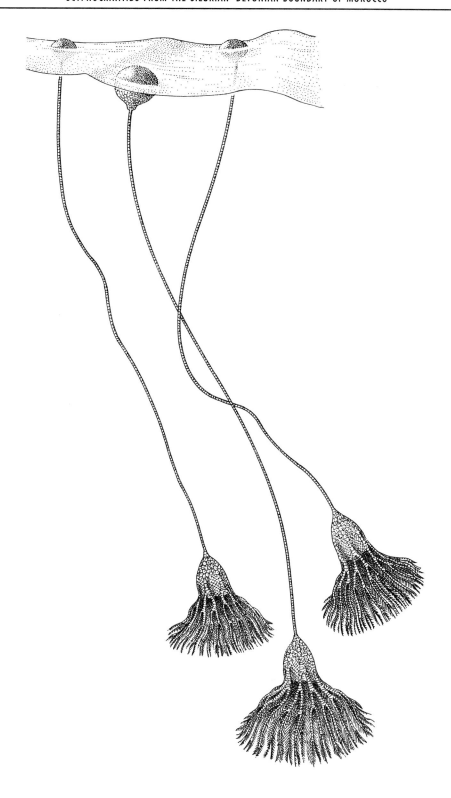

Fig. 113. Reconstruction of scyphocrinitids as pelagic, floating sea lilies. (After Haude *et al.* 1994.)

characteristics of the chambered bulbs, with their presumed buoyancy, as well as with their distribution, including the separate occurrence of crowns with broken stems and loboliths. Because it offers the most plausible explanation, it is widely accepted. The occurrence, at Haroum, of a large number of discrete lenses at the same stratigraphic horizon is difficult to explain by this paradigm. The lenses are separated from their neighbours by 2–10 m of marly clay and may have been formed by diagenetic processes (see section titled 'Sea Lilies in the Desert' above). A storm would not normally deposit floating animals neatly into lenses that thin out into clays or shales. However, such an occurrence seems more likely if the crinoids formed mats, possibly held together by their intermingled stems, or perhaps by some plant material contained in the crinoid colonies. This also explains the rarity of loboliths occurring *in situ* on most of the slabs, whereas one would expect the bulbs to be commonly preserved and attached to stems in bioherms growing from the bottom.

IMPORTANT COLLECTIONS IN EUROPE

Institut und Museum für Geologie und Paläontologie
    der Universität Göttingen, Germany
Natural History Museum, Basel, Switzerland
Naturmuseum Senckenberg, Frankfurt a.M., Germany

12

Lower Devonian Manlius/Coeymans Formation of Central New York, USA

CARLTON E. BRETT

A SHOWPIECE OF THE YALE PEABODY MUSEUM

In the late 1800s, a number of spectacular slabs of well-preserved fossil crinoids were discovered from the Lower Devonian thin-bedded limestones that had been excavated in small quarries for building and agricultural lime production in the township of Litchfield, Herkimer County, New York. The best-known localities were farm quarries in the hamlets of Jerusalem Hill and Days Corners (Fig. 114). A most spectacular, and now famous, slab with fossil crinoids was excavated by Charles Beecher and Charles Schuchert in the quarry of John Salisbury around the turn of the century. It was shipped in a series of crates weighing more than 2,100 kg and reassembled in the Yale Peabody Museum as one of the richest single-specimen slabs of fossil crinoids then known in North America. The slab, measuring 2.1 by 1.8 m, is presently displayed in a wall case at Yale University and has been the subject of several studies. In 1905, Mignon Talbot described the slab and its fossils, noting the occurrence of four species of crinoids. The most abundant are *Cordylocrinus plumosus*, followed by *Melocrinites* (now *Ctenocrinus*) *pachydactylus* and two new species: *Mariacrinus beecheri* (now *Ctenocrinus nobilissimus*) and *Thysanocrinus* (now *Ambicocrinus*) *arborescens*. With more than a thousand specimens of these four crinoids (some 887 calyces of *Cordylocrinus* alone), this slab contained more crinoids than had been found at that time in all the rest of the fossil-bearing beds of New York State.

The stratigraphy of the Helderberg Group, including beds containing the crinoids, was revised by Rickard

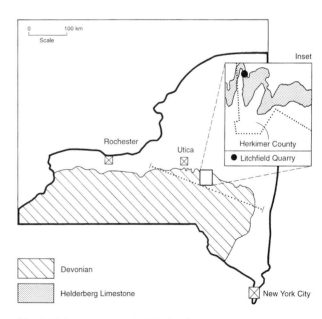

Fig. 114. Location map for Manlius/Coeymans crinoid locality near Litchfield, N.Y. Dotted line indicates approximate line of cross section shown in Fig. 115. (After Hicks & Wray in press.)

Hans Hess, William I. Ausich, Carlton E. Brett, and Michael J. Simms, eds., *Fossil Crinoids*. © 1999 Cambridge University Press. All rights reserved. Printed in the United States of America.

(1962), and depositional environments were considered in detail by Laporte (1967, 1975). Goldring (1923) discussed and illustrated all of the crinoids then known from the Litchfield area, including those from the famous Days Corners Quarry site. The type specimens and many other crinoids from the Litchfield area, illustrated by Goldring, are presently housed in the extensive collections of the New York State Museum in Albany.

However, relatively little subsequent work was undertaken on these spectacular fossil assemblages until 1988, when two students at Yale University, Jason Hicks and Charles Wray, made a detailed taphonomic and palaeoecological study of the famous Beecher slab in the Peabody Museum. The result of their work has been submitted for publication, and a good deal of what is noted about the fossil occurrence in this chapter is derived from that study.

The famous crinoid-bearing slabs from the Litchfield area are derived from thin, platy limestones of the Helderberg Group of Early Devonian age (Lochkovian, 405–408 million years old). In the Litchfield area, the lower two formations assigned to the Helderberg Group interfinger with one another (Rickard 1962). Crinoids are reported from transitional beds near the contact of the Olney Member of the Manlius Formation and the overlying massive crinoidal limestone of the Deansboro Member of the Coeymans Formation (Fig. 115).

SHALLOW LAGOONS AND CRINOIDAL SHOALS

The Manlius Limestone, in its type locality, is a particularly fine-grained, laminated, light grey weathering limestone. Portions of the Manlius Limestone display features such as mud cracks, small-scale rippling and very fine, wavy laminations, attributed to stromatolitic algal mats; these features are indicative of the wet–dry regime of the supratidal mud flat zone. These portions of the Manlius have generally been interpreted as intertidal to very shallow subtidal (Fig. 116). Lagoonal facies are also common within Manlius rocks and consist of thin, rather evenly bedded, light blue-grey laminated limestones separated by fossiliferous shaly partings; this facies is commonly referred to as ribbon-bedded limestone. Some of these layers display graded bedding indicative of storm deposition. Certain bedding planes are covered with low-diversity assemblages of brachiopods, bivalves, ostracodes and *Tentaculites*; the latter may display bimodal orientation suggestive of oscillating or wave activity. In parts of the Manlius Formation, repre-

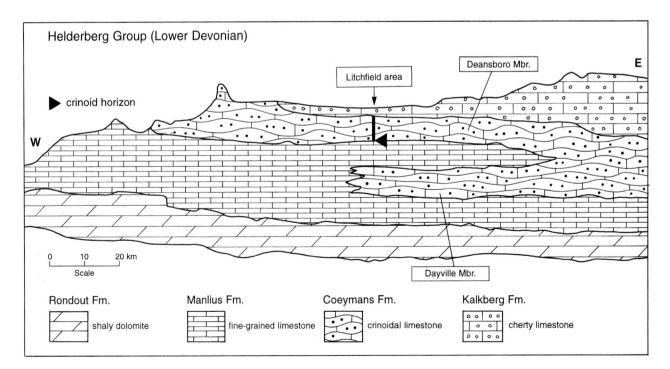

Fig. 115. Generalized stratigraphic profile for the Lower Devonian Helderberg Group of central New York; see Fig. 114 for line of cross section. (Modified from Rickard 1962.)

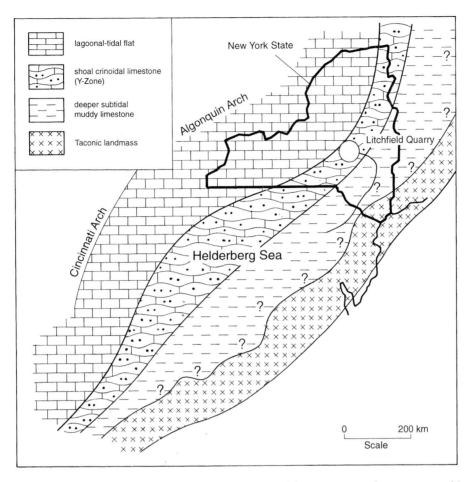

Fig. 116. Palaeogeography of Early Devonian Helderberg Group. Curved line emanating from position of Litchfield indicates northeastern limit of Devonian outcrop. (Modified from Hicks & Wray in press.)

senting the outer portions of the lagoon, cabbage head–like stromatoporoids and favositid corals may occur in large quantities, forming small patch reefs or, more commonly, biostromes of skeletal material or interbedded lime mud.

The Manlius Formation interfingers with and is eventually overlain by tongues of the Coeymans Limestone (Fig. 115). In contrast to the fine, platy, often thin-bedded to laminated weathering Manlius Limestone, the Coeymans is thick-bedded to massive skeletal limestone. It is composed primarily of the disarticulated plates and columnals of crinoids and cystoids. In addition, the Coeymans Limestone, in the central New York area, commonly possesses small, sausage-shaped calcitic holdfasts of the rhombiferan cystoid *Lepocrinites*. Complete crinoids and cystoids are rare within the Coeymans, which consists of skeletal debris that was probably reworked several times by storm and, possibly, normal fair-weather waves. There is relatively little mud in the Coeymans; rather, the echinoderm skeletal material is bound together by sparry, crystalline calcite cements. The absence of lime mud indicates that these skeletal deposits were winnowed by waves and/or currents that removed the fine-grained sediments.

GOOD PRESERVATION IN LOW SPOTS OF THE SEA FLOOR

The famous Yale slab and numerous smaller limestone slabs in the New York State Museum (Fig. 117) display exquisitely preserved crinoids, many of which preserve long sections of gently curving stems. On some bedding planes, stems and crowns occur in tangled masses, suggesting that these organisms were swept together by currents during the time of deposition. Further evidence for current orientation has been obtained in the studies

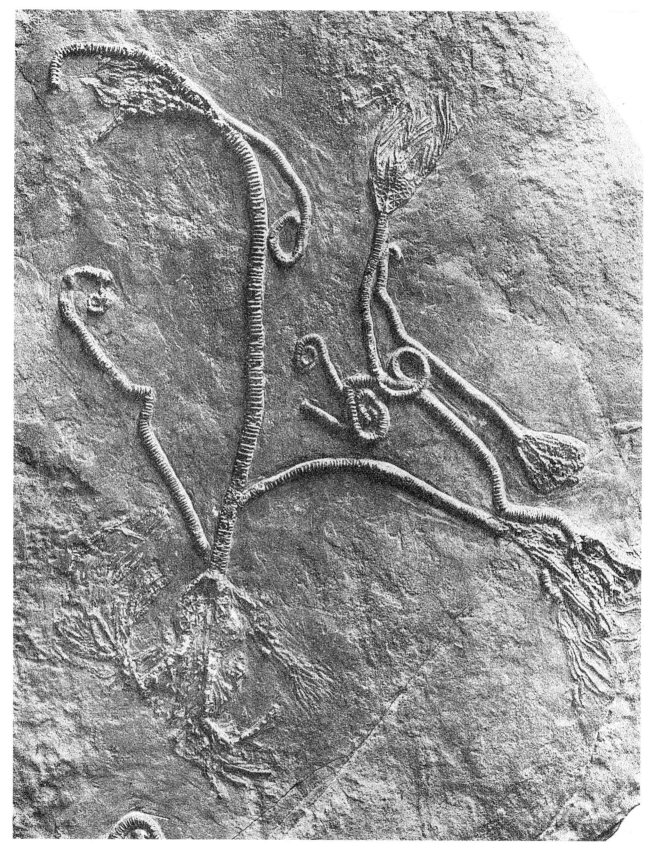

Fig. 117. Upper surface of slab with six nearly complete individuals of *Ctenocrinus pachydactylus*; note coiled distal stems. Manlius/Coeymans transitional limestone, Jerusalem Hill, Litchfield, N.Y. (New York State Museum; from Goldring 1923.) ×1.

of Hicks and Wray. They observed two prominent modes of orientation. One may represent the rolling of stems with the current. Another mode, at approximate right angles to the first, may represent complete individuals in which the stems were splayed out parallel to the current, with the crowns, slightly heavier, acting as drags or anchors.

The crinoids of the Yale slab appear to have been accumulated in a low spot on the sea floor or in a slight low-energy pocket where the remains became trapped during very local transport because of a locally lowered current velocity. Some of the slabs in the New York State Museum display complete crowns with outstretched arms, but they lack stems. Similar preservation has been observed in Triassic *Chelocrinus carnalli* beds (see Chapter 21). The absence of stems can be explained in one of two ways: either the crinoids died some brief period before their final burial and decay of ligaments had thus allowed the disarticulation of the crowns from the stems, or the crowns were autotomized. Modern crinoids may voluntarily cast off portions of their stems or arms during stress. This process, referred to as autotomization, could account for the stemless crowns. Perhaps the storm disturbances, which ultimately led to mass mortality and burial of the crinoids, stimulated the animals to cast off their stems. However, the intact nature of most crowns argues against this process, as it is the arms that are frequently cast off during autotomization in extant crinoids. Most accumulations on some bedding planes may represent the other side of this story – that is, the stems that were separated from the crowns and that were then moved separately because of slightly different hydrodynamic properties.

Crinoids on the Manlius bedding planes were buried in fine, yellowish, slightly dolomitic, carbonate-rich siltstone. This material resembles sediments in the mud-cracked supratidal facies of the Manlius that may have been washed from the shoreline during times of heavy hurricanes. Long-distance transport of the crinoids seems unlikely because the delicate pinnulate arms of many specimens are intact. However, local transport is evident from the orientation and aggregation of crinoid clusters. Once the remains had been aggregated, they were rapidly buried in the fine silt and not exhumed later. Early diagenetic cementation of the carbonate may have protected the crinoids from severe compaction. However, later diagenetic effects, particularly dissolution under pressure, have detracted somewhat from their appearance.

PREDOMINANCE OF LONG-STEMMED CRINOIDS WITHOUT PERMANENT HOLDFASTS

Crinoids on the Manlius slabs include seven species, heavily dominated by pinnulate camerates and the small cladid *Lasiocrinus*. These crinoids predominantly possessed long, relatively flexible-appearing stems with no permanent holdfast (Fig. 119). *Ctenocrinus* is the largest of the camerate crinoids; the slightly smaller *Ctenocrinus nobilissimus* is similarly adapted (Fig. 118). These forms had complex ramulate and pinnulate arms. Cowen's (1981) functional study of these melocrinids suggested that the evolution of heavily ramulate arms was a strategy for more effective filtering of a particular volume of water for suspended plankton. In any event, it is reasonable to postulate that these crinoids were adapted to the filtration of relatively small food particles by leeside suspension feeding in a manner similar to that of many modern isocrinids, which likewise possess highly efficient space-filling pinnulate arms (Figs. 235, 236). *Ctenocrinus* had a long, heteromorphic stem that appears to have had considerable flexibility (Fig 117). These stems typically formed distal coils and perhaps in some cases formed loops that helped to support the upright portion of the stem. These loops would have lain on the sea floor with a vertical portion of the column rising from them, rather like a coiled rattlesnake (the type specimen of *Acanthocrinus rex*, Fig. 124, can serve as a comparison). The long stems, sometimes approaching 1 m in length, served to elevate the large crowns into positions of higher current strength for feeding.

Two other, somewhat more common crinoids occupied a lower tier, typically 30 cm or less above the sea floor. *Cordylocrinus* (Fig. 120) belongs to the group of hapalocrinids that, relatively early in their evolutionary history, developed whorled radicular cirri on the stem. *Cordylocrinus* and its relatives appear to have borne runner-like portions of the stem on the sea bottom in the manner of modern isocrinids. Presumably, therefore, these crinoids may have fed from relatively low levels within the water column. They possessed sparsely branched but delicately pinnulate arms, which were presumably utilized in filtration-mesh feeding. The wider gaps of these arms may have been adapted for the interception of suspended particles somewhat larger than those of the longer-stemmed *Ctenocrinus*.

Likewise, the small but comparatively long-stemmed (up to 20 cm) cladid *Lasiocrinus* (Fig. 119) with its densely branched arms was presumably adapted for filtration-mesh feeding on relatively larger particles. Like

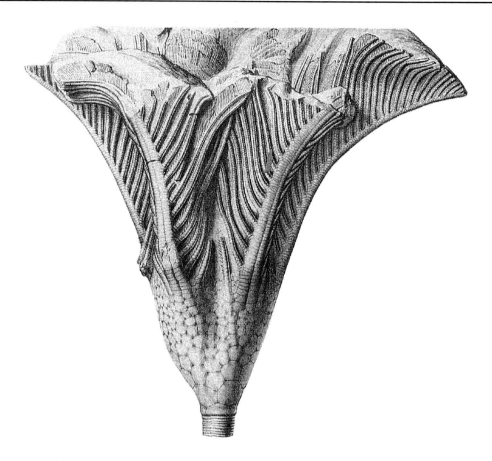

Fig. 118. *Ctenocrinus nobilissimus*, posterior view of Hall's type specimen. Note highly ramulate arm trunks. Manlius/Coeymans Limestone, Litchfield, N.Y. (American Museum of Natural History; from Goldring 1923.) ×1.

Cordylocrinus, Lasiocrinus possessed whorls of flexible radicular cirri on the stem, which may have aided in temporary attachment and adjustment of position on soft and shifting substrates of carbonate silt and fine sand. This small cladid appears to have been a generalist and perhaps an opportunistic species. It lived in large numbers in shallow lagoonal settings that were not inhabited by a high diversity of other echinoderms. *Lasiocrinus* tends to occur in clumps of up to hundreds of individuals on certain bedding planes in the fine-grained ribbon rock facies of the Manlius Limestone, where few other crinoids or other fossils are common. *Lasiocrinus* is present, but less common, in the more diverse camerate-dominated assemblages, such as the *Cordylocrinus* and *Ctenocrinus* associations.

The Manlius assemblages are heavily dominated by crinoids. However, a relatively low diversity of brachiopods, primarily the spiriferid *Howellella*, and small rhynchonellids occur on some of the crinoid-bearing slabs. In addition, large, smooth ostracods and the annulated tubes of *Tentaculites* may be common. Small ramose bryozoans and pteriniids (wing oyster bivalves) are typically associated. Only a few other echinoderms, aside from the four to five crinoids, occur in these assemblages. Most notable among these is the small mitrate (carpoid echinoderm) *Anomalocystites*. Some beds in the crinoid-bearing Manlius facies show evidence of early submarine cementation to form hardgrounds. These slabs may bear abundant encrusting edrioasteroids. These are small (0.5–1 cm in diameter) discoidal echinoderms that resemble fixed starfish with a web of inter-ambulacral plates between the rays. Edrioasteroids apparently filter-fed in areas of cleaner water near the substrate.

Finally, although not common in the crinoid-rich assemblages of the Manlius transition beds, the rhombiferan *Lepocrinites* does occur rarely. However, it is the dominant echinoderm in the adjacent crinoidal shoal deposits of the Coeymans Formation. These echinoderms appear to have been adapted to a high-energy environment and a substrate of shifting skeletal sands. Their particular attachment strategy is quite unique.

occur in considerable numbers on some bedding planes in the platy ribbon-bedded facies. This small crinoid may have become adapted to the conditions of the lagoon.

Presumably, the lime mud was transported from part of the lagoon into the lower-energy environments to the west and northwest of the Coeymans depositional belt (Fig. 116). The Coeymans Limestone may display cross-stratification and graded bedding. All this suggests that the skeletal debris accumulated under high-energy, turbulent water conditions close to or at normal fair-weather wave base. These skeletal sand and gravel deposits have been interpreted as offshore shoal facies that accumulated in the zone commonly referred to as the Y Zone or, in other words, the region at which fair-weather waves commonly disturb the sea floor. The development of a skeletal shoal or bar complex probably sheltered a shallower, more onshore zone or lagoons and muddy carbonate tidal flats. This is the environment represented by the Manlius Limestone. Although the lagoon was sheltered and was thus an area for the accumulation of fine-grained sediments on a day-to-day basis, occasional storms washed skeletal debris onshore and swept it into the shallow lagoon.

Crinoid beds are best developed in an area of interfingering between the Manlius and Coeymans facies.

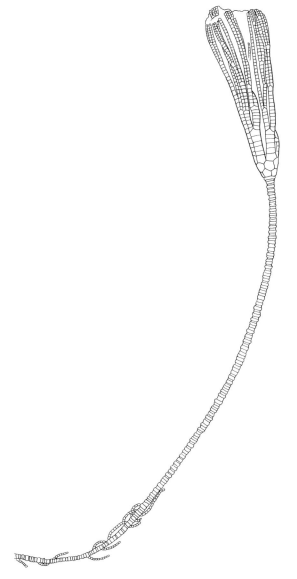

Fig. 119. *Lasiocrinus scoparius*. Complete specimen with crown and stem with whorls of radicular cirri along the distal end. Manlius/Coeymans transition at Jerusalem Hill near Litchfield, N.Y. (American Museum of Natural History; redrawn from Goldring 1923.) ×2.7.

They possessed a long, cylindrical or sausage-shaped body at the end of their short, stocky stems. This solid piece was apparently inserted into the loose skeletal sand or silt and served as an anchoring device.

A SHOAL-MARGIN ASSEMBLAGE

Crinoids are not common in most of the Manlius facies, although specimens of the small cladid *Lasiocrinus* may

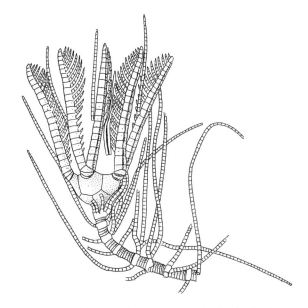

Fig. 120. *Cordylocrinus plumosus*. Note whorls of radicular cirri spaced regularly along the stem. Manlius/Coeymans Limestone, Days Corners near Litchfield, N.Y. (American Museum of Natural History, Yale University Collection; redrawn from Goldring 1923.) ×3.

These probably represented the inner margin of the skeletal shoal. Here, water was sufficiently well aerated and agitated to support dense stands of suspension-feeding crinoids but also quiet enough to prevent dislodgement and transport of the weakly rooted crinoids. Intermittent storms, however, disrupted the delicate crinoid gardens and dumped fine carbonate silt onto the slightly decayed remains on the sea floor. As in several other shoal-margin assemblages, the loose tethering of the Manlius crinoids may have made them particularly vulnerable to such episodic events.

IMPORTANT COLLECTION IN THE UNITED STATES

New York State Museum, Albany

13

Lower Devonian Hunsrück Slate of Germany

HANS HESS

ROOFS AND TILES

Visitors to this part of Germany cannot fail to notice the dark grey slate roofs and walls in the picturesque towns along the rivers Rhine and Moselle. The Hunsrückschiefer belongs, together with the Lower Jurassic Posidonia Shale and the Upper Jurassic Solnhofen Lithographic Limestone, to those sediments that have been exploited since ancient times for building purposes. The discovery of a large number of world-famous fossils in these strata is, therefore, due to the observation and skilled manual work of quarrymen preparing the slabs and tiles. Fossils wrested from the dark grey Hunsrück Slate are mostly small and, therefore, may not be quite so spectacular as those from some other sites, but they open a fascinating window onto the development of life in these early times.

DECLINE OF A CRAFT

The Hunsrückerschiefer is exposed in the Rheinisches Schiefergebirge between Koblenz, Trier and Mainz. The best localities for well-preserved fossils are the communities of Bundenbach and Gemünden. The slates were widely quarried in former times (slates for roofs, etc.); and during the 19th and present centuries, more than 600, mostly small, pits were exploited. Today, only one quarry remains open in the main fossiliferous region of Bundenbach. Unlike other famous *Lagerstätten*, the Hunsrück Slate has not undergone detailed sedimentological examination with modern techniques.

THICK SEDIMENTS FROM THE REMAINS OF AN OLD CONTINENT

The slates are of Early Devonian age and were deposited from the Late Siegenian (Late Pragian) Stage and ceased towards the end of the Early Emsian Stage (age about 390 million years). The sediments were deposited in a number of offshore basins separated by swells. Downwarping allowed the Hunsrück Slate to reach thicknesses that were estimated to be as much as 4,000 m but were probably considerably less (see Bartels *et al.* 1998). The Hunsrückschiefer proper (also called 'Dachschiefer' for its use as tiles for roofs) was laid down during the early Lower Emsian. A main belt of Hunsrückschiefer runs southwest to northeast for about 150 km (Fig. 121); smaller areas occur west and southeast of Koblenz. To the north and west the facies of the slates passes into sandy sediments with a shallow-water fauna dominated by brachiopods; the crinoids here are mostly sturdier (Schmidt 1942). To the south the thickness of the Hunsrückschiefer is reduced to about 200 m, and to the east open marine sediments were laid down (Mittmeyer 1980). Thus, deposition of the Hunsrückschiefer occurred in subsiding basins that were separated by swells

Hans Hess, William I. Ausich, Carlton E. Brett, and Michael J. Simms, eds., *Fossil Crinoids*. © 1999 Cambridge University Press. All rights reserved. Printed in the United States of America.

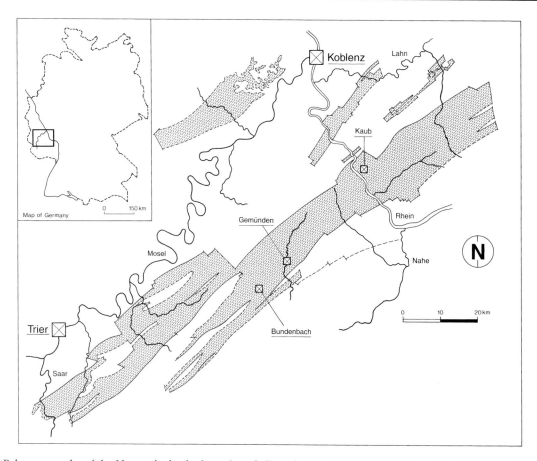

Fig. 121. Palaeogeography of the Hunsrückschiefer facies (stippled) in the Rhenish Massif. (Redrawn from Mittmeyer 1980.)

with reduced and partly sandy sedimentation. Mass occurrences of pelagic tentaculitoids in certain layers suggest the influence of an open ocean; water depth in the offshore setting of the slates has been estimated to be 40–60 m (Bartels & Blind 1995). This is considerably less than a depth of more than 200 m derived from sedimentary structures and trace fossils (Seilacher & Hemleben 1966), but is more in line with the occurrence of well-developed eyes of arthropods and vertebrates (Stürmer & Bergström 1973); see the section titled 'A Wealth of Fossils').

The Hunsrückschiefer, with a thickness of up to 1,000 m in the Bundenbach area, was deposited from clay and silt swept in by rivers from land situated to the northwest. The sedimentation rate varied in the Hunsrück Basin. The mudstone (made up of clay and silt) was transformed into slate under pressure during later Variscan mountain-building and thereby compressed to a tenth of its original thickness. Unfortunately, the bedding planes and, therefore, most fossils do not lie parallel to the cleavage planes of the slates.

PYRITE AND X-RAYS

The fossil remains are usually covered by a pyritic surface layer. When burial was rapid, the fossils, such as the delicate crinoids, were commonly completely replaced by pyrite, in which case preservation is excellent. Pyritized fossils can be examined in 0.5- to 1-cm-thick slabs by X-ray technique before preparation (Fig. 122); see Bartels and Blind (1995) for a discussion of pyritization and X-ray technique. This is of considerable help because the fossils cannot easily be separated from the fine-grained sediment matrix. Museum collections give the impression that intact fossils are the rule. However, the Hunsrück Slate contains mostly disarticulated remains, intact specimens being more common for sturdily built organisms, such as trilobites and their moulted carapaces. Due to the dissolution of thin calcareous shells (brachiopods, lamellibranchs, gastropods and cephalopods), such organisms were not well preserved, so that the number of fossils gives a biased picture of the original communities. Also, because the slates be-

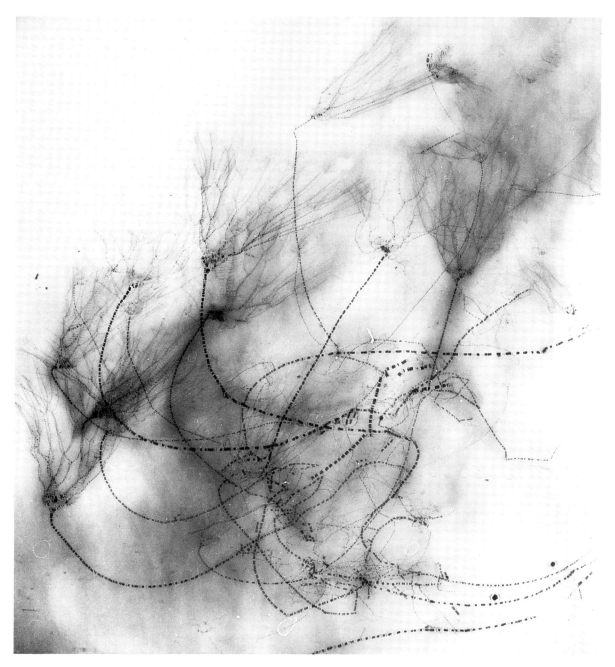

Fig. 122. X-ray photograph of a group of *Parisangulocrinus zeaeformis* with radicular cirri. Hunsrückschiefer, Bundenbach. (Staatliches Museum für Naturkunde Stuttgart; original photograph by W. Stürmer, whose contributions to the X-ray photography of Bundenbach fossils are outstanding; courtesy M. Urlichs.) ×1.

came famous for their arthropods and echinoderms, quarrymen ignored less spectacular fossils. It must be realized that excellently preserved fossils are confined to localized patches, mainly around Bundenbach and in some other localities from the Rhine valley to the middle part of the Hunsrück, where they are restricted to a few horizons. Soft-body preservation is quite exceptional.

A WEALTH OF FOSSILS

To date, at least 264 different species and subspecies of animal fossils have been described from the Hunsrück Slate, which also includes 6 species of plants and 46 species or subspecies of land-based spores (Bartels *et al.* 1998). In the main depositional basin (localities of Kaub, Bundenbach and Gemünden), echinoderms are concentrated in the southwestern part around Bundenbach, whereas the northeastern part is rich in brachiopods. Corals and trilobites with well-developed eyes are in sediments from the more shallow environments as well as from the central basin areas. This suggests moderate water depths of considerably less than 200 m (Mittmeyer 1980; Bartels 1995). Breaks in sedimentation in the shallow-water areas indicate sporadic emergence. However, remains of land plants (*Psilophyton*) and spores are found only rarely.

Animal fossils include rare sponges and jellyfish, corals and, more commonly, conularians, brachiopods, gastropods (frequent in silty and sandy layers), lamellibranchs and cephalopods (*Orthoceras* and early ammonoids). Traces of worms are abundant in places. Certain horizons are filled with tiny tentaculitoids, pelagic inhabitants of the open ocean to the south. The main attractions, as demonstrated in numerous museum and private collections, are the arthropods (trilobites and others) and echinoderms, as well as the rather rare vertebrates. About 50 species of asterozoans have been described. Other echinoderms, such as holothurians (mostly isolated sclerites and parts from the calcareous mouth ring), echinoids, homalozoans, cystoids and blastoids are present, but only crinoids are very common, especially as disarticulated ossicles. The vertebrates include flattened jawless and other fishes, which were living on or near the bottom in the Bundenbach area. From other regions numerous remains of sharks have been reported. For a complete overview see Bartels *et al.* (1998).

THE DELICACY OF MOST CRINOIDS

According to Bartels *et al.* (1998), the 63 crinoid species from the Hunsrückschiefer belong to 30 genera. All the main Palaeozoic groups are represented, most commonly with the cladid genera *Codiacrinus*, *Imitatocrinus* and *Parisangulocrinus*, the disparid genera *Calycanthocrinus* and *Triacrinus* and the camerates *Hapalocrinus* and *Thallocrinus*; the flexible *Eutaxocrinus* is somewhat less common (Südkamp 1995). Crinoids vary from heavily branched forms with numerous arms to *Triacrinus* with only five long arms (Fig. 129). Some species have long anal tubes with protective spines; on others such spines are missing. This is also true of the cup (see Fig. 124 for a cup with spines). According to Schmidt (1934), almost half of the species show some development of spines. Animals that might have fed on the crinoids appear to be uncommon. The snail *Platyceras* has been found on the cup of only two species. Stem lengths vary considerably, but the stems are generally rather long in comparison with the crown (Figs. 122, 125–127); those of *Eutaxocrinus* may reach a length of more than 1 m. Most of the crinoid species are unique to the Hunsrückschiefer, whereas these strata share about half of the genera with the sandy Lower Devonian facies. It appears that the different crinoids occupied separate niches by collecting food at different elevations above the sea floor. Most crinoids in the Hunsrücker Slate are lightly built with small cups, in contrast to the sturdy animals in the nearby contemporaneous sandy facies. This points to a rather sheltered life on the muddy bottom, protected from regular wave action. Even under higher magnification, the Hunsrückschiefer crinoids do not readily show detailed structures of stem and crown, but they beautifully illustrate the different lifestyles of an extraordinary assemblage. We have therefore chosen, for discussion in the following section, a set of complete individuals and groups with different modes of attachment.

ATTACHMENT STRATEGIES ON A MUDDY SEA FLOOR

Crinoids attached themselves with radicular cirri or roots, commonly to hard objects lying on the muddy bottom, such as shells of *Orthoceras* (Figs. 123, 126), brachiopod valves (Fig. 128) or corals (Fig. 125). In many instances, crinoids attached themselves to stems, some of them lying on the sea floor, of other crinoids. Juvenile individuals particularly preferred the stems of adult individuals (Figs. 125, 127). Crinoids that were rooted in the muddy sea floor by radicular cirri growing

Fig. 123. (*On facing page*) *Hapalocrinus elegans* (upper right, with 10 pinnulate arms) and several *Parisangulocrinus zeaeformis*, a form without pinnules, growing out of the body chamber of an empty *Orthoceras* shell. On the outer surface of the shell, small crinoid roots are preserved whose stems and crowns were presumably torn off the shell while it was scraping or rolling along the bottom before burial. Hunsrückschiefer, Bundenbach. (Geologisch-Paläontologisches Institut, Johann Wolfgang Goethe-Universität, Frankfurt a.M.; figured by Seilacher 1961; courtesy K. Vogel.) ×0.6.

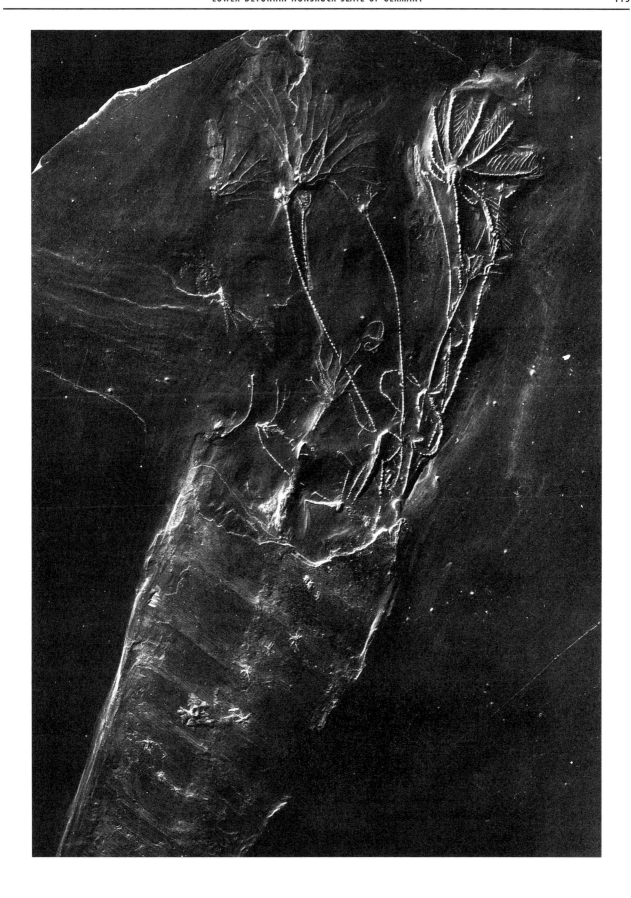

from the distal end of the stem appear to be exceptional (Fig. 129). Rare discoidal holdfasts are found only in the Eifel region, where hard substrates occur. The rare *Acanthocrinus rex* used the strategy of a coiled distal stem for fixation, as shown by the famous type specimen (Fig. 124). It is interesting that the same species may vary considerably in the development of its stem (length, occurrence of radicular cirri). In a few instances, crinoids were attached to rarely occurring sponges, rooted in the soft sediment by means of root tufts. By virtue of its attachment to the sponge, a specimen of *Dictenocrinus semipinnulatus* figured by Südkamp (1992) seems to have needed only a short stem (3 cm) for elevating the crown high enough into the current, whereas other specimens of this species have much longer stems. The calceocrinid *Senariocrinus maucheri* (Fig. 30) has a rudimentary, tapering stem. The anal sac is articulated to the cup and, therefore, could be moved like an arm, perhaps to release waste material downcurrent. Schmidt (1934) thought that *S. maucheri* with its rudimentary stem was free-living and could even swim. As already discussed, we think it more probable that this animal lived on the muddy bottom, used the stem for attachment and raised the crown for feeding.

BOTTOM LIFE WITH CURRENTS AND OCCASIONAL MUD SLIDES

Hunrückschiefer fossils represent mainly sessile and vagile benthic organisms and include numerous trilobite tracks. This is proof of a rich bottom life on the muddy sea floor. The Hunsrück Slate was deposited in several environments (Bartels *et al.* 1998). The shallower parts of the submarine fan above storm level basis were fully oxygenated; the shelly bottom fauna was therefore not pyritized. Density currents from channel distributories reached the lower parts of the fan; the current-transported sediment was initially oxygenated and supported an infauna. The classic fossil sites of Bundenbach and Gemünden are preserved in interchannel areas with oxygenated bottom water that allowed the establishment of bottom communities. Occasional density currents buried the animals, many of them where they lived, and the thick sediment layer protected them from scavengers. The iron-rich sediment rapidly became anoxic beneath the surface, leading to conditions that promoted pyritization and further inhibited any scavengers.

Disarticulated specimens and oriented fossils as well as ripple marks and flow marks point to the occasional presence of rather strong bottom currents (Koenigswald 1930; Seilacher 1960). Rapid burial must have occurred in the case of completely preserved crinoids as well as the many current-oriented intact asterozoans, whereas disarticulated specimens must have lain on the sea floor for some time or represent previously buried and decayed specimens exhumed and disarticulated by bottom currents. The mobile asterozoans seem to have been swept into somewhat deeper water, where the crinoids were anchored; in some instances the crinoid stems served as traps. It has been observed that the root part of crinoid stems is commonly on a lower bedding plane than the crown (Fig. 125). Larger groups may pierce up to 3 cm of slate that were compressed from 20–30 cm of mud (Bartels 1995). Such cases indicate rapid burial of the crinoids by a considerable amount of sediment from a turbidity current or a tempestite event. Arthropods (mainly trilobites) are preserved in all possible positions, including sideways, indicating that they came to rest in pasty mud (Bartels 1995). In addition, groups of crinoids of all age classes have been found, including individuals with a size of only a few millimetres, the young ones often attached to adult animals of the same or some other species (Figs. 125, 127). Such specimens were buried at or near the place of living. However, other crinoids have been detached and transported for at least a short distance before burial. Opitz (1932) pointed out that the majority of crinoids are found with broken stems. Intact specimens, as shown in our figures, are quite exceptional. The burial events with preservation of soft tissues at certain horizons were local in extent; their distribution reflects considerable relief on the sea floor within the basin (Bartels *et al.* 1998). This is in line with local differences in the composition of the Hunsrück fossils, which derive from different communities.

Fig. 124. (*On facing page*) *Acanthocrinus rex*. The specimen – certainly one of the most beautiful crinoids ever found – belonged to the collection of the Preussische Geologische Landesanstalt (Geologisches Landesmuseum), Berlin; it was lost during World War Two. Visible to the right of the stem are one larger and four small specimens of the blastoid *Pentremitidea medusae*. The shell at the coiled end of the stem may have belonged to a brachiopod. Hunsrückschiefer, Kaub. (Photograph S. Dahint from Jaekel's 1895 original lithographic plate.) ×0.8.

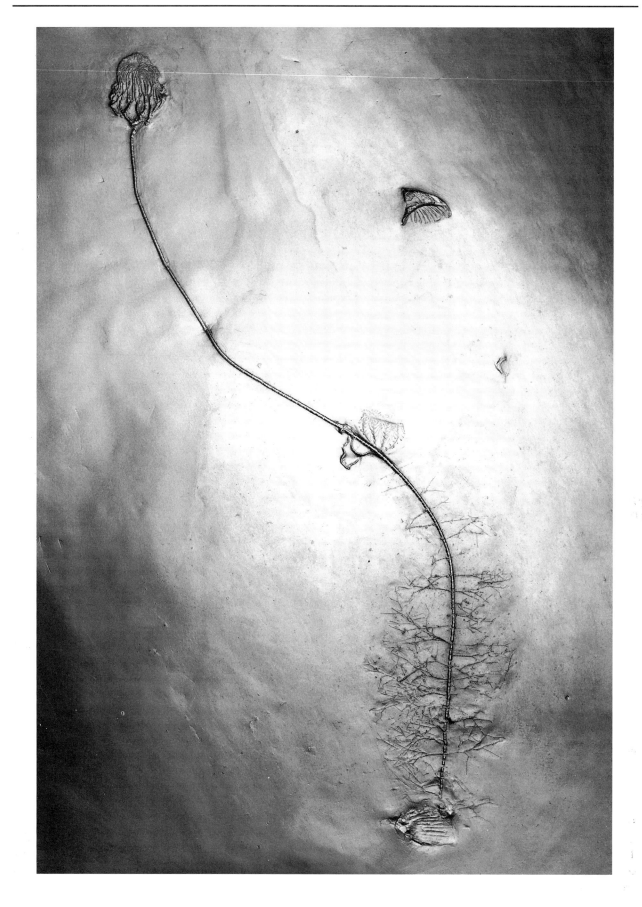

Fig. 125. (*On facing page*) *Eutaxocrinus stürtzii* attached to a rugose coral; fixed to the middle part of the stem is a juvenile *Hapalocrinus elegans*. The slab has been prepared from below, the stem was broken at a distance of about a third from the crown, and this proximal part is in a somewhat higher sediment layer. It must be assumed that the distal part of the stem of the *Eutaxocrinus* with its radicular cirri was lying on the sea floor while the crown and the proximal part of the stem were elevated into the water. The stem must have been broken during burial. The animal appears to have initially grown on the coral and formed its dense network of radicular cirri only later. Hunsrückschiefer, Bundenbach. (C. Bartels Collection; Deutsches Bergbau Museum, Bochum; photograph A. Opel; courtesy C. Bartels.) This photograph as well as those in Figs. 126–129 are of specimens treated with mineral oil; they therefore appear lighter than that in Fig. 123 and many Hunsrückschiefer fossils figured previously. ×0.5.

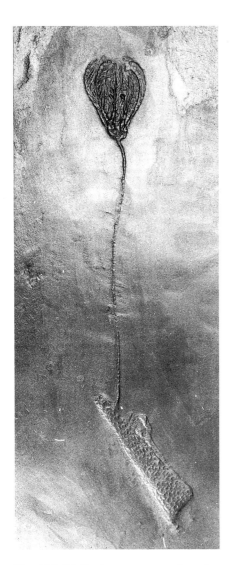

Fig. 126. *Thallocrinus procerus* anchored to the shell of a small orthocone cephalopod, which is overgrown with epizoic tabulate corals. Hunsrückschiefer, Bundenbach, Eschenbach-Bocksberg pit. (G. Brassel Collection; preparation and photograph C. Bartels; courtesy C. Bartels.) ×0.8.

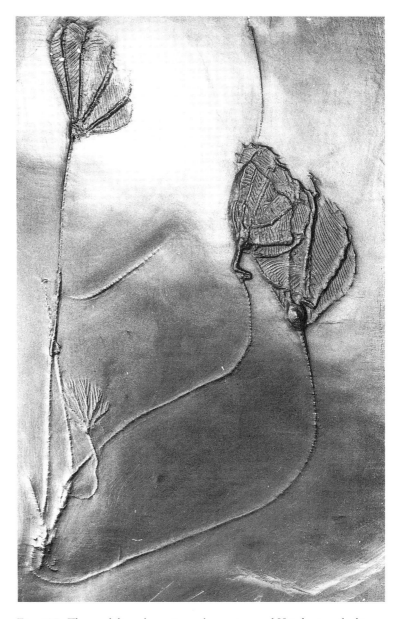

Fig. 127. Three adult and one juvenile specimen of *Hapalocrinus frechi*, anchored to the stem of a fifth, now-lost crinoid. Hunsrückschiefer, Bundenbach, Eschenbach-Bocksberg pit. (G. Brassel Collection; photograph C. Bartels; courtesy C. Bartels.) ×0.6.

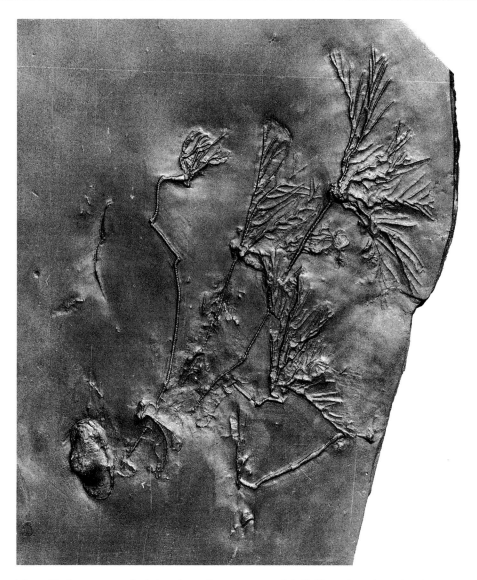

Fig. 128. Group of three *Rhadinocrinus* sp. (right) and one *Bactrocrinites jaekeli* (left), anchored to a brachiopod shell; stem and cup of another specimen of *Bactrocrinites* are visible at lower right. Hunsrückschiefer, Bundenbach, Eschenbach-Bocksberg pit. (G. Brassel Collection; preparation and photograph C. Bartels; courtesy C. Bartels.) ×0.9.

IMPORTANT COLLECTIONS IN GERMANY

Bayerische Staatssammlungen für Paläontologie und historische Geologie, Munich. This contains a large collection of Ferdinand Broili with many specimens described from the Hunsrückschiefer; in addition, important private collections (Maucher, Rievers) were deposited here.

Humboldt University, Berlin. This has taken over the collection of the former Reichsstelle für Bodenforschung with specimens collected by F. Kutscher before World War Two. Many of W. E. Schmidt's originals are deposited here.

Museum am Besucherbergwerk Bundenbach (started in 1993). The collection of the community shows a cross section of the local Hunsrückschiefer fauna.

Natur-Museum Senckenberg, Frankfurt a.M. This important museum contains a large part of the Opitz Collection, as well as the Rudolf and Emma Richter Collection. The classic X-ray films of Stürmer are

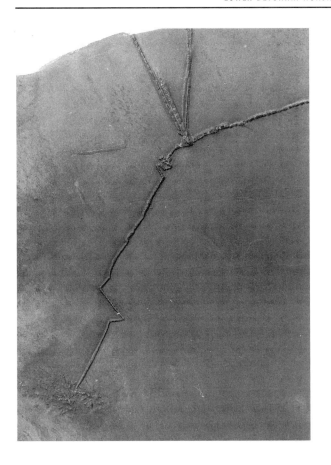

also preserved here, as will be in the near future G. Brassel's collection, one of the most important ones accumulated since World War Two.

Schlossparkmuseum Bad Kreuznach. This contains parts of the collection of Opitz as well as the Herold Collection with numerous type specimens. Opitz and Herold accumulated the most important collections made between 1920 and 1939.

University of Bonn, Geological Institute. This collection contains many of the beautiful asterozoans described by Lehmann (1957), and was accumulated between 1930 and 1959.

Deutsches Bergbau Museum, Bochum. This houses the unique Bartels Collection, with an almost complete documentation of the Bundenbach area and other, largely unknown Hunsrückschiefer sites in the southeast Eifel region.

Fig. 129 *Triacrinus koenigswaldi*, specimen with dense tuft of root radicular cirri at the end of the stem. This form has three basals and five long, unbranched arms. Hunsrückschiefer, Bundenbach, Eschenbach-Bocksberg pit. (Collection, preparation and photograph C. Bartels; Deutsches Bergbau Museum, Bochum; courtesy C. Bartels.) ×1.

14 Middle Devonian Windom Shale of Vincent, New York, USA

CARLTON E. BRETT

REDISCOVERY OF A LOST SITE

In the late 1800s and early part of the 1900s, large colonies of crinoids were excavated from a small stream bank and hillside exposures of the Middle Devonian Windom Shale near the town of Vincent (formerly called Muttonville), Ontario County, New York. These were described and illustrated in some detail by Goldring (1923). For many years this location was overgrown and the horizon that had yielded the crinoids was unknown, although it was presumed to be in the Lower Moscow Formation, possibly the Deep Run Shale. Careful prospecting in the vicinity of Vincent by the late James Nardi and the author of this chapter resulted in the rediscovery of the concretionary crinoid horizon within the Upper Moscow Windom Shale along Mud Creek, about 2 km from the original site (Fig. 130). Subsequent excavations have helped to elucidate the nature of the occurrence.

Crinoids in these colonies occur in the upper portion of the Middle Devonian (mid-Givetian) Hamilton Group, approximately 380 million years before present. Specifically, they are derived from layers of highly fossiliferous shale in the upper portion of the Windom Member of the Moscow Formation (Fig. 131).

As in the case of the Silica and Arkona crinoids (see Chapter 15), lenses of fossil debris may contain complete crinoids; those from the Windom Shale are typically associated with pods of stick bryozoans and crinoidal debris. These are lenticular, typically concretionary beds, ranging from a few millimetres to more than 1 cm in thickness. Some display internal laminae of mud, suggesting that they represent composite beds resulting from the stacking and time-averaging of skeletal debris through a series of storms. Intervening shales are com-

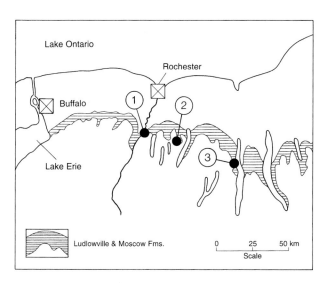

Fig. 130. Location map for Middle Devonian Windom Shale crinoid beds in the Finger Lakes region of western New York. (1) Fall Brook; (2) Vincent; (3) Kashong Creek. (Modified from Brett & Baird 1994.)

Hans Hess, William I. Ausich, Carlton E. Brett, and Michael J. Simms, eds., *Fossil Crinoids*. © 1999 Cambridge University Press. All rights reserved. Printed in the United States of America.

moderate to long crinoid stem fragments. They may also contain complete brachiopods, particularly large spiriferids (*Mediospirifer*) at their tops, preserved *in situ*; this mode of preservation suggests rapid but not instantaneous burial of the organism. However, the tops of one or two of the beds display clumps of articulated crinoids, attesting to rapid, probably live burial by storm-generated mud suspension clouds. The famous slabs illustrated by Goldring display completely articulated specimens with long stems bearing whorls of radicular cirri and pinnulate crowns.

A CAMERATE-DOMINATED CRINOID ASSOCIATION

As is typical of crinoid occurrences within Palaeozoic mudstones, the associated fauna consists largely of remains of thickets of stick-like branching bryozoans, particularly the ribbon-like cryptostome *Sulcoretepora*, to which some crinoids may have been attached. Also common are small mounds, up to 10 cm high, of crinkly, lettuce-like, fistuliporoid bryozoans and fenestrates. Common brachiopod associates include spiriferids, such as *Mucrospirifer* and *Mediospirifer*, atrypids, chonetids and strophomenids, such as *Protodouvillina*. The trilobites *Phacops* and *Monodechenella* are commonly associated with bryozoan–crinoid clumps. Some of the larger camerate crinoid stems display attached ball-like tabulate corals of the genus *Antholites*, which presumably benefitted by being elevated above the soft substrate into areas of favourable current velocity. Abundant platyceratid gastropods evidently lived coprophagously on their crinoid hosts.

About a dozen species of crinoids occur in the Upper Windom Beds. Most of these were relatively long-stemmed, with stem lengths ranging from about 15 cm to nearly 1 m. These long stems elevated the crowns well above the substrate into areas of lower turbidity and probably higher current strength.

The Windom crinoids were dominated by the rather peculiar camerate *Clarkeocrinus*, a form with a 20- to 30-cm stem bearing whorls of flexible cirri up to the level of the crown (Fig. 132). Those nearest the crown typically occur as tight coils, which may have served to protect the crown (Fig. 133). Otherwise, the calyx and arms are typical of the dolatocrinids. The flexible radicular cirri of *Clarkeocrinus* were probably adapted to penetration into soft muds. Radicular cirri could also be entangled about other objects, such as ramose bryozoans.

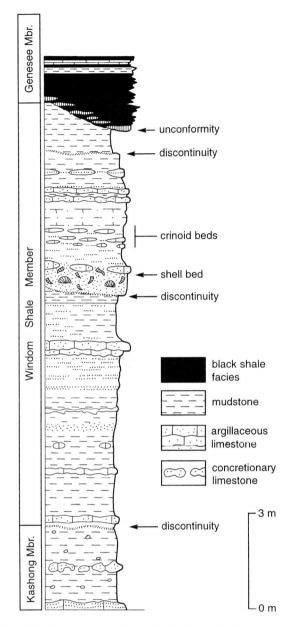

Fig. 131. Stratigraphic column for Windom Shale in Genesee and western Finger Lakes region of New York. (Modified from Brett & Baird 1994.)

posed of medium dark grey, somewhat calcareous and sparsely fossiliferous mudstone.

The taphonomy of upper Windom crinoid-rich beds has been discussed in some detail by Parsons *et al.* (1988). As in most other crinoid occurrences, there is a mixture of well-preserved fossils and disarticulated skeletal debris. Within the shelly layers, the majority of material consists of disarticulated and comminuted crinoidal and bryozoan debris. Many shell beds display

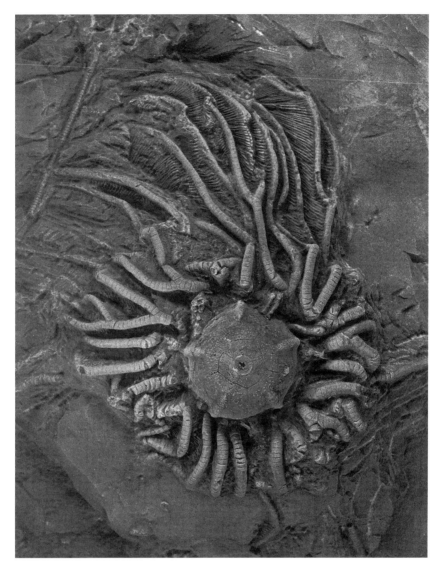

Fig. 132. Aboral view of crown of *Clarkeocrinus troosti* from concretionary portion of Middle Windom Shale near Vincent, N.Y. (J. Nardi, collector; University of Rochester Collections.) ×1.5.

Moreover, it is likely that *Clarkeocrinus* held a portion of its stem recumbent on the sea bottom, as is the habit of most modern stalked crinoids that possess whorled cirri.

Other associated crinoids used different types of attachment strategies. Notably, *Gennaeocrinus* (Fig. 134) had a long column with radicular cirri directed downward along the distal portion that may have served as a stiff propping root. This type of attachment structure was common in many of the camerate crinoids of Devonian and Mississippian age. It was presumably an adaptation to attachment in relatively soft, slightly shifting substrates. The peculiar triangular flanges on the proximal columnals of *Gennaeocrinus* stems (Fig. 134, arrows) are of unknown function; possibly they served to catch the current to aid in elevating the crown. *Megistocrinus* possessed a distally thickened stem with cemented areas that lay recumbent on the sea bottom and may have attached to bryozoans or other crinoids. Adults of both of these camerates had relatively large, globose calyces with broad tegmens and minor anal chimneys. *Gennaeocrinus* also had short spines on calyx plates. The many-branched, biserial, pinnulate arms probably formed a filtration device to capture fine suspended food.

Thylacocrinus (Fig. 135) had a globose, sac-like calyx, 20 unbranched pinnulate arms and a rather short pentagonal stem with a distally tapered coiled end. It evidently wound this terminal part of the stem around upright supports, including other crinoid stems. Hence, it might have occupied a higher position, taking advantage of the height of other crinoids or bryozoan colonies to elevate itself further above the substrate. Both *Gilbertsocrinus spinigerus* and *Acanthocrinus spinosus* had small box-like calyces with broad, flat tegmens and long spines on the calyx plates (Fig. 136). It has been suggested that such spines served to keep platyceratid gastropods off the crinoid crowns. However, this function is refuted by the fact that these crinoids actually served as hosts to the commensal snails more frequently than other non-spiny Devonian crinoids, possibly because of their expansive tegmens. We suggest that, rather than being a defence from the innocuous snails, the spines protected the crinoids and, incidentally, their guests from predators that may have been attracted by the sessile gastropods. *Gilbertsocrinus* had both biserial pinnulate arms and curious flexible, tubular tegminal appendages of unknown function (Fig. 159). These crinoids are very typical of muddy substrates. Both *Gilbertsocrinus* and *Acanthocrinus* possessed rather long, distally tapering stems without radicular cirri or cemented holdfasts but, commonly, with coiled distal tips. It is possible that they had become secondarily recumbent on the sea floor. The strange downward-directed spines on the radial plates (Fig. 136) could then have served to prop the crown up slightly off the bottom.

The cladids *Botryocrinus nycteus*, *Decadocrinus nereus* and *Atelestocrinus* sp. had relatively short stems with distal radicular cirri for anchorage. *Decadocrinus* had pinnulate arms and anal sacs that may have supported respiration. *Atelestocrinus* was unusual in (generally) having only four free arms.

Blastoids, such as *Devonoblastus*, possessed shorter, coiled stems. However, one notable blastoid, *Eleutherocrinus*, was stemless and rested its theca directly upon the substrate.

AN OFFSHORE MUD BOTTOM ENVIRONMENT

The crinoid-bearing concretions in the upper portion of the Windom Shale represent an offshore, shallow-marine, mud bottom environment. The environment was evidently unsuitable for the majority of rugose and tabulate corals, perhaps because of relatively high rates of sedimentation and/or turbidity. However, colonial organisms having upright branching skeletons with good sediment-shedding ability and mound-like colonies of fistuliporoid bryozoans appear to have flourished in this same setting, at least in local thickets and patches. A soft, muddy substrate, with fairly abundant skeletal debris, provided attachment sites, not only to the bryozoans but to a diversity of brachiopods and crinoids. The Windom Shale provides evidence of a tiered, moderate- to high-diversity community that contained a wide array of organisms and was dominated by suspension-feeding bryozoans, crinoids and brachiopods.

Detailed studies of the biostratinomy of upper Windom fossil beds indicate that this interval consists of a

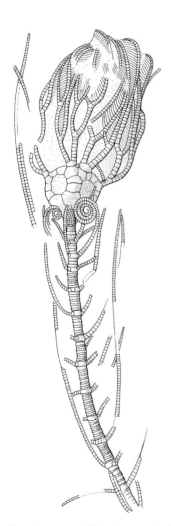

Fig. 133. *Clarkeocrinus troosti*. Note long, whorled, radicular cirri tightly coiled beneath calyx. Windom Shale, Vincent, N.Y. (New York State Museum; redrawn from Goldring 1923.) ×0.7.

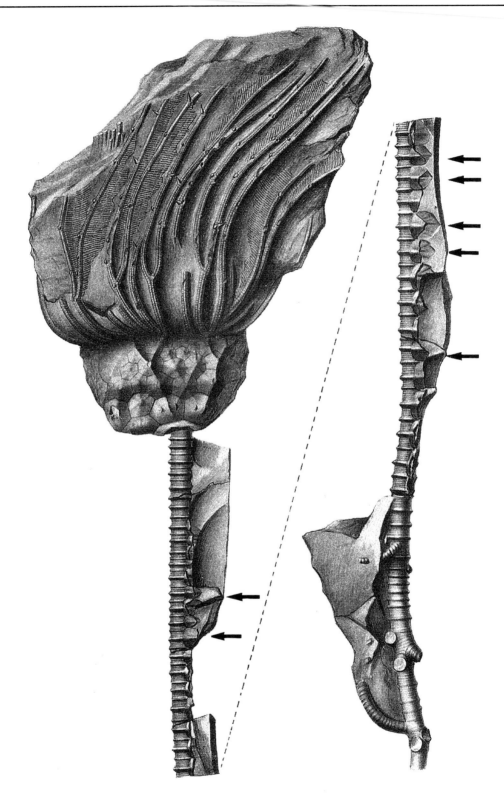

Fig. 134. Complete specimen of *Gennaeocrinus eucharis*. Note rows of triangular flanges on proximal stem (arrows). Windom Shale, Vincent, N.Y. (New York State Museum; from Goldring 1923.) ×1.

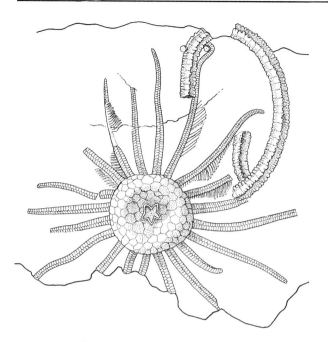

Fig. 135. *Thylacocrinus clarkei*. Aboral view of a well-preserved crown with pentagonal stem disarticulated, but lying close to the crown. Windom Shale, Vincent, N.Y. (New York State Museum; redrawn from Goldring 1923.) ×0.85.

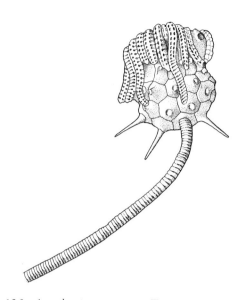

Fig. 136. *Acanthocrinus spinosus*. Type specimen showing spines directed downward, recumbent arms and a portion of the stem. Windom Shale, Vincent, N.Y. (New York State Museum; redrawn from Goldring 1923.) ×2.

series of shell-hash horizons that are laterally traceable and may represent much of the time span of the formation (Parsons *et al.* 1988). Conversely, the typically thicker (1–10 cm), nearly barren mudstone layers between the fossil hash beds appear to represent pulses of siliciclastic mud deposition. Background sedimentation rates were relatively low, although there may have been plentiful suspended sediment in the water. During these times, organisms lived, grew and died, and their remains accumulated to form slightly time-averaged skeletal hash beds. Debris of crinoids, bryozoans and other organisms may have provided substrates for the attachment of additional hard-substrate-adapted forms, including certain species of echinoderms, some bryozoans and pedically attached brachiopods. Nonetheless, the well-preserved fossils are patchy in their distribution, probably reflecting an original patchiness of the living communities; crinoid and bryozoan thickets were separated by areas of more barren, muddy sea bottom.

The influx of clouds of suspended sediment, as in the case of the Silurian Rochester Shale (see Chapter 10), abruptly blanketed the living communities. Uprooting and burial resulted in the mass death of crinoid communities over a wide region of western New York. Beds rich in *Clarkeocrinus*, at approximately the same level in the Windom Shale, have been traced for nearly 65 km in outcrop. The disruption and burial of the crinoid colonies suggest intermittent but relatively infrequent disturbance of the muddy sea floor by storms. Storm-generated currents may have been responsible for the uprooting of the crinoids, as well as for bringing in suspended fine-grained sediment to bury the recumbent animals. The occurrence of concretionary masses around the well-preserved crinoids indicates that early diagenesis of calcium carbonate cements took place within the muds that had buried the crinoid colonies.

The generation of bicarbonate and/or ammonia as a result of the decay of the organisms' soft parts within the sediment may have triggered the initial cementation of the muds. However, the development of relatively large concretions around the fossil layers is a common feature within the Windom at several horizons. These may reflect the tops of small-scale sedimentary cycles. Minor periodic sea level rises or drier climatic events may have suspended siliciclastic sedimentation for a sufficiently long period to allow bicarbonate cements to build up within the pore waters of the muddy sediment. Over a period of time, probably on the order of a few thousand years, the cements gradually hardened the mud layers around the crinoids, encasing them in a semi-protective jacket that helped to inhibit further deformation of the crinoids by compaction as overlying sediment accumulated.

In summary, we visualize the Middle Devonian Windom Shale depositional environment as a muddy, relatively shallow, offshore sea floor. On the basis of the occurrence of intermittent storm effects, we suggest that the depths may have been of the order of 50–70 m, well below normal wave base, but vulnerable to more distal storm waves. This seems to be a common theme with many of the best-preserved fossil crinoid assemblages.

IMPORTANT COLLECTION IN THE UNITED STATES

New York State Museum, Albany

15 Middle Devonian Arkona Shale of Ontario, Canada, and Silica Shale of Ohio, USA

CARLTON E. BRETT

The two related fossil assemblages discussed in this chapter occur in possibly age-equivalent Middle Devonian mudstones in southern Ontario and Ohio (Fig. 137). The Arkona Shale is exposed along the banks of the Ausable River and in its tributary streams in the vicinity of Arkona, Lambton County, Ontario, Canada. The Silica Shale is exposed in the quarries of the Medusa Cement Company and the New Genstar Cement Company, both north and south of Centennial Road in the town of Silica, Lucas County, Ohio. These mudstones are considered to be in the lower part of the Hamilton or Traverse Group and are of Early Givetian age, approximately 385 million years before present.

LIMESTONE LENSES WITH BRYOZOANS, BRACHIOPODS, TRILOBITES AND COMPLETE CRINOIDS

The thin skeletal limestone lenses within the Arkona and Silica Shales carry a moderately diverse fauna of bryozoans, brachiopods, gastropods, bivalves and trilobites, as well as crinoids and blastoids. Approximately 40–50 species can be found with considerable effort. The Silica fossils are described in a richly illustrated volume compiled by Kesling and Chilman (1975). Among the most common fossils are the stick-like bryozoans, *Sulcoretepora*, as well as *Fenestella*. Brachiopods include abundant, small, concavo-convex chonetids and *Mucrospirifer*. The trilobite *Phacops* is also commonly associated and has been found in clumps of articulated individuals in the Silica Shale of Ohio; it has become the landmark of these strata and is a highly valued collector's item. The surrounding mudstones are quite sparsely fossiliferous, but do occasionally contain isolated specimens of chonetids, *Mucrospirifer* and other brachiopods. Scattered, pyritized specimens of small bivalves (nuculids), goniatites and bactritids also occur.

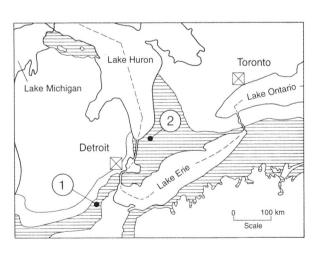

Fig. 137. Location map for Devonian Arkona and Silica Shales in southern Ontario and Ohio. (1) Silica; (2) Arkona.

Hans Hess, William I. Ausich, Carlton E. Brett, and Michael J. Simms, eds., *Fossil Crinoids*. © 1999 Cambridge University Press. All rights reserved. Printed in the United States of America.

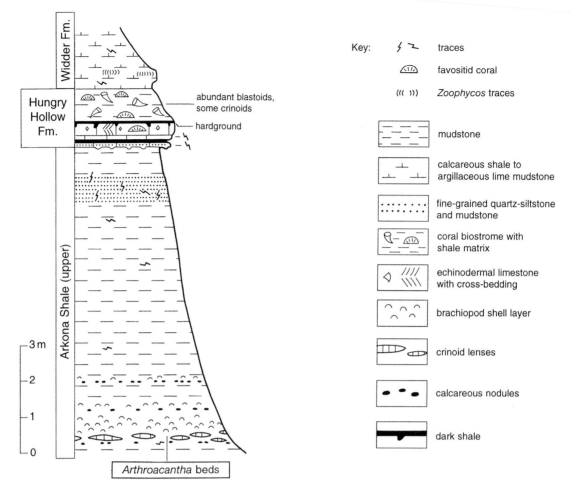

Fig. 138. Stratigraphic section of Middle Devonian Arkona Shale, exposed in a brick pit on the north side of the Ausable River near Arkona, Ontario. (Modified from Landing & Brett 1987.)

Ostracodes are common on some parting planes in the mudstone.

Complete crinoids in both areas appear on the top of thin lenses of skeletal debris of crinoids, bryozoans and occasional brachiopods. These lenses are included in soft, slightly petroliferous, bluish-grey mudstone (often termed shale, although it lacks fissility). Surrounding mudstones are nearly barren of fossils. However, crinoids and other fossils occur in great densities at the interface between some lenticular pods of skeletal debris and the overlying sparsely fossiliferous mudstone (Figs. 138, 139). In some cases, completely articulated crinoids are also found at the margins of lenses on bedding planes that are otherwise obscure. That is, they occur at the boundary between one mudstone and another, with only a very thin scattering of skeletal debris and/or complete crinoids marking this interface. In at least one horizon within the Arkona Shale, crinoids are also associated with the top of a thin, somewhat concretionary siltstone layer within the mudstone.

The lenses of skeletal debris form an apparent substrate beneath the crinoid colonies. In both cases, debris consists of broken crinoid stems (individual columnals and very short sections), plates of the crown and even partly articulated cups. Brachiopod shells are usually whole valves, although many are disarticulated. Bryozoans, which are predominantly of the narrow blade-like cryptostome *Sulcoretepora*, tend to be fragmented into short segments no more than 1 cm in length. In contrast, crinoids, blastoids and other delicate fossils found at the interface between the skeletal lenses and the overlying mudstone may be perfectly preserved with a considerable length of stem and radicular cirri, carrying complete crowns with delicately pinnulate arms and, in many cases, attached platyceratid gastropods. The pinnulate arms on the camerate crinoids are usually stretched out or splayed, although in some cases the arms appear to have been drawn together. The perfect preservation of these crinoids suggests a complete lack of transport and nearly immediate burial by the clay-rich sediment.

SPINY CRINOIDS AND PLATYCERATID SNAILS

The crinoid assemblages are of relatively low diversity and are dominated by one or two species of camerates. In both the Silica and the Arkona Shales, the most common crinoid is the simple, monocyclic camerate *Arthroacantha*. This crinoid had a stem approximately 20–30 cm long, with whorls of radicular cirri towards the distal end, and a relatively large crown with pinnulate arms (Fig. 140). Its most unique feature is the presence of small articulated, movable spines on the cup plates and of larger spines on the axillaries (hence the name 'spiny joints'). A majority of these crinoids have platyceratid gastropods, such as the spiny *Platyceras dumosum*, attached to the tegmen. Another, relatively common crinoid in the Silica Shale is the spiny camerate *Gilbertsocrinus*. This crinoid is unique in having peculiar, snake-like tegminal appendages that were originally mistaken for arms; in the famous Mississippian species *Gilbertsocrinus tuberosus* (see Chapter 18) the appendages are pendent and dominate the crown, with the delicate, pinnulate, biserial arms being inserted between them. *Gilbertsocrinus ohioensis*, the more common species in the Silica Formation, has much stronger arms and only weak, string-like appendages. The function of the appendages, which are hollow, remains something of an enigma. The box-like cup carries on its base long spines that are directed downward. The large, flat tegmen of these camerates also commonly supported com-

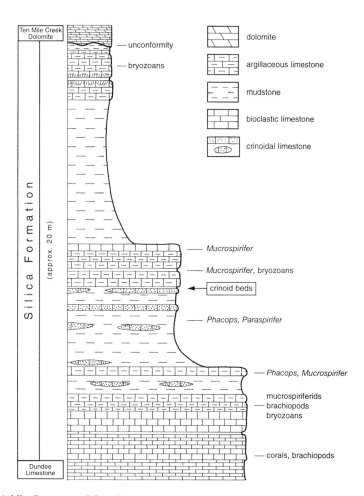

Fig. 139. Section of the Middle Devonian Silica Formation exposed in quarries near Silica, Ohio. The thickness of the Silica Formation is approximately 20 m. (Modified from Kesling & Chilman 1975.)

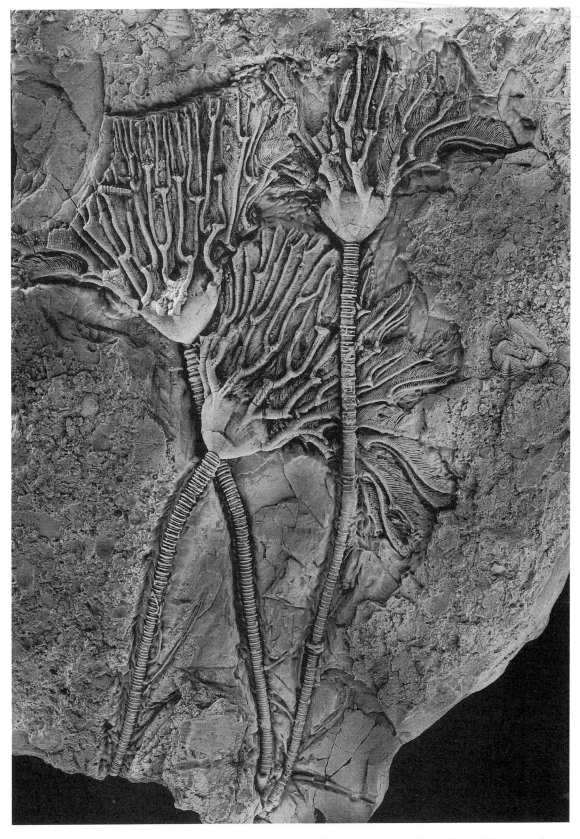

Fig. 140. *Arthroacantha carpenteri*. Silica Formation, Medusa Cement Quarry, Sylvania, Ohio. (Hess Collection; photograph S. Dahint.) ×1.

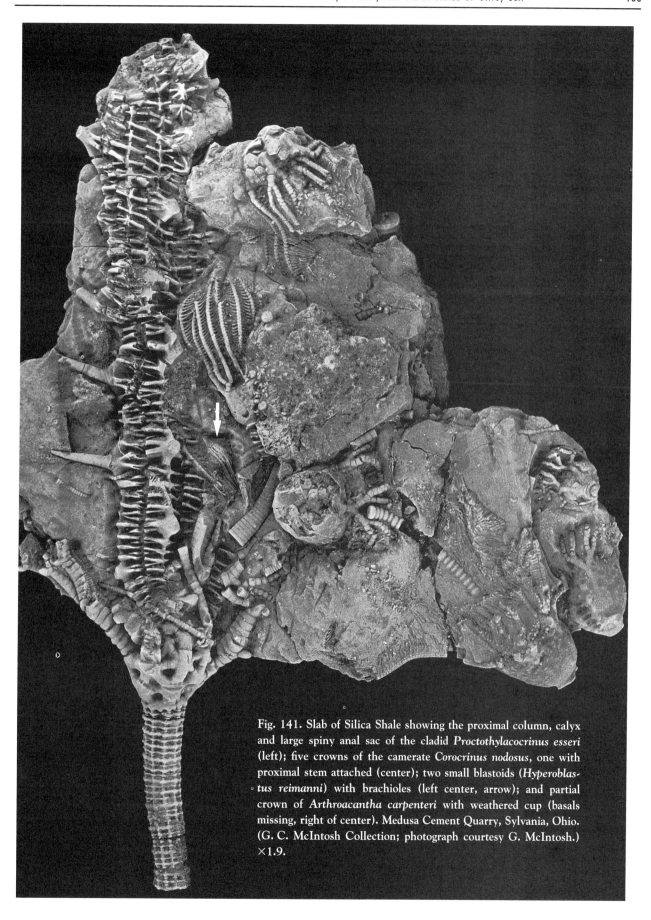

Fig. 141. Slab of Silica Shale showing the proximal column, calyx and large spiny anal sac of the cladid *Proctothylacocrinus esseri* (left); five crowns of the camerate *Corocrinus nodosus*, one with proximal stem attached (center); two small blastoids (*Hyperoblastus reimanni*) with brachioles (left center, arrow); and partial crown of *Arthroacantha carpenteri* with weathered cup (basals missing, right of center). Medusa Cement Quarry, Sylvania, Ohio. (G. C. McIntosh Collection; photograph courtesy G. McIntosh.) ×1.9.

mensal platyceratid gastropods. In the Arkona and Silica Shales another camerate crinoid, *Corocrinus nodosus*, is abundant in certain assemblages (Fig. 141) and, again, commonly has attached platyceratid snails. Associated specimens of the small cladid *Decadocrinus*, *Proctothylacocrinus esseri* (Fig. 141) and the flexible *Synaptocrinus* are also found. The blastoid *Hyperoblastus* is commonly associated with the crinoids at Sylvania, Ohio, and is typically preserved with stems and brachioles (Fig. 141).

A MUDDY SEA FLOOR

Both the Arkona and Silica Shales were deposited in relatively offshore, deeper water, characterized by muddy bottoms. Some evidence of winnowing and fragmentation of fragile fossils and of erosive sole marks on the bases of some coarsely skeletal limestone lenses suggests intermittent activity of storm-generated currents on the sea bottom. However, the crinoids lived predominantly below the effect of storm waves. Clays deposited in this offshore setting were ultimately derived from the erosion of eastern siliciclastic sources, uplifted in the Acadian orogenic belt along the present eastern seaboard of North America. The sparsely fossiliferous nature of most of the mudstone suggests rather inhospitable conditions associated with a soft, possibly soupy substrate; and the occurrence of a diminutive, pyritized fossil in some of these shales indicates low-oxygen conditions at least at and below the surface of the sediment. During most of the time, relatively little sediment accumulated. However, considerable mud was deposited during brief pulses associated with the development of mud-rich slurries, possibly winnowed from upslope areas of flocculated clays by storm waves and transported into this setting by storm generated gradient currents.

Most of the benthic organisms in these assemblages were apparently adapted to soft-substrate attachment or support. The brachiopods, for example, display large bearing surfaces or elongate ski-like wings in the *Mucrospirifer* brachiopods. The crinoids such as *Arthroacantha* possessed flexible radicular cirri that may have permitted either attachment to other objects (including other crinoids) or temporary anchoring to soft substrate. *Gilbertsocrinus* was tethered by a distal coil that could be wrapped around bryozoan stalks, other crinoids or positioned on the sea floor in a series of coils resembling a coiled snake. In this crinoid, and perhaps in *Arthroacantha*, a portion of the stem may have been borne horizontally as a runner on the sea bottom, as is noted for some modern isocrinids (see Chapter 29). Blastoids and some associated small crinoids were anchored to thickets of bryozoans by either coils or small discoidal holdfasts. The occurrence of crinoids and other animals preferentially around pods of skeletal debris suggests a process of taphonomic feedback in which an armouring of crinoidal skeletal material from the sea bed permitted attachment of new types of epifaunal organisms in a tiered ecological succession. However, it is clear that most of the crinoids could anchor in soft muds as well as attach themselves to loose piles of skeletal debris.

IMPORTANT COLLECTION IN THE UNITED STATES

University of Michigan, Museum of Paleontology, Ann Arbor

16

Lower Mississippian Hampton Formation at LeGrand, Iowa, USA

WILLIAM I. AUSICH

COLOURFUL CRINOIDS

Colonies and colours provide the fascination of LeGrand, Iowa, crinoids. The invertebrate palaeontology halls of most U.S. museums display a slab of buff-coloured dolomite covered with crinoid crowns. The slab may be up to 2 m across with small, perfectly preserved crowns and a tangle of stems forming an eye-catching display (Figs. 142, 143). Most of these museum display pieces are from the original large crinoid colony discovered during the preceding century. Crinoids occur in extensive, thin lenticular beds representing the original distribution of living crinoid colonies.

The buff-coloured dolomite provides an attractive background for these crinoids, but their aesthetic appeal nearly hides the most remarkable aspect of this fauna: the crinoids themselves are preferentially coloured. Preservation of this fauna was so good that certain species retain distinctive coloration. The earth tones that shade these fossils undoubtedly do not reflect living coloration, but the species-specific nature of this remarkable preservation does suggest some type of primary vital effect. However, in contrast to situation with other crinoids (see Chapter 4), the chemistry of the pigments responsible for the coloration has not yet been examined.

STRATIGRAPHY OF THE LEGRAND CRINOIDS

LeGrand crinoids are from the Maynes Creek Member of the Hampton Formation in north-central Iowa (Mississippian, approximately 355 million years old). The Maynes Creek is composed of approximately 20 m of fossiliferous buff- to brown-coloured dolomite interbedded with chert. These rocks were deposited in a fairly shallow-water epicontinental setting (Laudon 1931). The member immediately beneath the Maynes Creek Member is an oolitic unit. The Hampton Formation is Kinderhookian (Tournaisian 2) in age (Anderson 1969). Hampton crinoids occur in the centre of lenticular beds.

CRINOID COLONIES RECOVERED THROUGH QUARRYING

James Hall was the first to collect crinoids at LeGrand, Iowa, in 1858. However, nearly another quarter century passed before serious collections were made at LeGrand (see Laudon & Beane 1937), and the first LeGrand specimens were not described until 1890.

Many thousands of crinoids have been collected, up

Hans Hess, William I. Ausich, Carlton E. Brett, and Michael J. Simms, eds., *Fossil Crinoids*. © 1999 Cambridge University Press. All rights reserved. Printed in the United States of America.

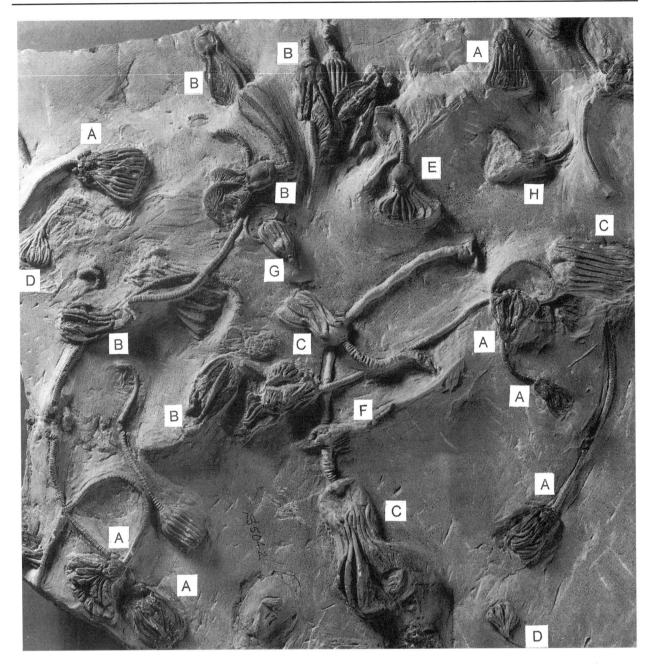

Fig. 142. Bedding surface with numerous LeGrand crinoids. (A) *Rhodocrinites kirbyi*; (B) *Strimplecrinus inornatus*; (C) *Platycrinites symmetricus*; (D) *Aphelecrinus elegantulus*; (E) *Orophocrinus conicus*; (F) *Orophocrinus fusiformis*; (G) *Abatocrinus macbridei*; (H) *Cribanocrinus watersianus*. (Reprinted by permission of the National Museum of Natural History, Washington, D.C.) ×0.6.

to now all from a relatively narrow stratigraphic zone. Crinoid colonies were entombed in a relatively soft, fine-grained dolomite. The initial colony was collected between 1874 and 1890 during a time when quarrying was done by hand. Quarrymen recognized the value of these unusually fossiliferous rocks and took care to preserve them.

B. H. Beane of LeGrand renewed the collection of these crinoids in 1931 with the discovery of a small colony of *Rhodocrinites kirbyi*, which had more than 200 specimens in 1 m². In 1934, Beane discovered another colony. Quarry operations blasted into the margin of the colony (Laudon & Beane 1937) and pieces were strewn throughout the rubble of the blast. The bulk of the new colony remained in place and was exposed for 6 m along the quarry face. With patience, much work and the close co-operation of the quarry operators, Beane excavated the remainder of this colony through 1937. This

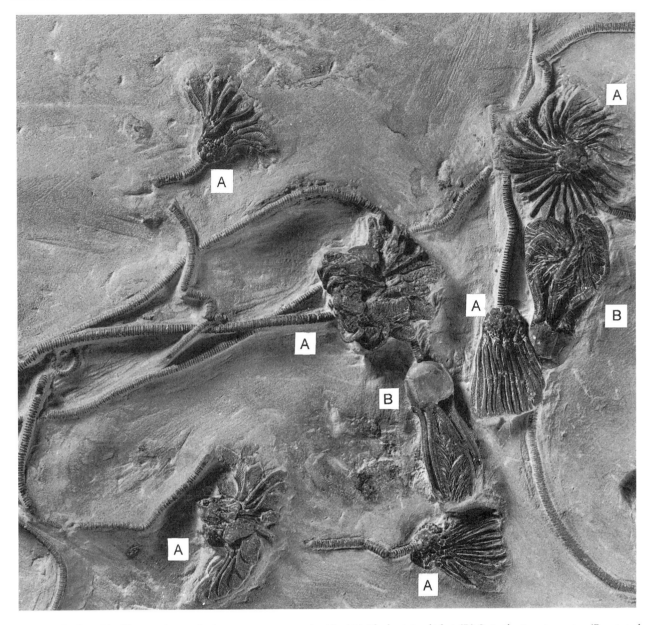

Fig. 143. LeGrand bedding surface with the more common crinoids. (A) *Rhodocrinites kirbyi*; (B) *Strimplecrinus inornatus*. (Reprinted by permission of the National Museum of Natural History, Washington, D.C.)

final colony yielded several thousand of the best crinoid specimens recovered from this site. Beane collaborated with Lowell R. Laudon to publish a monograph on the LeGrand crinoid fauna in 1937.

THE CRINOIDS OF LEGRAND

Presently, 41 species assigned to 25 genera are known from the LeGrand crinoid beds. In addition, blastoids, asteroids and echinoids also belong to this fauna. Advanced cladids are the most diverse group, but the camerates, diplobathrids and monobathrids are the dominant faunal elements. The most common crinoid in this fauna is the diplobathrid camerate *Rhodocrinites kirbyi* (Figs. 55, 142, 143). Other common crinoids include the monobathrids *Aorocrinus immaturus*, *Strimplecrinus inornatus* and *Platycrinites symmetricus* (Figs. 142, 143) and the flexible *Taxocrinus intermedius*. Although relatively uncommon, pinnulate cladids are reasonably diverse in this fauna (11 species) and provide an early glimpse of these forms, which dominated crinoid faunas by the Late Mississippian.

Species-specific coloration discussed by Laudon and Beane (1937) includes the following: *Rhodocrinites kirbyi*, very dark colour; *R. nanus*, very light colour for genus; *R. nanus glyptoformis* and *Cribanocrinus watersianus*, midway between *R. kirbyi* and *R. nanus*; *Strimplecrinus inornatus* and *Dichocrinus hammondi*, dark colour; and *D. delicatus*, white. Other coloration distinctions not mentioned by Laudon and Beane (1937) are also present among LeGrand species.

The LeGrand fauna has received little modern study. However, its remarkable preservation provides insight into crinoid palaeoecology. LeGrand crinoids lived in dense colonies, some monospecific and others very diverse (Fig. 142). Colony size varied greatly, from small circumscribed patches 1 m in diameter to colonies with a diameter approaching 10 m. These colonies also appear to have been established in depressions on the sea floor, which is in apparent conflict with living stalked crinoids that prefer exposed areas where currents are locally enhanced.

Blastoids lived among crinoids in the larger colonies. Although most LeGrand crinoids are rather small, interspecific differences in both filtration fan density and stem length are apparent, suggesting that these crinoids partitioned food resources in the same way as geologically younger crinoids (Ausich 1980) (see Chapter 18). Finally, the larger colonies must have been established for a relatively long time because juveniles and adults are preserved together.

IMPORTANT COLLECTIONS IN THE UNITED STATES

Beloit College, Beloit, Wisconsin (the Beane Collection)
Field Museum of Natural History, Chicago, Illinois
National Museum of Natural History, Smithsonian Institution, Washington, D.C.

17 Lower Mississippian Burlington Limestone along the Mississippi River Valley in Iowa, Illinois, and Missouri, USA

WILLIAM I. AUSICH

CRINOIDS BY DOCTOR'S PRESCRIPTION

Charles Wachsmuth was an ill man. This German native immigrated to the United States to pursue a mercantile career and eventually settled in Burlington, Iowa, in an attempt to find a better climate. In his early 30s and with his health continuing to fail, his doctor prescribed fresh air and exercise. According to Keyes (1897, p. 13), the doctor suggested 'that the collecting of fossils, which abounded in the rocks of the neighborhood, would soon provide an incentive for sufficient exercise'. Thus, this resident of Burlington began to collect fossils from country underlain by perhaps the largest accumulation of crinoidal remains anywhere in the world, the Lower Mississippian Burlington Limestone. The study of fossil crinoids became a consuming interest for Charles Wachsmuth for the next 30 years. With the encouragement of Alexander Agassiz (Harvard University), Wachsmuth's collecting zeal matured to scientific curiosity. Wachsmuth accumulated a fantastic collection of fossil crinoids, which was eventually sent to the Museum of Comparative Zoology. Soon thereafter, Wachsmuth began to develop a second collection, which became even better than the first. Most of Wachsmuth's scientific work was completed in collaboration with Frank Springer. Springer was a native Iowan who developed an interest in natural history and fossils at an early age. Springer began a law practice in Burlington, Iowa, and also began collaborative studies on crinoids with Wachsmuth. In 1873 Springer moved his practice to New Mexico and became established as one of the foremost lawyers in this developing area of the United States. However, he returned during the summers to Burlington and continued his research on crinoids with Wachsmuth. Wachsmuth and Springer published numerous scientific papers on fossil crinoids, with the most influential being the 'Revision of the Palaeocrinoidea' (1880–1886) and 'The North American Crinoidea Camerata' (1897). From this collaboration, Springer became the foremost North American crinoid worker. In 1906 he retired from the law profession and devoted himself to crinoid research. He amassed magnificent echinoderm collections (in combination with Wachsmuth's second collection), which are now housed in the National Museum of Natural History. He did much to comprehensively describe North American crinoids, and he provided much of the fundamental morphological, stratigraphic and systematic knowledge on which our current understanding of crinoids is based.

CLEAR WATERS OF A LOWER MISSISSIPPIAN CARBONATE RAMP

If the Mississippian is the 'Age of Crinoids', it is largely because of the Burlington Limestone. Approximately 345 million years old, this crinoidal limestone is as

Hans Hess, William I. Ausich, Carlton E. Brett, and Michael J. Simms, eds., *Fossil Crinoids*. © 1999 Cambridge University Press. All rights reserved. Printed in the United States of America.

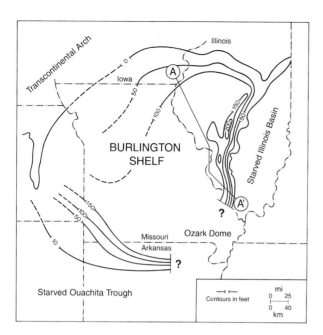

Fig. 144. Palaeogeographic reconstruction of midcontinental United States during deposition of the Burlington Limestone (Middle Osagean), with isopach contours showing thickness of the Burlington Limestone. Figure 145 is a cross section from A to A'. (Modified from H. R. Lane 1978.)

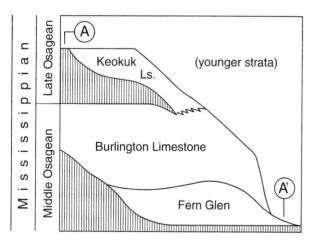

Fig. 145. Cross section from the Burlington Limestone shelf into the Illinois Basin from NNW to SSE. The relative disposition of the Fern Glen Formation, Burlington Limestone and Keokuk Limestone is indicated; vertical lines are times and places where no rocks are present. The line of cross section, A to A', is given in Fig. 144. (Modified from H. R. Lane 1978.)

much as 50 m thick, covering much of Missouri and parts of Illinois and Iowa (Fig. 144). Dott and Batten (1976) estimated that the Burlington is composed of 300×10^{10} m^3 of crinoidal limestone;[1] and Ausich (1997) used the Burlington as the primary example of the regional encrinites type of deposit, an expansive and stratigraphically thick deposit of principally crinoidal limestone. The Burlington Limestone is Middle Osagean (Tournaisian 3) in age and was deposited as a carbonate shelf on the western flank of the Illinois Basin (Fig. 145). H. R. Lane (1978) subdivided the Burlington shelf into four facies, with the main facies (the Burlington Limestone) the principal crinoid-bearing one. It is composed largely of undolomitized crinoidal limestones. During deposition the sediment was an unconsolidated, mobile, sand-to-gravel substratum, and, apparently, the main shelf was deposited well within storm wave base. The episodically mobile, coarse-grained, poorly sorted sediment did not provide a suitable substratum for many of the typical Mississippian benthos, such as bryozoans, brachiopods or corals. Instead, crinoids and blastoids with rooted holdfasts, either terminal cirri or cirri positioned along the column in the distal stem, were well suited for this substratum composed largely of disarticulated crinoidal remains. The result was nearly total domination of an entire shelf margin by pelmatozoan echinoderms.

BATOCRINID MEADOWS

Bassler and Moodey (1943) listed more than 400 species of crinoids from the Burlington Limestone. Although many of these species have since been judged to be junior synonyms and many more await systematic revision, this is undoubtedly the single most diverse crinoid fauna from any time or place. Similarly, the Burlington Limestone contains the most diverse fauna of blastoids known, with at least 17 genera in the Burlington fauna. This is approximately 20% of all known blastoid genera.

Most crinoid specimens collected from the Burlington Limestone are camerate crinoids from the family Batocrinidae – for example, *Abatocrinus*, *Azygocrinus*, *Dizygocrinus*, *Eretmocrinus*, *Eutrochocrinus* (Fig. 38), *Macrocrinus* and *Uperocrinus* (Fig. 37). Other important elements of the Burlington camerate fauna are members of the Rhodocrinitidae, Actinocrinitidae, Coelocrinidae, Dichocrinidae and Platycrinitidae. Most commonly they are preserved as calyces disarticulated from both the arms and stem. This taphonomic style of preservation occurred, presumably, because the sediment was episodically mobile and well within storm wave base; thus, complete crinoid crowns were rarely buried permanently. Most buried crinoids were probably exhumed by a subsequent storm after connective tissue decay, which disarticulated and scattered the skeletal elements of the arms and stem. The rigidly sutured calyx of camerates would be less susceptible to this sort of taphonomic destruction and would be left behind as an articulated residue of the crown.

In addition to camerates, the Burlington crinoid fauna also contains numerous disparids, cladids and flexibles. Because there is no way way this extensive fauna can be treated comprehensively here and because camerates are the conspicuous crinoids of this fauna, we will discuss a few of the most common and a few of the most unusual Burlington camerates to illustrate the morphological variety among Lower Mississippian camerate crinoids. Specifically, consider the morphological disparity among *Agaricocrinus inflatus*, *Azygocrinus rotundus*, *Camptocrinus praenuntius*, *Dorycrinus missouriensis*, *Eretmocrinus remibrachiatus*, *Eucladocrinus pleuroviminus* and *Strotocrinus glyptus*. In addition, species of the genera *Gilbertsocrinus*, *Macrocrinus* and *Platycrinites* (Fig. 17), discussed in Chapter 18, also occur in the Burlington Limestone.

Agaricocrinus inflatus is representative of this abundant Lower Mississippian coelocrinid genus (Fig. 146). *Agaricocrinus* typically has a flat or deeply excavated base, a high tegmen but no anal tube, an anal opening along the side of the tegmen and very robust arms. This Burlington species is characterized by a very high, broad tegmen and very substantial arms with fine, long pinnules.

Azygocrinus rotundus (family Batocrinidae) is a typical camerate crinoid (Figs. 147, 148). It is commonly preserved as a ball of calcite with the plate sutures difficult to distinguish and 19–26 holes – at opposite poles are holes for the column attachment and anal tube, and around the 'equator' there are 17–24 arm openings. Where complete, *A. rotundus* has an anal tube nearly as long as the arms, and stout, biserially pinnulate arms. More commonly, *A. rotundus* is preserved without arms and is characterized by a subspherical calyx, flat calyx plates, 1–5 interbrachials not in contact with the tegmen and a relatively short anal tube (Lane 1963a).

Camptocrinus praenuntius is one of the unusual dichocrinids, with a very simplified crown and a highly modified, planispirally coiled column (Fig. 149). The aboral cup of *C. praenuntius* mimics that of a cladid by having the arms nearly free above the radials. The aboral cup is composed nearly exclusively of the basal and radial circlets, but it has 10 biserially pinnulate arms. In *Camptocrinus* the columnals are modified into elliptical shapes

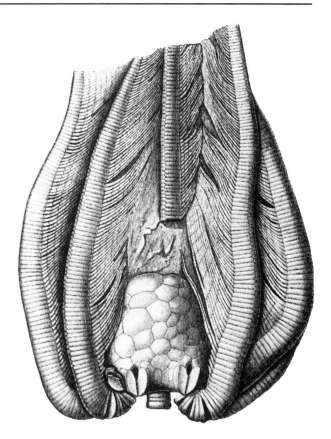

Fig. 146. *Agaricocrinus inflatus*. Burlington Limestone. Side view with some arms missing. (From Wachsmuth & Springer 1897.) ×1.

Fig. 147. *Azygocrinus rotundus*. Burlington Limestone. Side view of calyx, equator with small arm openings. (From Wachsmuth & Springer 1897.) ×1.

Fig. 148. *Azygocrinus rotundus*. Burlington Limestone. Complete crown; anal tube not visible. (From Wachsmuth & Springer 1897.) ×1.

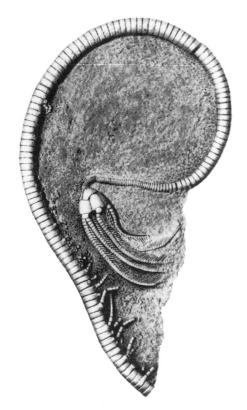

Fig. 149. *Camptocrinus praenuntius*. Burlington Limestone. (From Wachsmuth & Springer 1897.) ×1.

with an articular ridge (synarthrial articulation). The articular ridges are aligned parallel along the stem, so that a planispiral coil occurs in the stem (Fig. 15). Distal and medial columnals have radicular cirri, which presumably acted as a holdfast and also enclosed the crown in some species. The column immediately beneath the crown bends, reversing the coiling direction by inserting wedge-shaped columnals. It is quite probable that this small, unusual camerate lived recumbent along the bottom, similar to the Silurian myelodactylids (Fig. 100), but its palaeoecology requires further study.

Eretmocrinus remibrachiatus is another member of the family Batocrinidae. This species is characterized by highly modified distal arms (Fig. 150). The 18–20 arms of this species begin as normal biserially pinnulate arms, but in the distal half or third they lose their pinnules, widen and become flattened. Other members of this genus and a few species of *Dizygocrinus* display this modification, but its expression is extreme in *E. remibrachiatus*. The function of these modified arms is not known. Did the arms function for the animal when they were open or when they were closed? When open, these arms would have formed a fairly solid barrier to current flow that may have conferred feeding advantages; alterna-

Fig. 150. *Eretmocrinus remibrachiatus*. Burlington Limestone. Aboral (dorsal) view of complete crown. (From Wachsmuth & Springer 1897.) ×1.

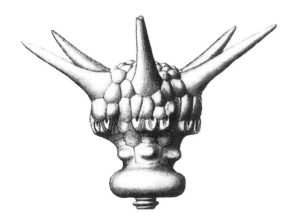

Fig. 151. *Dorycrinus missouriensis*. Burlington Limestone. Side view of calyx lacking arms, tegminal spines well preserved. (From Wachsmuth & Springer 1897.) ×1.

tively, when the arms were closed, they may have made a formidable protective roof over the animal.

Eucladocrinus pleuroviminus also has highly modified arms (Fig. 23). On examination of the calyx and stem, this platycrinitid looks very much like a typical member of *Platycrinites* (see discussion of *P. hemisphaericus* in Chapter 18), with an aboral cup composed almost exclusively of basals and radials, arms nearly free above the radials and a helically twisted column. Upon closer inspection, however, the columnals of this crinoid are seen to be very exaggerated ellipses, and the radial facets are nearly vertical (Ausich & Kammer 1990). This latter character gives away the most distinctive feature of this crinoid, its highly modified arms. Instead of arms, each radial supported a thick, tubular arm trunk that divided once. These 10 arm trunks bore alternating biserial armlets that formed a highly modified and unusual filtration fan. This arm system may have been largely immobile, so that the arms were fairly permanently oriented into a filtration fan.

Unlike the batocrinids discussed here, *Dorycrinus missouriensis* lacked a long anal tube extending distally from the centre of the tegmen (Fig. 151). The anus of this coelocrinid is situated along the side of the tegmen just above the arm openings (not visible in Fig. 151). Five long spines, which are simply modified tegmen plates, dominated the tegmen of this crinoid, rather than the anal tube as in batocrinids. These spines are long enough to extend through closed arms, and this and other spinose adaptations of crinoids have been interpreted to be anti-predatory in nature (Meyer & Ausich 1983; Signor & Brett 1984).

Strotocrinus glyptus is an actinocrinitid that has an extreme number of fixed brachials and attenuation of the margin where the arms become free. The width of the calyx approaches a maximum at 12 cm, and as many as 150 free arms may extend from this margin (Fig. 152). The proximal portion of the calyx is normal in appearance, but above the secundibrachials the fixed brachials

Fig. 152. *Strotocrinus glyptus*. Burlington Limestone. Side view of calyx with insertion for arms. (From Wachsmuth & Springer 1897.) ×1.

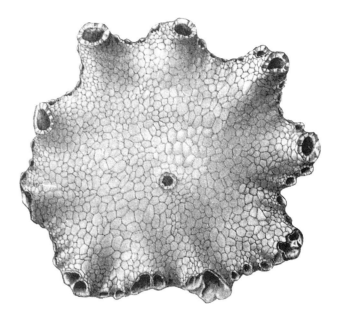

Fig. 153. *Strotocrinus glyptus*. Burlington Limestone. Oral view of tegmen; broken anal tube in centre and arms attached around periphery. (From Wachsmuth & Springer 1897.) ×1.

and interradials become a nearly irregular mosaic. The tegmen is flat and composed of numerous very small plates and a small, nearly centrally located anal tube (Fig. 153).

IMPORTANT COLLECTIONS IN THE UNITED STATES

Field Museum of Natural History, Chicago, Illinois

Museum of Comparative Zoology, Harvard University, Cambridge, Massachusetts

National Museum of Natural History, Smithsonian Institution, Washington, D.C.

NOTE

1. Approximately 1 million times the volume of the largest Egyptian pyramid!

18 Lower Mississippian Edwardsville Formation at Crawfordsville, Indiana, USA

WILLIAM I. AUSICH

A 'WARTY TOAD'

In 1842 Horace Hovey was collecting 'encrinites' along the banks of Sugar Creek, north of Crawfordsville, Indiana. He was collecting in response to an advertisement offering $5 per bushel for 'encrinites.' In addition to 'encrinites', this nine-year-old boy discovered a calcareous 'warty toad', the first crinoid calyx (a specimen of *Abatocrinus grandis*; see Fig. 154 for a complete crown) from the now-famous Crawfordsville crinoid beds.

Perhaps it is fitting that the first crinoid calyx discovered at Crawfordsville was found by someone who was fossil collecting for profit, because the magnificent Crawfordsville crinoids have been highly prized fossils ever since. Crawfordsville is one of the richest accumulations of exquisitely preserved crinoids in the world, and it has attracted numerous scientists, amateur collectors and professional collectors for the 150 years since the 'warty toad' was discovered. As in the latter half of the 19th century, today Crawfordsville crinoids are actively being studied by palaeontologists and sought by professional collectors.

DELTA SHED FROM THE EMERGING APPALACHIAN MOUNTAINS

Crinoids are known from a number of stratigraphic intervals in the greater Crawfordsville area, principally in

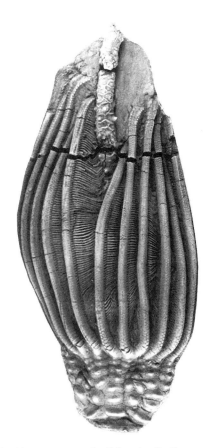

Fig. 154. *Abatocrinus grandis*. Edwardsville Formation, Crawfordsville, Indiana. Anal tube projecting above arms. This is the very first crinoid calyx found at Crawfordsville. (Reprinted by permission of the National Museum of Natural History, Washington D.C.) ×1.

Hans Hess, William I. Ausich, Carlton E. Brett, and Michael J. Simms, eds., *Fossil Crinoids*. © 1999 Cambridge University Press. All rights reserved. Printed in the United States of America.

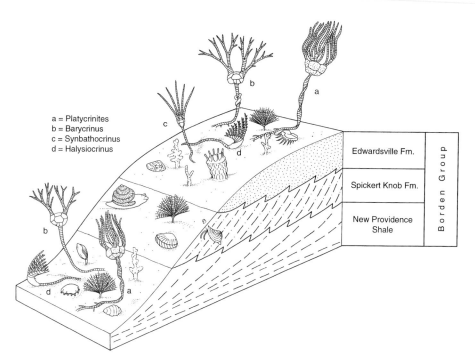

Fig. 155. Diagrammatic cross section through the Borden deltaic complex in southern Indiana and north-central Kentucky with characteristic communities, drawn very schematically on each part of the delta. (Modified from Ausich et al. 1979.)

Montgomery County, Indiana. However, they are most abundant at two localities – Corey's Bluff north of Crawfordsville and along Indian Creek southwest of Crawfordsville. Strata at both of these localities are now considered to be within the Edwardsville Formation (Borden Group) (approximately 340 million years old), with the Indian Creek locality in the uppermost part of the formation (most previous literature placed the Indian Creek Beds within the Ramp Creek Formation). In central and southern Indiana, the Edwardsville Formation is Late Osagean in age (Visean). The Borden Group was deposited as a broad, low-relief delta that was one of the westward sediment pulses eroded from the newly formed Appalachian Mountains. The Borden delta is divisible into three basic parts – the subaqueous delta platform, delta slope and prodelta (Fig. 155). The Edwardsville Formation records the subaqueous delta platform in Indiana and yields numerous crinoid localities, including those in the Crawfordsville area.

CRINOID MINING

Crawfordsville crinoids were collected during six major periods. From 1832 to 1859 they were collected principally by residents interested in acquiring a teaching collection for the newly established Wabash College in Crawfordsville and developing natural history cabinets (see Van Sant & Lane 1964 for a history of Crawfordsville collection). In 1859 Lyon and Casseday described four new species from the cabinet of O. W. Corey, a Crawfordsville resident. The description of these crinoids, *Abatocrinus grandis Cyathocrinites multibrachiatus*, *Gilbertsocrinus tuberosus* and *Onychocrinus ramulosus*, ushered in the second period of collecting (1859–1875). The publication of information about these crinoids attracted collectors, palaeontologists and their agents to Crawfordsville to collect and quarry crinoids. In 1864 F. H. Bradley initiated systematic quarrying methods, used by many later; and various preparation techniques were developed. The period from 1875 to 1887 was dominated by land purchases for collecting and by more crinoid quarrying. During this period, D. A. Bassett developed preparation techniques that are widely considered the best (Van Sant & Lane 1964). Bassett prepared exquisite large slabs with fine detail of preparation and with many complete individuals – arms, calyx, column and holdfast. Large-scale quarrying continued from 1887 to 1906, and this collecting (Fig. 156) was funded by Frank Springer, who eventually amassed the largest collection of Crawfordsville crinoids (Van Sant & Lane 1964).

After 1906 little or no fossil collecting occurred in the Crawfordsville area until Gary Lane opened two small excavations (1964–1965) to study fossil distribution and the palaeoecology of these strata (Lane 1973). The final period of serious collection began in the early 1980s, providing magnificent crinoid specimens for amateurs and museums around the world (Figs. 157, 158).

THE CRAWFORDSVILLE FAUNA

More than 63 crinoid species assigned to 42 genera are known from Crawfordsville. (This takes into consideration the recent taxonomic revisions by T. W. Kammer and W. I. Ausich, in which numerous species were placed into synonymy). All major groups of Lower Mississippian crinoids are present, including the diplobathrid camerates, monobathrid camerates, disparids, primitive cladids, advanced cladids and flexibles.

A remarkable feature of these Late Osagean crinoids is their size. The crowns of many of these crinoids are quite large in comparison, especially, to those of younger faunas. Crowns greater than 10 cm are common for several species in this fauna, which adds further to their universal appeal. For example, crowns of the primitive cladid *Barycrinus rhombiferus* measure up to 15 cm or more, and the primitive cladid *Pellecrinus hexadactylus* may reach heights of more than 12 cm. Other especially large crinoids included the advanced cladids *Springericrinus magniventrus*; the monobathrid camerates *Abatocrinus grandis*, *Actinocrinites gibsoni*, *Agaricocrinus americanus* and *Paradichocrinus polydactylus*; and the flexibles *Onychocrinus exsculptus* (Fig. 56) and *O. ramulosus*.

Some of the most abundant and well-known Craw-

Fig. 156. Quarrying operations at the Sugar Creek, Crawfordsville, Indiana, site during the summer of 1906. Another photograph from the same time was published by Van Sant and Lane 1964. (Reprinted by permission of the National Museum of Natural History, Washington, D.C.)

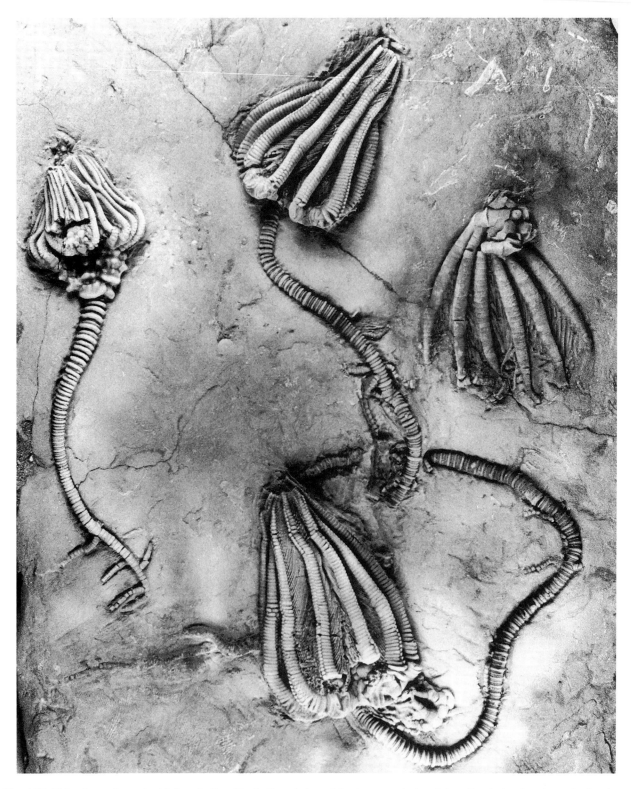

Fig. 157. Slab of complete crinoids from Indian Creek. From left to right these are *Actinocrinites gibsoni*, *Agaricocrinus splendens* (two specimens) and *Scytalocrinus decadactylus*. (Reprinted by permission of the National Museum of Natural History, Washington, D.C.) ×1.

Fig. 158. Complete crinoids from Indian Creek, uppermost Edwardsville Formation. From left to right these are *Macrocrinus mundulus* (upper left corner), *Agaricocrinus splendens*, *Platycrinites saffordi* (the large complete specimen) carrying the gastropod *Platyceras*, *Taxocrinus ungula* (inside the stem of *P. saffordi*), probably a juvenile of *Taxocrinus ungula*, and *Scytalocrinus decadactylus*. (Siber Collection, Saurier Museum Aathal, Switzerland; photograph S. Dahint.) ×0.6.

fordsville crinoids are *Cyathocrinites multibrachiatus* (Fig. 47), *Dizygocrinus indianensis*, *Gilbertsocrinus tuberosus*, *Histocrinus coreyi*, *Onychocrinus exsculptus*, *Pachylocrinus aequalis*, *Platycrinites hemisphaericus*, *Scytalocrinus robustus* and *Taxocrinus colletti* (Table 4). Perhaps the most important aspect of this fauna is its high diversity of numerous well-preserved specimens, which allows for a full understanding of the morphology of a wide array of crinoids. This provides the baseline knowledge for understanding details as well as the range of structures of Lower Mississippian crinoids – for example, anal sacs, tegmens and stems.

Gilbertsocrinus tuberosus may, by default, be considered a typical Late Osagean diplobathrid camerate. As a

Table 4. *Taxonomic Distribution of the More Common Crawfordsville and Indian Creek Crinoids*

DIPLOBATHRID CAMERATES
Gilbertsocrinus tuberosus (C)

MONOBATHRID CAMERATES
Abatocrinus grandis (C)
Actinocrinites gibsoni (C) (IC)
Dizygocrinus indianensis (C) (IC)
Macrocrinus mundulus (C) (IC)
Platycrinites hemisphaericus (C)
Platycrinites saffordi (IC)

DISPARIDS
Halysiocrinus tunicatus (IC)

PRIMITIVE CLADIDS
Barycrinus rhombiferus (C) (IC)
Barycrinus stellatus (C) (IC)
Cyathocrinites iowensis (C) (IC)
Cyathocrinites multibrachiatus (C) (IC)

ADVANCED CLADIDS
Abrotocrinus coreyi (C) (IC)
Abrotocrinus unicus (C) (IC)
Histocrinus coreyi (C) (IC)
Hylodecrinus gibsoni (C)
Lanecrinus depressus (C) (IC)
Pachylocrinus aequalis (C)
Scytalocrinus robustus (C) (IC)

FLEXIBLES
Onychocrinus exsculptus (C)
Onychocrinus ramulosus (C)
Taxocrinus colletti (C) (IC)
Taxocrinus ungula (IC)

Abbreviations: C, Crawfordsville; IC, Indian Creek.

diplobathrid camerate, it has infrabasals, basals and radials, fixed brachials in the calyx and biserial arms; but beyond these generalities, it is a most unusual crinoid. *Gilbertsocrinus tuberosus* is a medium to large crinoid with a broad, flat tegmen. Tubular, bifurcating extensions of the tegmen grew interradially from the tegmen margin (Fig. 159). The arms are narrow threads, rarely preserved, that hung beneath these tegmen extensions. The calyx is deep and wide, with the spinose basal plates the lowest calyx plates visible in side view. The infrabasal circlet is hidden in a basal concavity and is nearly covered by the proximal-most columnal. In addition to

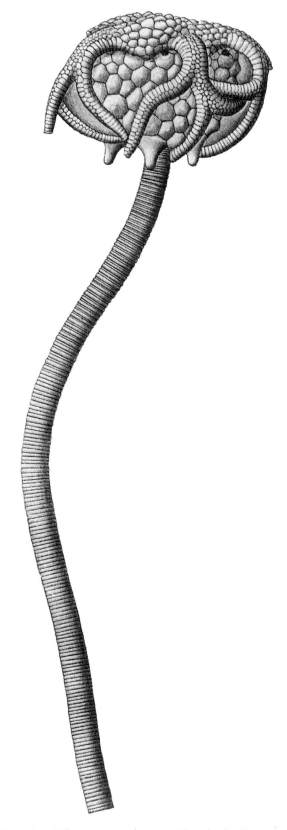

Fig. 159. *Gilbertsocrinus tuberosus*. Edwardsville Formation, Crawfordsville, Indiana. (From Wachsmuth & Springer 1897.) ×1.

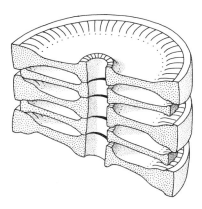

Fig. 160. Reconstruction of *Gilbertsocrinus tuberosus* stem showing the radical changes to the articulation on columnal facets. These changes greatly increased the potential flexibility of these stems. (Redrawn from Riddle *et al.* 1988.) ×8.

Fig. 161. *Macrocrinus mundulus*. Edwardsville Formation, Crawfordsville, Indiana. (Hess Collection; photograph S. Dahint.) ×1.8.

the peculiar tegmen and arms, the stem is unusual in its high degree of flexibility, which is commonly preserved looped back over itself. This high degree of flexibility is the result of highly modified columnal facets. The crenularium for intercolumnal articulation is on a small raised area immediately surrounding the lumen called the perilumen (Riddle *et al.* 1988), rather than being broad and extending to the periphery of the facet (Fig. 160). With this arrangement, the mechanics of the articulation is altered and greatly increases the column flexibility (Baumiller & Ausich 1996). Stems may be as long as 65 cm, and the holdfast is a relatively small bulbous feature that was apparently anchored into the sediment (Lane 1963b).

Macrocrinus mundulus is a typical medium to small monobathrid camerate crinoid (Fig. 161). It has a truncate, conical calyx, with basal plates visible in side view and articulating with the stem. Fixed brachials and interradial plates are present, and the second primibrachial is axillary. At the top of the calyx, 16–20 free arms emerge that are biserially pinnulate and unbranched. The tegmen is high, conical and dominated by a long, conical, centrally located anal tube with a terminal anal opening. The arms and the anal tube are approximately the same length. The stem of *M. mundulus* may be as much as 20 cm long. The distal half of the stem tapers, and radicular cirri for the holdfast are along the distal quarter of the column.

Platycrinites hemisphaericus is probably the most common camerate at Crawfordsville (Fig. 158 shows *Platycrinites saffordi*, the species occurring at Indian Creek). It is the monobathrid camerate crinoid that mimics an inadunate. The calyx is basically the aboral cup, and the arms are basically free above the radials. Both radial and basal plates are large, and distinctive, coarse nodes form the plate sculpturing. Free arms divide as many as four times in each ray and are biserially pinnulate. Columnals of *Platycrinites* are highly modified into an elliptical shape with a fulcral ridge nearly paralleling the long axis (a synarthrial articulation) (Fig. 162). The articular

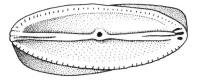

Fig. 162. Articular facet of *Platycrinites* columnal showing articular ridge along its maximum width. (Redrawn from Ubaghs 1978.) Approx. ×8.

ridges on opposite facets of an individual columnal are slightly offset from one another, which results in a helically twisted stem (Riddle 1989). As in *Gilbertsocrinus tuberosus*, modifications of the *Platycrinites* column altered the mechanics of the articulation and increased the flexibility of the stem (Baumiller & Ausich 1996). This helically twisted stem reached lengths of up to 30 cm. The distal holdfast had modified columnals with radicular cirri growing from the narrow ends of the columnal. This produced the appearance of a sparse bottle brush, and the distal stem probably lay along the sea floor as a runner-type holdfast similar to that of living isocrinids.

A typical cladid is *Scytalocrinus robustus* (see Figs. 157 and 158 for a species of *Scytalocrinus* from Indian Creek). Only part of the infrabasals are visible in side view, and both the basals and radials are large plates. The arms are free above the radial plates, and only one large, axillary primibrachial is present. The arms divide only this one time, and they are pinnulate with broadly rounded, cuneate, uniserial brachials. The stem length of *Scytalocrinus robustus* can reach up to 40 cm.

Many disparid crinoids seem unusual, but this is especially true of the catillocrinids. The catillocrinid at Crawfordsville is *Eucatillocrinus bradleyi* (Fig. 163). This crinoid has multiple fine, non-branching arms articulated to the radial plates. Centrally a very long, thick anal sac is present; and when the arms were in a closed posture, the arms fit into grooves along the proximal anal sac. Plating of the aboral cup is unusual because radials vary in size. Three radials are very narrow and have only one arm (one radial also supports the anal sac). However, two radials are very large, and each gives rise to approximately 20 arms. The proximal stem is differentiated from the middle column and called the proxistele. The proxistele is composed of extremely thin and wide columnals, and more typical columnals are present in the middle column, or mesistele. Intuitively, the very thin columnals of the proxistele would seem to be an adaptation for increased flexibility, but the opposite is true. Proxisteles constructed similarly to those of *E. bradleyi* were interpreted to be extra rigid portions of the stem, based upon preserved postures (Baumiller & Ausich 1996). The total stem length of *E. Bradleyi* is as much as 35 cm.

Finally, flexibles are represented by *Forbesiocrinus wortheni* (Fig. 27) and *Taxocrinus ungula* (Fig. 158), typical flexibles. Flexibles generally have poorly sutured calyx plates. The subclass name, Flexibilia, refers to the general flexible appearance of the crown due to typical

Fig. 163. *Eucatillocrinus bradleyi*. Edwardsville Formation, Crawfordsville, Indiana. Note massive anal sac and very fine arms, which are mainly broken away. (Reprinted by permission of the Field Museum of Natural History, Chicago.) ×3.

crushing during compaction. *Forbesiocrinus wortheni* is a large crinoid; the calyx may reach up to 10 cm in height. Characteristic of flexibles, three infrabasal plates are present. In *F. wortheni*, the infrabasal plates are rather short, and radial and basal plates are large. Several primibrachials are present before the first bifurcation, above which the arms branch several times. The distal arms curve or coil in toward the centre of the crown, which is a typical arm posture among the Flexibilia. As in *Eucatillocrinus bradleyi*, the proximal stem is differentiated into a proxistele with extremely thin and wide columnals. The calyx, arm and stem construction of *Taxocrinus ungula* from Indian Creek is similar to that of *F. wortheni*. However, *T. ungula* can be differentiated by its distinct column of anal plates in the posterior, higher basal circlet and arms that are more rounded and distinct from one another.

COMPLETE CRINOIDS: THE SEED FOR THE IDEA OF TIERING

The number and diversity of crinoids from Crawfordsville are nearly unbelievable, but even more amazing is the fact that complete crowns with attached stems are so abundant. With careful quarrying and care of preparation, like that of Bassett, numerous complete individual crinoids – with a crown and stem complete to the holdfast – have also been collected (Fig. 164). Of the approximately 63 species of crinoids recognized from Crawfordsville, nearly half (29) are known from complete or nearly complete specimens, and a few species are represented by multiple complete individuals. Such wonderfully preserved fossils provide data that are rarely available, and such a *Lagerstätte* commonly leads to more general interpretations about the behaviour, ecology or physiology of fossil organisms.

In this case, the exceptional Crawfordsville crinoids led Lane (1963b) to recognize the ecological significance of differing stem lengths among crinoids. In 1980, Ausich developed a more comprehensive model for ecological niche separation among crinoids, again by careful consideration of Crawfordsville crinoids. In a sea floor setting, suspended food particles move horizontally across the substratum (from a crinoid individual's perspective). This resource can be partitioned among different crinoid species in two ways: (1) different species can be elevated to different levels (tiers) above the sea floor, thereby filtering food from different parcels of water, and (2) even within a single tier crinoids can filter food particles of different sizes. In a community of crinoids with a range in stem length and a range in filtration fan densities and ambulacral groove widths, a series of non-overlapping or only slightly overlapping niches can be established, thereby making it possible to separate different species ecologically (Ausich 1980).

Furthermore, understanding niche differentiation among crinoids led to the development of the general tiering model (Ausich & Bottjer 1982; Bottjer & Ausich 1986). Tiering is a general ecological structural ordering defined as 'vertical subdivision of space by the organisms in a community' (Bottjer & Ausich 1986). Entire epifaunal suspension-feeding communities are tiered, as are infaunal suspension-feeding communities and infaunal deposit-feeding communities.

Tiering is now widely recognized to have been an important aspect of the structure of many benthic marine communities throughout geological history. Tiering was especially important for the numerous crinoid communities discussed in this book (e.g., Figs. 80, 107, 155). It is important to remember, however, that this concept is rooted in the detailed collection and preparation of crinoids by D. A. Bassett during the late 19th century.

OTHER LATE OSAGEAN CRINOID FAUNAS

Late Osagean crinoids are present throughout the eastern United States. Crawfordsville and Indian Creek contain subaqueous delta platform faunas, and faunas from the delta platform are present elsewhere, such as Monroe County, Indiana. Delta slope facies supported few crinoids, but at the toe of the delta in the prodelta

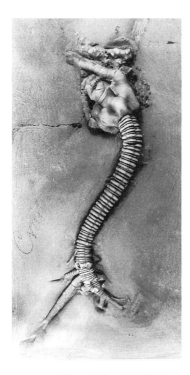

Fig. 164. *Barycrinus stellatus*. Edwardsville Formation, Crawfordsville, Indiana. Example of a complete specimen. (Reprinted by permission of the National Museum of Natural History, Washington, D.C.) ×1.

facies, crinoids also flourished locally (Ausich et al. 1979; Kammer 1984). Prodelta faunas occur in the New Providence Shale in southern Indiana and Kentucky. Farther south, where the toe of slope along this basin margin received less sediment, carbonate build-ups developed in the Fort Payne Formation of Kentucky and Tennessee. These deeper-water carbonate build-ups associated facies also supported significant crinoid faunas. West of the Illinois Basin, the extensive Burlington–Keokuk carbonate ramp developed (see Chapter 17). The latest phase of this ramp was the Keokuk Limestone, composed principally of crinoids. Crinoids continued to flourish on this western margin until the close of the Osagean, even after Borden siliciclastics (the Warsaw Formation) encroached onto this platform. Both the Keokuk Limestone and the lower part of the Warsaw Formation are Late Osagean in age and approximately coeval with the Edwardsville Formation, New Providence Shale and Fort Payne Formation, completing a regional perspective of facies and crinoid assemblages around this epicontinental basin.

IMPORTANT COLLECTIONS IN THE UNITED STATES

Field Museum of Natural History, Chicago, Illinois
National Museum of Natural History, Smithsonian Institution, Washington, D.C.

19

Upper Pennsylvanian LaSalle Member, Bond Formation of Central Illinois, USA

WILLIAM I. AUSICH

COMPLETE PENNSYLVANIAN CRINOIDS

As one drives across the seemingly featureless till plains of central Illinois, geology and fossils typically do not come to mind. This is corn and soybean country. The ground is a rich, black soil; rocks are not part of the picture. Remarkably, however, rich Palaeozoic fossiliferous beds lie beneath these glacial deposits. The most famous is the Mazon Creek soft-bodied *Lagerstätte*, but these Pennsylvanian strata have also yielded an exceptional crinoid fauna from the LaSalle Limestone. By the standards of the faunas described in this book, this locality may not seem exceptional, but it is an exceedingly important fauna because it was one of the first Pennsylvanian crinoid faunas described principally from *complete crowns* (Fig. 165). Most crinoid faunas younger than the Middle Mississippian are dominated by cladid inadunates; and especially in the Pennsylvanian, crinoids are rarely preserved with arms. The majority of Pennsylvanian crinoids are found only as bowl- and saucer-shaped isolated aboral cups, so LaSalle crinoids offer an important glimpse of Late Palaeozoic cups with articulated arms.

CYCLICAL SEDIMENTATION

The Pennsylvanian of the midcontinental United States was represented by cyclical sedimentation and is well known for extensive and economically important coal deposits. Individual cycles are called cyclothems (in England, the Coal Measures), and each cyclothem has alternating marine and non-marine facies. In Illinois, cyclothems are composed of substantial portions of both marine and non-marine facies, whereas to the east non-marine facies dominated and to the west marine facies dominated. A typical, complete Illinois cyclothem is composed of 10 units: (1) fluvial channel sandstone; (2) grey shale; (3) non-marine limestone; (4) underclay; (5) coal; (6) grey shale; (7) marine limestone; (8) black shale; (9) marine limestone; and (10) a marine grey shale. The LaSalle crinoids occur in shaly pockets of the underlying LaSalle Limestone, at the base of this cyclothem unit (Fig. 166). The process or processes that drove sea level changes to produce Pennsylvanian cyclothems has received a tremendous amount of attention recently. Some of the most likely potential causes are glacio-eustatic changes in sea level, other climatic change and crustal deformation associated with the formation of the super-continent Pangaea.

The LaSalle Limestone Member is Missourian, upper Middle Pennsylvanian in age (Stephanian, approximately 300 million years old). It is the lowest limestone member of the Bond Formation, which is part of the middle formation of the McLeansboro Group. The LaSalle Limestone is a fossiliferous limestone that ranges from 4 to 8 m in thickness.

Hans Hess, William I. Ausich, Carlton E. Brett, and Michael J. Simms, eds., *Fossil Crinoids*. © 1999 Cambridge University Press. All rights reserved. Printed in the United States of America.

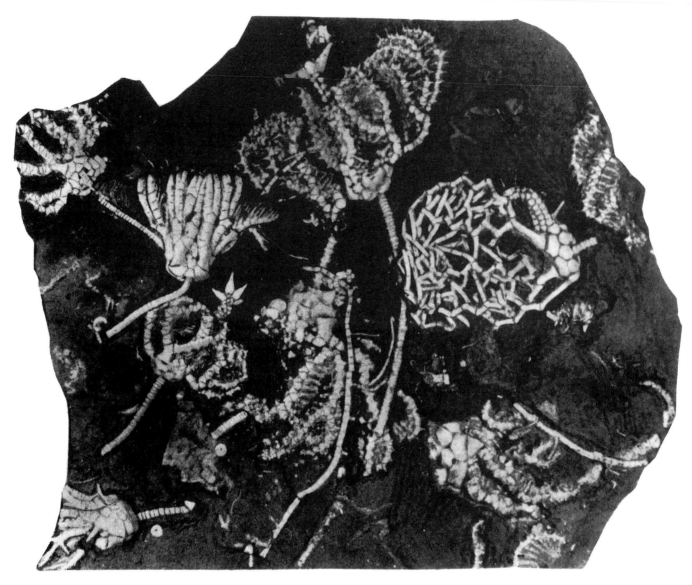

Fig. 165. Complete crowns from the LaSalle Limestone Member of the Bond Formation, Ocoya, Livingston County, Illinois. Crinoids are *Brabeocrinus christinae* (five specimens), *Clathrocrinus clathratus* (middle, at right), *Exocrinus wanni* (compact crown left of middle) and *Stenopecrinus* cf. *S. planus* (lower left corner). (Figs. 165 and 167–174 are from Strimple & Moore 1971; reprinted by permission.) ×1.5.

A CLADID-STYLE ARMS RACE

The LaSalle crinoid fauna is composed of 38 species assigned to 32 genera; however, it is dominated by cladids. Only one disparid, one camerate and two flexibles are part of this fauna; consequently, 34 species (assigned to 28 genera) of cladids dominate.

This degree of cladid dominance is typical of Pennsylvanian crinoid faunas. Whereas the design of aboral cups is restricted in these cladid faunas, a great diversity of arm types is present. In older faunas much of the arm-type diversity was present because different subclasses or orders had different arm types (see Ausich 1980). In Pennsylvanian faunas, a rich arm-type diversity evolved among just cladids, which was apparently a response to broaden the spectrum of available suspension-feeding niches. The LaSalle fauna contains crinoids with atomous, ramulate, rectangular uniserially pinnulate, cuneate uniserially pinnulate and biserially pinnulate arms.

The single disparid, *Kallimorphocrinus lasallensis*, is typical of disparids in that it has a few atomous arms and some radials bearing more than one arm (Fig. 167). This arm style is unique among LaSalle crinoids. Similarly, LaSalle flexibles have relatively few arms with

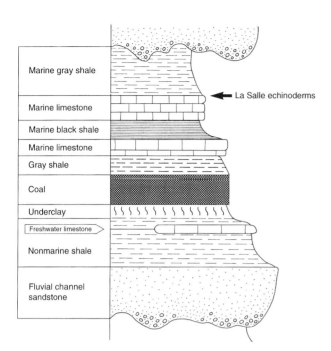

Fig. 166. Idealized Illinois cyclothem.

Fig. 167. *Kallimorphocrinus lasallensis*. LaSalle Limestone Member, Livingstone County, Illinois. ×8.25.

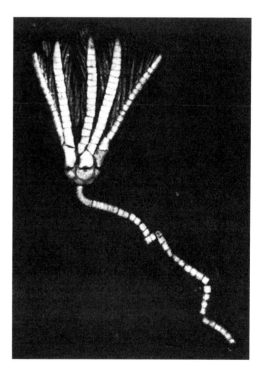

Fig. 168. *Apographiocrinus typicalis*. LaSalle Limestone Member, Livingstone County, Illinois. ×1.5.

ramulate branching, again unique in this fauna. Both the atomous arms of disparids and the ramulate arms of flexibles have fairly wide ambulacral grooves and filtration fans that are not dense in comparison to pinnulate arms.

The standard arms of cladids are uniserially pinnulate, with either rectangular or cuneate brachials. Prior to the Middle Mississippian, nearly all biserial pinnulate arms known were among the camerates; however, after the Middle Mississippian demise of most camerates, cladids developed biserial arms.

Typical cladid arms (uniserial pinnulate) are also present among members of the LaSalle fauna. Among the simplest of these is *Apographiocrinus typicalis*, which has 10 arms with rectangular to slightly cuneate brachials (Fig. 168). *Exocrinus wanni* is also relatively small, with rectangular to slightly cuneate brachials, but its arms divide as many as three times in an exotomous pattern to yield 30 total arms (Fig. 165). Its brachials are equidimensional to high, in contrast to the very low brachials of *Haeretocrinus wagneri*, which is a much larger crinoid. Arm branching of *H. wagneri* is similar, with up to 30 arms with cuneate brachials and exotomous branching (Fig. 169).

Where brachials become strongly cuneate, the arms can take on a zig-zag appearance, and variations on that theme, from simple to extreme are present in the LaSalle fauna. The low cuneate uniserial arms of *Microcarinus conjugulus* have a slight zig-zag appearance (Fig. 170), but the similar brachials with alternating spines in *Brabeocrinus christinae* produce a striking zig-zag arm (Fig. 165), which would have made a more dense filtra-

Fig. 169. *Haeretocrinus wagneri*. LaSalle Limestone Member, Livingstone County, Illinois. ×0.9.

Fig. 170. *Microcaracrinus conjugulus*. LaSalle Limestone Member, Livingstone County, Illinois. ×4.5.

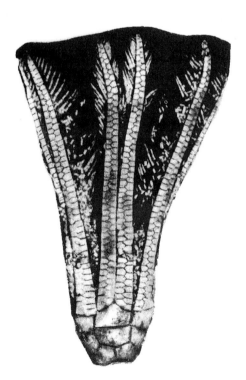

Fig. 171. *Erisocrinus typus*. LaSalle Limestone Member, Livingstone County, Illinois. ×0.9.

Fig. 172. *Endelocrinus tumidus spinosus*. LaSalle Limestone Member, Livingstone County, Illinois. ×2.2.

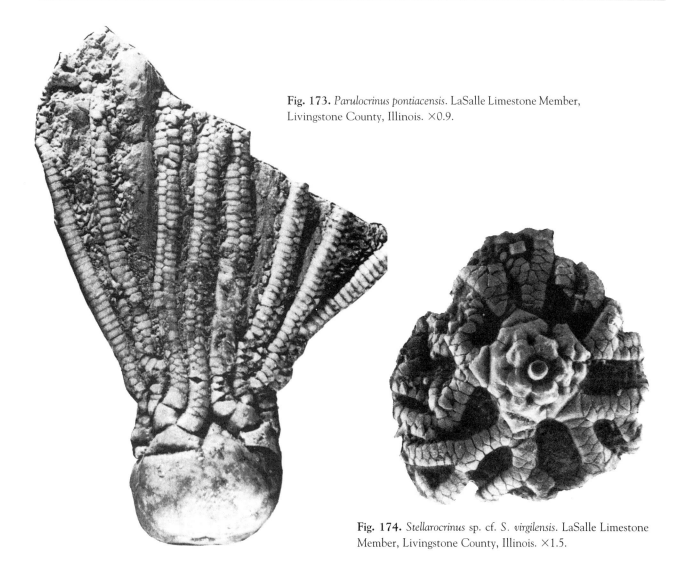

Fig. 173. *Parulocrinus pontiacensis*. LaSalle Limestone Member, Livingstone County, Illinois. ×0.9.

Fig. 174. *Stellarocrinus* sp. cf. *S. virgilensis*. LaSalle Limestone Member, Livingstone County, Illinois. ×1.5.

tion fan. The coarse, lacy appearance of the arms of *Clathrocrinus clathratus* is a result of extreme development of the cuneate brachials (Fig. 165). In this crinoid the cuneate brachials are very high and the main axis of the arm takes approximately 60° bends at each articulation. The branches in the pinnule position were considered small armlets by Strimple and Moore (1971). Very unusual biserially arranged cover plates with pores roof the ambulacral grooves. *Clathrocrinus clathratus* is a most unusual and beautiful crinoid.

Finally, the biserial pinnulation characteristic of camerate crinoids earlier during the Palaeozoic is present among LaSalle cladid crinoids. Examples include *Erisocrinus typus* with 10 arms (Fig. 171), *Endelocrinus tumidus spinosus* with 10 arms and spinose axillary first primibrachials (Fig. 172), *Parulocrinus pontiacensis* with 15 very camerate-like-appearing arms (Fig. 173), and *Stellarocri-*

nus sp. cf. *S. virgilensis* with 20 arms (free biserial arms divide once) that coil inward distally (Fig. 174).

This wide diversity of arm branching styles produced a wide range of filtration densities, despite the fact that nearly all crinoids were cladids. As discussed in Chapter 18, the resultant filtration fans did not serve to increase efficiency; rather, they were adapted to different types of particle capture, probably mostly for size differentiation. This arm diversity offered a wide range of suspension-feeding niches to support the diverse LaSalle crinoid fauna.

IMPORTANT COLLECTIONS IN THE UNITED STATES

Illinois Geological Survey, Urbana
University of Iowa, Iowa City

Permian

HANS HESS

PALAEOZOIC TWILIGHT

The Permian crinoids are the last to represent all four Palaeozoic subclasses – the Camerata, Flexibilia, Disparida and Cladida (see Simms & Sevastopulo 1993). With the extinction of 30 families, the Camerata, Flexibilia and Disparida disappeared altogether at or near the end of the Permian in the greatest extinction of all time. This crisis, 245 million years ago, nearly wiped out the crinoids as a whole (Figs. 3 and 61), but, fortunately, some survived to form in the Triassic the stem group of the articulates. Compared with older Palaeozoic systems, crinoids were fairly rare throughout most of the Permian, and assemblages, as described elsewhere in this book, are exceptional. However, because the Permian was a critical time in crinoid existence and evolution, a brief chapter on Permian crinoids is justified.

CLASSIC CRINOID SITES IN TIMOR AND AUSTRALIA

Until World War Two, the most diverse and abundant Permian crinoid fauna known was the one from the Island of Timor in Indonesia. It was discovered, exploited and described by Johannes Wanner in a series of outstanding papers during the first half of the twentieth century. In one of his later monographs Wanner (1937) listed 320 species belonging to 100 genera, most of them new and unique to Timor. The total fauna includes other echinoderms (echinoids and blastoids) as well as gastropods, bivalves, brachiopods and corals. On the basis of the massive skeleton of many of the forms, Wanner concluded that the animals lived in a warm, shallow and partly agitated sea. Their mode of preservation indicates a certain amount of transport before burial. The Timor crinoids, usually isolated cups, have a large variety of cladids and flexibles, but disparids and camerates are rare. Stem fragments of more than 10 mm diameter are common, whereas the cups, on which most species are based, show much smaller attachments for the stem. These cups are generally quite small, with a diameter of 1 or 2 cm, at the most. In addition, many of the Timor cladids exhibit evolutionary features that demonstrate a trend toward simplification of an earlier, more complex structure (Lane & Webster 1966). The suppression of one or two arms (*Indocrinus*, *Sundacrinus*, *Tribrachyocrinus*) is a significant feature among these Permian crinoids. Some have only one arm (Fig. 35), or the arms are completely absent. The reduction of arms goes hand in hand with a high variability and many (pathological?) deformations. A number of the crinoids have asymmetric cups. In the flexibles *Prophyllocrinus* and *Proapsidocrinus*, the arms could be enclosed within prolonged radials, as in the Jurassic cyrtocrinids *Phyllocrinus* and *Apsidocrinus*; see Fig. 33 (Wanner 1924). In the cladid *Timorechinus*, the arms could be received between projecting ribs of the balloon-like anal sac. In the

Hans Hess, William I. Ausich, Carlton E. Brett, and Michael J. Simms, eds., *Fossil Crinoids*. © 1999 Cambridge University Press. All rights reserved. Printed in the United States of America.

strange, stemless genus *Timorocidaris* (first mistaken for an echinoid spine), the skeleton is reduced to a single, extremely variable piece that carried at the most three arms on a small neck. Even by Timor standards this crinoid was hugely successful, as judged by the incredible number of about 110,000 cups of the main species, *Timorocidaris sphaeracantha*, collected at Basleo (Wanner 1940). Such stemless and unattached crinoids with reduced arms are unknown from later strata (the unattached Middle Jurassic *Ailsacrinus* has well-developed arms; see Chapter 24), but in many respects the Timor crinoid fauna shows similarities to the cyrtocrinids from the Jurassic and Lower Cretaceous of Europe (see Chapter 3, section titled 'Mesozoic Cyrtocrinid Assemblages'). This raises some doubts as to the classification of the Timor crinoids as reef forms as postulated by Wanner (1924, 1940). However, Wanner (1924) also noted striking similarities with the Eifel region, where Middle Devonian crinoids thrived in forereef shelf areas.

The Upper Permian (most likely Guadalupian) age of the Timor crinoids (Erwin 1993) has been questioned by Webster (1987, 1990) on the basis of the similarities between the Australian and the Timor crinoids. On Timor, the crinoids occur in yellowish to reddish mudstones, but also in tuffs and tuffaceous siltstones; these are interbedded with highly fossiliferous, partly crinoidal limestones. This so-called Maubisse Formation also includes basaltic rocks. These strata were displaced from the north to the present Timor area and may have been part of the West Australian geosyncline (Webster 1990). Accordingly, the crinoid-bearing beds of western and eastern Australia were deposited on the shelf at the southern margin of this sea (Fig. 175).

Of the roughly 100 crinoid species described from Australia, about half were from the eastern (including Tasmania) and half from the western part of the continent, with no species common to both regions. Because some crinoid genera occur in eastern and western Australia and some species are closely related, migration pathways around the northern margin or across the central interior of the continent must have been open during part of the Permian (Webster 1990). The Permian

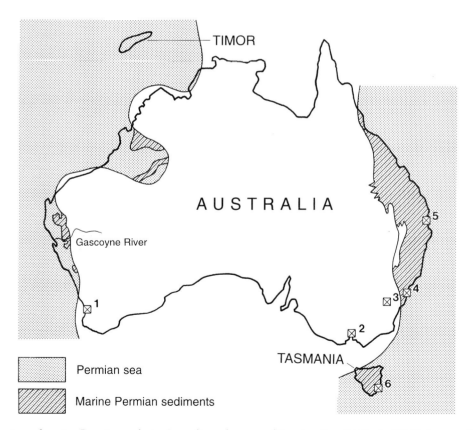

Fig. 175. Distribution of marine Permian rocks in Australia and extent of Permian Sea. (1) Perth; (2) Melbourne; (3) Canberra; (4) Sydney; (5) Brisbane; (6) Hobart. (After Teichert 1951; Willink 1979.)

strata of Australia are dominated by clastic sediments (sandstones, siltstones and shales) and limestones are rare. About half of the Australian Permian crinoids are from limestones or calcareous mudstones, and the other half from sandstones and siltstones. Many of the crinoids, embedded in fine-grained sandstones or mudstones with sandy layers, are preserved as iron oxide replacements of the original calcite. The crinoid faunas from Australia are the only cooler water faunas from the Palaeozoic and have a much lower diversity than the Timor crinoids. They include forms reaching a very large size, with specialized arm structures.

CALCEOLISPONGIIDS: A PERMIAN SUCCESS STORY

The most outstanding forms of Australian crinoids are the calceolispongiids, the genera being *Calceolispongia* and *Jimbacrinus*. These are large dicyclic cladid crinoids with bulky, commonly spinose basals, slender stems and unbranched, uniserial arms. The amazing *Calceolispongia* is characterized in a number of species by strongly hypertrophic basal plates, which are drawn out into long, horn-shaped processes resembling slippers (hence the first part of the name of the genus, from the Latin *calceolus*) (Fig. 176). These plates were originally misinterpreted as sponge remains (hence the second part of the name). Wanner named this form *Dinocrinus* on the basis of some plates from Timor. This appropriate name was, however, proposed one year after the name '*Calceolispongia*' was assigned and is therefore a junior synonym. It is interesting that many of the eastern Australian species of this genus have no significant thickening of the basals (Willink 1979). Teichert (1949, p. 3) characterized this crinoid as follows: 'It is as if Nature had not been content with producing one of the oddest crinoids that ever existed: individuals of each species vied with each other to excel in oddities and individual peculiarities. This is all the more remarkable since the cup of *Calceolispongia* is of the simplest design; number and general size and arrangement of the cup plates never vary, and anomalies are rare. Yet the shape and ornamentation of individual plates is subject to variation without limit. As a result the degree of variability in general appearance of the cups is such as is rarely found and perhaps never surpassed in invertebrates.'

In western Australia, 13 species of *Calceolispongia* follow each other in a stratigraphic sequence of about 2,000 m of Artinskian strata, estimated by Teichert (1949) to be approximately 6 million years. During the evolution of this genus, the basal plates increased in bulk 400 times. The last species of this phylogenetic series, *Calceolispongia robusta*, is a giant among crinoids, with a cup diameter of 130 mm (Fig. 176). With the exception of *Calceolispongia*, the western Australian crinoid fauna may be regarded as an extremely impoverished Timor assemblage. Well-preserved calcitic specimens of *Jimbacrinus bostocki* have been found in recent years in the Artinskian of the Gascoyne River in western Australia (Fig. 177), where they occur in a fine-grained sandstone. In more or less complete specimens, the stem is sharply bent just below the crown. Teichert, who figured one such specimen (1954, Pl. 13, Fig. 1), believed this to be the result of muscular contraction at the death of the animals. However, the stem of these crinoids may have been bent in life, with the stem as runner along the substrate. Bent-stem flexible crinoids

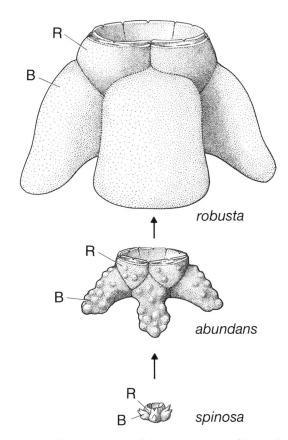

Fig. 176. Three examples of the size increase of Lower Permian *Calceolispongia* species in successive strata of western Australia. Side views of *C. spinosa* (Callytharra Series, Sakmarian), *C. abundans* (Middle Wandagee Series, Artinskian) and *C. robusta* (Upper Wandagee Series). Key: B, basals; R, radials. (After Teichert 1949.) ×0.5.

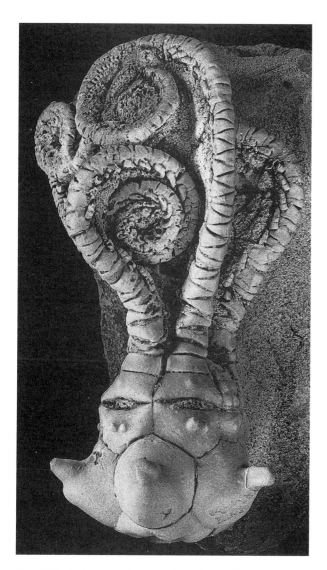

Fig. 177. *Jimbacrinus bostocki*. Artinskian, Gascoyne River, western Australia. Note swollen basal plate with spine and enrolled atomous arms with uniserial, wedge-shaped brachials. (Hess Collection; photograph S. Dahint.) ×1.6.

suggested that the basals helped to stabilize the cup on the substrate and to lift the crown above the bottom. Teichert, who wrote a monograph on the genus *Calceolispongia* in 1949, thought that the stem was too weak to support the crown and may have been without any function. This may have been a reasonable possibility for species with very large basals such as *C. robusta*, but this theory is difficult to accept for the species with rather long stems and radicular cirri. Other possibilities have been discussed by Willink (1979), such as a baffle function of the protruding basals inducing an eddy effect in currents. Thickening of the basals may thus have been an adaptation to life in strong currents; as a matter of fact, most species with very large basals occur in sandstones (a common substrate for Australian crinoids) and are disarticulated, whereas species with less developed basals occur in mudstones and are commonly preserved completely. The articulations of the arms indicate that these could be coiled quite tightly, perhaps another adaptation to stronger currents, but they also conferred an advantage against predation (Fig. 177).

WIDE DISTRIBUTION OF THE REMAINS OF PERMIAN CRINOIDS

The best-known Permian crinoid faunas are those from Timor and Australia, but Permian crinoids have also been reported from North America, Bolivia, Europe (Sicily) and Tunisia, as well as from Russia (Urals), China, India and Thailand. The Russian crinoids form a link between Timor and North American faunas and contain advanced poteriocrines and indocrinids. A rich fauna of Lower Permian crinoids with 37 genera and 47 species, mostly cladids, has been described from southern Nevada (Lane & Webster 1966; Webster & Lane 1967). A number of these silicified fossils are preserved as more or less complete crowns and were collected from two biohermal lenses of coarse-grained, well-bedded, skeletal limestone composed mainly of disarticulated crinoids. This fauna is dominated by advanced poteriocrines, but the flexibles *Trampidocrinus* and *Nevadacrinus* with their wedge-shaped columnal beneath the crown appear to be an adaptation for a stem running along the substrate. Lane and Webster assumed that the rather robust Nevada crinoids adapted to life in agitated water.

have been reported from the Lower Permian of Nevada (see the next section). The Early Permian (probably Artinskian) limestones of Callytharra Springs in western Australia have furnished a crinoid fauna composed of a range of species feeding at different levels. Unfortunately, the fossils occur mainly as disarticulated cups, with only a very few crowns (Webster 1987).

What could have been the reason for the excessive growth of the basals of many of these forms? It has been

21

Triassic Muschelkalk of Central Europe

HANS HAGDORN

WITCHES' MONEY AND CHICKEN LEGS: THE RESEARCH HISTORY OF *ENCRINUS LILIIFORMIS*

Long before the advent of scientific palaeontology, common fossils were connected with superstition and legends of popular belief (Abel 1939). In Lower Saxony the abundant columnals of the Muschelkalk sea lily *Encrinus liliiformis* were called *Sonnenräder* (sun wheels), and in Thuringia and Hessia they were called *Bonifatiuspfennige* (St. Boniface's pennies) because the saint who baptised the German tribes was said to have cursed all the pagan money, which turned into stone. In southwestern Germany, *Encrinus* columnals were called *Hexengeld* (witches' money). According to a legend from Beuthen in Upper Silesia, in 1276 St. Hyacinth's rosary broke when he was praying at a fountain, and the beads dropped into the water. The saint prayed for them to multiply, and since then the fountain has been producing rosary beads: columnals of a diverse Middle Muschelkalk crinoid association (Hagdorn et al. 1996).

Therefore, it is not astonishing that columnals of *Encrinus liliiformis* were among the first crinoid remains described in the scientific literature. In his monograph, *De natura fossilium* (1546), Georgius Agricola from Chemnitz in Saxony introduced the name 'Encrinos lilgenstein', which means stone lily. However, he used this name for *Chladocrinus* columnals from the Lias of Hildesheim. For the cylindrical columnals of *Encrinus liliiformis*, Agricola coined the names 'Trochites' (wheel stones, a translation of their trivial German name, *Rädersteine*) for single columnals and 'Entrochus' for pluricolumnals. A hundred years later, Fridericus Lachmund, in *Oryctographia Hildesheimensis* (1669), illustrated columnals, cups and cup elements (*Pentagonus*) as well as a fragmentary crown, the arms of which he compared to chicken legs. Misinterpreting Agricola, he transferred the name 'Encrinus' to these fossils. From that time onward, the name 'Encrinus' became attached to this common and earliest recognized crinoid crown. In 1719 the Hamburg physician Michael Reinhold Rosinus figured complete crowns and elements of crown and stem that he regarded as fragments of some kind of starfish. Other 18th-century authors explained them as vertebrae of sea animals, marine plants, corals or parts of 'Jew stones' (sea urchins). Finally, the complete animal was correctly reconstructed by Johann Christophorus Harenberg in 1729. *Encrinus* specimens were found and described from many sites in Germany, such as the classic sites of Hildesheim, Erkerode, Göttingen (all in Lower Saxony), Crailsheim, Schwäbisch Hall and Neckarwestheim (in Württemberg).

Before the introduction of binominal scientific nomenclature, Harenberg's *Lilium lapideum* (stone lily) was the most common name for the fossil. Until 1840, *Encrinites fossilis* Blumenbach 1802, *Encrinites trochitiferus* Schlotheim 1813 and *Encrinites moniliferus* Miller 1821 were used in parallel, but eventually Lamarck's name, *Encrinus liliiformis* 1801, was formally established as the valid species name (comp. ICZN 1962; opinion 636).

Hans Hess, William I. Ausich, Carlton E. Brett, and Michael J. Simms, eds., *Fossil Crinoids*. © 1999 Cambridge University Press. All rights reserved. Printed in the United States of America.

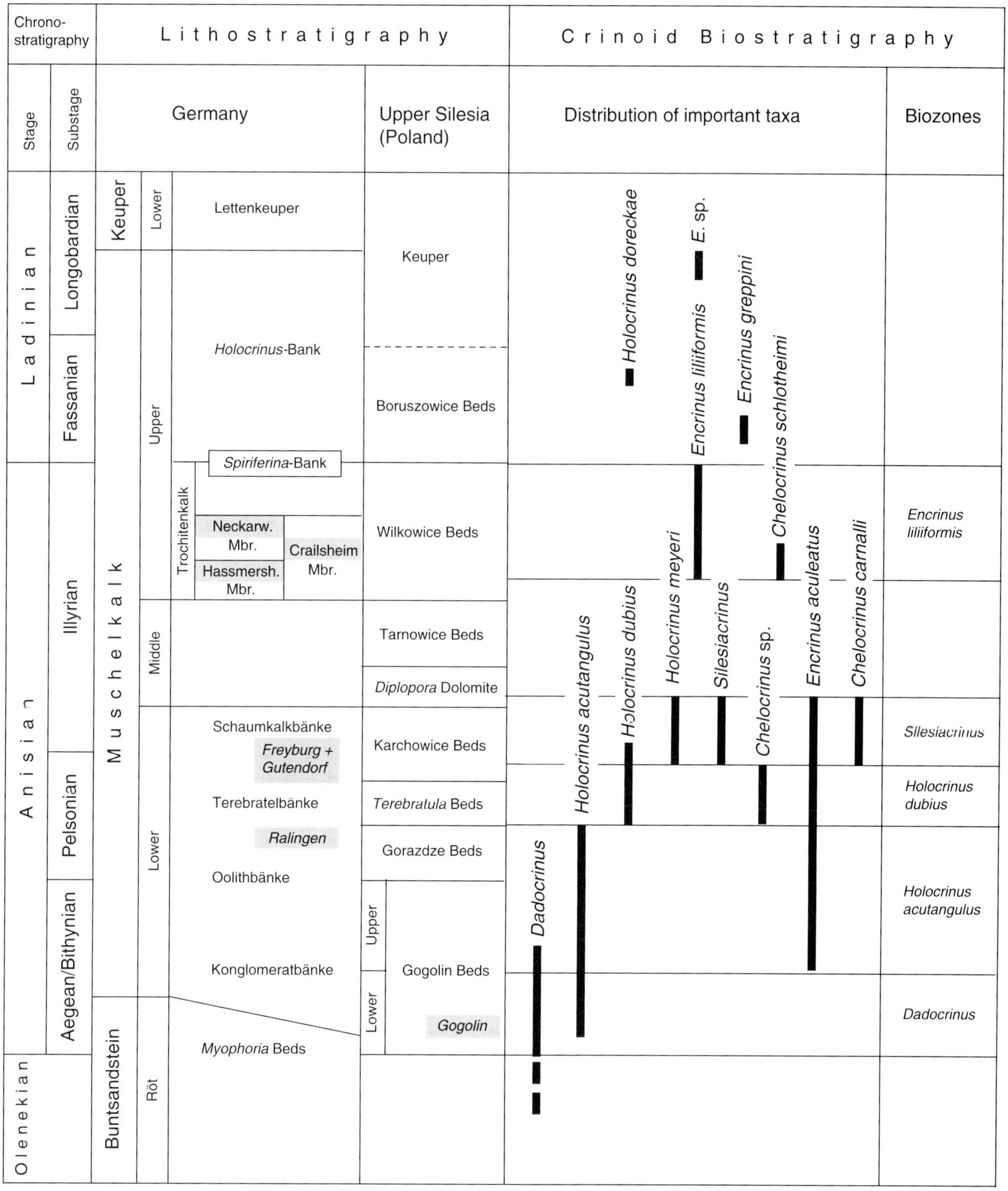

Fig. 178. Stratigraphic table of the Muschelkalk with crinoid *Lagerstätten*. Biostratigraphy of Middle Triassic crinoids; distribution of some important crinoid species.

THE MUSCHELKALK SEA

The middle Triassic Muschelkalk (Fig. 178) forms the central group of the tripartite Germanic Triassic beginning with mostly continental red beds of the Buntsandstein, followed by the marine Muschelkalk carbonates and evaporites, which in turn are overlain by the multi-coloured Keuper claystones, sandstones and evaporites deposited under changing terrestrial, playa lake or even marine conditions. The Muschelkalk deposits cover an area of Europe between the French Massif Central, Fennoscandia and the Russian Platform (Fig. 179). Towards

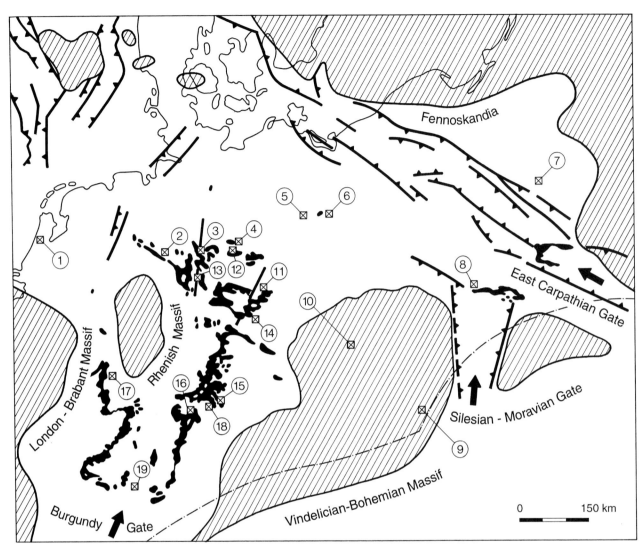

Fig. 179. Palaeogeographic map of the Germanic Basin during Middle Triassic with Muschelkalk outcrops (black) and position of crinoid *Lagerstätten*. Key: 1, Amsterdam; 2, Örlinghausen; 3, Hildesheim; 4, Erkerode; 5, Berlin; 6, Rüdersdorf; 7, Warsaw; 8, Gogolin; 9, Vienna; 10, Prague; 11, Freyburg/Unstrut; 12, Wittmar; 13, Göttingen; 14, Gutendorf near Weimar; 15, Crailsheim; 16, Neckarwestheim; 17, Ralingen-Kersch; 18, Schwäbisch Hall; 19, Basel. (Palaeogeography according to Ziegler 1982; from Hagdorn 1991.)

the North Sea Central Swell, the Muschelkalk sediments grade into terrestrial red beds. To the south, the basin was separated from the western branches of the Tethys Sea by the Vindelician–Bohemian Massif. During the Muschelkalk transgression, marine faunas immigrated into the Germanic Basin. The first transgression, in Early Anisian times, brought elements from the Asiatic faunal province via the northern Palaeotethys branch through the East Carpathian Gate into the eastern and central part of the Germanic Basin (Kozur 1974). During Pelsonian and Early Illyrian times (upper part of Lower Muschelkalk) the eastern part of the Germanic Basin was closely connected with the Alpine realm via the Silesian–Moravian Gate. Therefore, the middle and upper parts of the Lower Muschelkalk in Upper Silesia belonged to the Austroalpine faunal province with abundant and diverse stenohaline benthos composed of crinoids, echinoids, brachiopods, corals and hexactinellid sponges. Towards the west, faunal diversity generally decreased due to elevated salinity. Later in the Early Illyrian, the Germanic Basin became isolated by the lowering of sea level, and only euryhaline faunas lacking crinoids survived this salinity crisis. Along the subsidence centres, conditions leading to evaporite development did not allow any benthic life. Towards the end of Anisian times, renewed transgression from the

Tethys induced faunal exchange documented in the Upper Muschelkalk by *Encrinus liliiformis*. This incursion connected the western part of the basin with the western Tethys via the Burgundy Strait. Lower and Upper Muschelkalk crinoid faunas are closely related, although they do not have any species in common. A detailed scenario of crinoid and echinoid immigration and spreading over the Muschelkalk Basin was reconstructed by Hagdorn (1985) and refined by Hagdorn and Gluchowski (1993).

During Anisian and Early Ladinian times, a first rapid speciation among the encrinids and holocrinids took place, and this is best documented in Muschelkalk sediments. The second radiation occurred during latest Ladinian and Early Carnian. Consequently, a parastratigraphic biozonal scheme based on distinct isolated crinoid and echinoid ossicles has been proposed by Hagdorn and Gluchowski (1993) for the eastern part of the Muschelkalk Basin (Fig. 178). This scheme can also be applied to the western Tethys and its marginal basins (southern and northern Alps, Bakony, Hungarian Muschelkalk in the Mecsek Mountains).

The Muschelkalk sediments were deposited in a warm and shallow carbonate-dominated sea with considerable terrestrial influx, indicated by a clastic marginal sand facies and by thinly bedded limestones, becoming intercalated with mudstones (marls) and shales in the central part of the basin. On the tops of regional or local elevations, thickly bedded skeletal and oolitic limestones give evidence of extremely shallow water. Excellent exposures in cliff-like escarpments along valleys and small gorges, as well as extensive quarrying, have allowed high-resolution bed-by-bed litho- and eco-stratigraphy. During the past few years, this has been interpreted in terms of sea level fluctuations and sequence stratigraphy.

Unlike the contemporaneous rocks of the Alpine Triassic, the Muschelkalk is rich in *Lagerstätten* (Seilacher 1993) resulting from smothering of epibenthic organisms *in situ*, thus representing a snapshot of the sea floor. Some of these *Lagerstätten* have yielded articulated skeletons of crinoids, echinoids, ophiuroids, asteroids and even holothurians.

ENCRINUS LILIIFORMIS IN WÜRTTEMBERG'S CRINOID GARDENS

Since Quenstedt's collecting trips to the Gaismühle on the river Jagst northwest of Crailsheim during the 1850s, the Crailsheim Trochitenkalk has become as famous for *Encrinus liliiformis* as the North German localities. Extensive quarrying for building stones has yielded hundreds of specimens. From the facies patterns, Vollrath (1958) reconstructed a regional shoal of thickly bedded crinoidal limestones situated several tens of kilometres offshore from the Vindelician Massif. Towards the basin centre, this 16-m-thick limestone sequence (Crailsheim Member) grades into single crinoidal beds interbedded with mudstones and micritic limestones (Hassmersheim and Neckarwestheim Members) (Fig. 184). Most of the crinoidal beds can be used as stratigraphic marker horizons.

THE NECKARWESTHEIM SEA LILY BED

One of these marker beds is the Trochitenbank 6 (*pulcher* Biozone). At Neckarwestheim near Heilbronn, approximately 70 km west of Crailsheim, it has yielded well-preserved and complete *Encrinus* specimens. In several excavations, Linck (1954, 1965) recovered big slabs that consisted of three units:

(a) a 5- to 10-cm-thick micritic limestone that was extensively burrowed (*Balanoglossites*) and had an eroded surface; this firmground was patchily cemented and subsequently bored (*Trypanites*);
(b) a thick (up to 20 cm) graded skeletal and crinoidal limestone with large intraclasts eroded from unit a; the surface has megaripples;
(c) an ochre-coloured marl (up to 10 cm) with silt and pellet-filled feeding traces that are also present between the closed crinoid arms of the unit below and may even penetrate the tegmen.

The crinoids – adult specimens with a length of up to 160 cm – are distributed irregularly on the upper surface of unit b and covered by unit c. The mudstones below and above Trochitenbank 6 contain thin shelly layers of soft-bottom infauna and are devoid of crinoids. Some of the crinoids have holdfasts still attached to an intraclast or to a valve of the big oyster-like terquemiid *Enantiostreon* or *Newaagia*. The stems of about half of the specimens end with ruptures in their distal or central parts (Fig. 180). Secondary callus growth on these 're-generated terminal stems' (Linck 1954) indicates that the animals survived the traumatic event that broke their stem. Linck's (1954) reconstruction of uprooted, yet still upright and floating crinoids implies buoyancy of the crown, which is most improbable. The presumed lifestyle of such 'terminally regenerated' animals is further described in the section titled 'Encrinus liliiformis: A Gregarious Sea Lily with Intraspecific Tiering'. A certain number of the Neckarwestheim Encrinus specimens are preserved without their arm tips. The reason for this is not known, because regenerated arms, which would indicate ecological stress or partial predation some time prior to final burial, have rarely been recovered. Further outcrops of this bed demonstrate that this Encrinus habitat covered several hundreds of square kilometres.

From this sedimentological and preservational evidence, a habitat between wave base and storm wave base may be assumed. Currents provided the suspension-feeding community with nutrients. Occasional storms (which also caused megaripples) affected populations in a limited area. Such storms repeatedly caused the crinoid remains that had previously accumulated to disintegrate. The animals killed during a final event were uprooted, sank to the bottom and were covered by mud. Subsequently, crinoids and other epibionts from unaffected areas resettled suitable anchoring grounds. Such

Fig. 180. Complete specimen of *Encrinus liliiformis* from Neckarwestheim, arms closed. Note regenerated stem, distally rounded. The length of these crinoids reached 160 cm. (Paläontologisches Museum, Zürich; courtesy H. Rieber.) ×0.4.

a scenario explains the fairly uniform size in these crinoid populations. The final preservation of the Neckarwestheim crinoid bed is due to increased subsidence, a long period without a severe storm and an increasing background sedimentation, as documented by the succeeding mudstones (Hagdorn & Simon 1984).

THE CRAILSHEIM CRINOID BIOHERMS

From the Neckarwestheim occurrence, Linck (1965) concluded that the 16-m-thick crinoidal limestones around Crailsheim (Crailsheim Member) accumulated from complete animals floating onshore from deeper parts of the basin. He assumed that *Encrinus* occupied vast areas, but was rather patchily distributed, similar to the Neckarwestheim occurrence. However, Linck did not take into account that only the very top of the Crailsheim Member can be time-correlated with Trochitenbank 6 of Neckarwestheim.

Unequivocal evidence for the autochthony of *Encrinus liliiformis* in the Crailsheim Member came from the discovery of bioherms with encrinid roots preserved *in situ* (Hagdorn 1978) (Fig. 181). Such structures also occur in the Hassmersheim Member, which was deposited in the deeper parts of the southwestern German Trochitenkalk carbonate ramp (Aigner 1985; Hagdorn 1991). It is characterized by a cyclical change of four skeletal beds and intercalated mudstones devoid of crinoids. The skeletal beds with epifaunal suspension-feeding communities and bioherms are interpreted as parts of high-frequency sequences with flooding surfaces on their tops (Aigner & Bachmann 1992; Hagdorn & Ockert 1993). The calcareous mudstones with infaunal suspension and deposit-feeding communities, preserved in tempestite beds and gutters, represent the relative highstands. Towards the shallow ramp, the mudstones wedge out in the skeletal limestones of the Crailsheim Member (Fig. 182). Bed-by-bed lithostratigraphy has allowed the tracing of these marker beds from the Hassmersheim to the Crailsheim Member with their changing facies and communities. The offshore–onshore zonation of infaunal to epifaunal communities corresponds to the spectrum of community replacement in the high-frequency sequences (Hagdorn & Ockert 1993).

Bioherms are most abundant (up to 1 per 100 m²) in a 2-m-thick sequence of crinoidal limestones interbedded with sheets of marl in the Crailsheim Member (*Encrinus* Platten). Bioherms also occur in the thick, oolitic

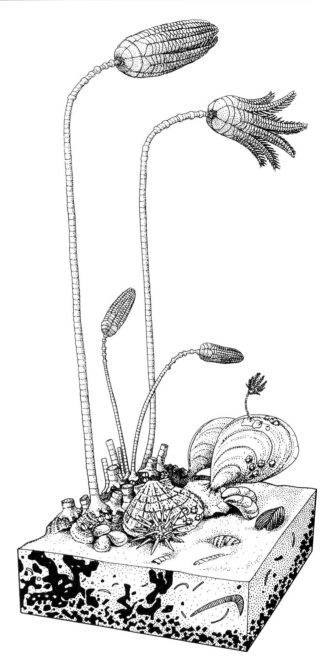

Fig. 181. Reconstruction of *Encrinus liliiformis* on an encrinid–terquemiid bioherm. Crinoids of different age are attached to the oyster-like terquemiids *Enantiostreon* and *Newaagia* (left) with holdfasts; juvenile individuals are attached to the mussel *Myalina* (right, fixed by byssus threads to a *Newaagia* shell). Large stems of *Encrinus* are broken near the base. (Redrawn from Hagdorn 1991.)

shallow-water limestones but are less common there. This may be due to frequent sediment reworking, diminishing their preservation potential. The bioherms have a diameter of 2–3 m and a thickness of up to 1.8 m.

Fig. 182. Trochitenkalk with disarticulated *Encrinus* ossicles (mainly columnals) from the flank of an encrinid–terquemiid bioherm. Upper Muschelkalk, Crailsheim Member, Mistlau near Crailsheim. (Muschelkalkmuseum, Ingelfingen; photograph H. Hagdorn.) ×1.2.

Their framework was built primarily by terquemiid false oysters and is commonly encrusted with small *Placunopsis*, spirorbid worm tubes and foraminifera (Fig. 181). *Encrinus* roots encrust the framework most densely towards the bioherm tops and contribute to it. Bioherms are extensively bored by phoronids (*Talpina, Calciroda*), acrothoracican barnacles and algae. According to Schmidt (1992), the bioherms originated in the photic zone at a maximum depth of 50 m. Monaxon spicules indicate the presence of frame-binding sponges; algal or bacterial mats have not been recovered (Hüssner 1993). Articulate brachiopods and byssate bivalves were flexibly attached to the bioherm; the bivalves are commonly encrusted by a single or by small groups of *Encrinus* holdfasts (Fig. 183). In bioherms situated towards the deeper water on the carbonate ramp (Hassmersheim Member), crinoids decrease in number and brachiopods (*Coenothyris*) become dominant (Fig. 184).

Near Schwäbisch Hall, approximately 30 km west of Crailsheim, a large bioherm complex in an intermediate position on the carbonate ramp persisted through two sedimentary cycles, with periods of lateral expansion during transgression and restriction during regression, when terrigenous mud was prograding. It has yielded 120 *Encrinus* specimens with a crown length ranging from 6 to 120 mm. The juvenile crinoids with their entirely flexible stem were still attached to the bioherm framework, whereas larger specimens were found uprooted in the bioherm flanks, either directly on top of a shell bed with *in situ* brachiopod and bivalve colonies or in the mudstones covering this bed.

As a rule, complete *Encrinus* specimens are most abundant close to a bioherm on the muddy tops of shell beds. Low-tier brachiopod colonies in life position also occur there; these grew on a shelly bottom devoid of mud. Such communities appear to have been smothered by bottom backflows of mud avalanches originating from higher areas of the carbonate ramp (Aigner 1985).

Fig. 183. Juvenile individual of *Encrinus liliiformis* attached to the mussel *Myalina*, discoid holdfast and distal stem of a slightly larger specimen. The bivalve covering the small specimen was deposited at the base of a thin bed of crinoidal limestone from a bioherm flank. Upper Muschelkalk, Crailsheim Member, Mistlau near Crailsheim. (Museum für Geologie und Paläontologie, Tübingen; photograph W. Wetzel.) ×1.

Complete *Encrinus* specimens also occur inside metre-thick, oolitic, crinoidal limestones deposited in extremely shallow water. During lateral shift of the skeletal material, single crinoids were buried by the moving sediment. Due to early cementation, their crowns were not laterally compressed, but remained completely closed, resembling corn cobs.

In southwestern Germany, *Encrinus liliiformis* settlements persisted through two ceratite zones, or six high-frequency sequences on the shallow carbonate ramp of the Crailsheim shoal. Towards the deep ramp, the crinoids are restricted to the transgressive crinoidal marker beds and disappear in the interbedded mudstones that were dominated by soft-bottom communities during sea level lowstands. With rising sea level, the epifaunal communities with crinoids expanded laterally towards the deep ramp until they were covered again by mud deposited during low sea level of the next sequence.

ENCRINUS LILIIFORMIS: A GREGARIOUS SEA LILY WITH INTRASPECIFIC TIERING

The Crailsheim and Neckarwestheim *Encrinus Lagerstätten* demonstrate that this crinoid was living between extremely shallow water above wave base and deeper water below storm wave base (Fig. 184). Larvae settled in the bioherms close to the adults, preferentially on their basal stems. Thus, they succeeded in finding solid anchoring ground slightly above the muddy sea floor, where they were less endangered by smothering. From this, a short larval period and gregarious behaviour of *Encrinus* may be inferred. After a pioneer settlement on winnowed shell beds during omission periods, bioherm frame builders (terquemiids, encrinids) were able to compensate for increasing sedimentation rates by upward growth. In the *Lagerstätten* just discussed, *Encrinus liliiformis* is the only crinoid species. Its size distribution in the Crailsheim and Schwäbisch Hall bioherms indicates intraspecific gradual tiering between juveniles, occupying low-level tiers, and adults, occupying high-level tiers up to 160 cm above the sea floor. Unlike other crinoids, *Encrinus* did not reach its maximum size in a short time by accelerated juvenile growth.

The crown of larger individuals was elevated by a stem that was rigid in the distal and middle parts but flexible in the proximal growth zone (Seilacher et al. 1968); thus, current pressure could be compensated for by passive orientation. Nevertheless, in an environment with occasional storms, rupture of the distal stem uprooted many individuals (Fig. 180). Post-rupture secondary growth of the terminal stem by sealing the axial canal indicates that these animals were able to survive for some time, possibly with part of the stem along the sea floor. However, in such a position they were in increased danger of being buried. Unlike the cirriferous holocrinids, encrinids were unable to become reattached after their stem broke. For successful long-term settlement, they needed solid substrates as provided by the bioherm framework. Individuals settling on flexibly attached bivalves like *Myalina* (Fig. 183) lost their attachment when growing up because the bivalve byssus was not able to anchor such large passengers against stronger current pressure. *Myalina* shell fragments, marginally overcrusted by the crinoid holdfasts, indicate that such crinoids shared the fate of those with broken stems.

The taphonomic record of isolated cups proves that *Encrinus liliiformis* had tightly sutured plate circlets forming a robust cup. Bending of the arms by active muscular movement was restricted to the radial and axillary facets and to the uniserial proximal part of the arms with straight muscular articulations. The oblique muscular articulation between the first pinnular and brachial allowed the pinnule to bend away from the arm. *Encrinus* was able to unfold a filtration fan with the arms at an angle of about 45° to the long axis. Between the arms

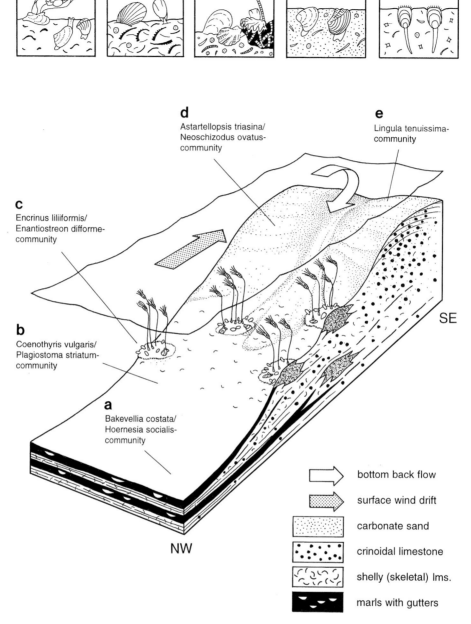

Fig. 184. Facies model and fossil community zonation on the southwestern German Trochitenkalk carbonate ramp during the *atavus* Biozone with crinoid bioherms. The profile covers an area extending approximately 90 km from northwest to southeast along the river Jagst, northeast of Stuttgart. Hassmersheim Member on the deep ramp (left), Crailsheim Member on the shallow ramp (right). According to Aigner's (1985) hydrodynamic model, fair-weather alongshore currents (SW–NE) provided nutrients. Storm-induced onshore wind-drift currents of surface water piled up skeletal debris that was continuously reworked in shallow water. Bottom backflows were responsible for the smothering of habitats in deeper water. (Redrawn from Hagdorn 1991.)

were dense pinnular combs, but the arms could not recurve into a parabolic fan (Fig. 48). Pectinate ornaments on the outer side of the pinnular food grooves were interpreted by Jefferies (1989) to be evidence of food specialization. He argued that this crinoid was not able to form a real filtration fan, but rather used its dense pinnule system as a microfilter for extremely small plankton, creating its own currents. However, this hypothesis does not explain the lack of other crinoids that filtered normal-sized plankton in the southwest German

Trochitenkalk Sea. The extraordinary ecological success of *Encrinus liliiformis* may be due to its intraspecific tiering.

Encrinus liliiformis Lagerstätten in central and northern Germany and in France have not been studied in such detail. Bioherms of the Crailsheim type, occurring in most *Encrinus* localities, indicate similar habitats. In northwestern Germany another encrinid, the 20-armed *Chelocrinus schlotheimi*, occurs together with *Encrinus liliiformis* at the base of the Trochitenkalk (Gelbe Basisschichten Member). In an assemblage from Wittmar (Lower Saxony), many specimens had their stems broken a few millimetres below the cup (Hagdorn 1982). Sealing of the axial canal or rounding of the stump indicates that the stemless crowns survived for some time on the sea floor, presumably with their oral side facing upward, as with the stemless *Agassizocrinus* or *Ailsacrinus* (Fig. 206).

CHELOCRINUS CARNALLI: VIOLET FLOWER NESTS

The Schaumkalk ('foamy limestone') derives its name from dissolved oolites. This makes it a soft, porous and easily worked building stone that has been quarried since the Middle Ages. Buildings as famous as the romanesque Naumburg Cathedral with its great statues have been constructed from local Schaumkalk. These marker beds belong to the basin-wide expansions of a shallow-water oolitic and skeletal sequence characterizing the upper part of the Lower Muschelkalk in central and eastern Germany. In deeper parts of the basin, oolitic beds are interbedded with thinly bedded, marly Wellenkalk ('wavy limestone'). Again, this interplay was governed by regional tectonics and sea level fluctuations. The quarries of Zscheiplitz and Schleberoda (near Freyburg an der Unstrut, Saxony-Anhalt) and Gutendorf (near Weimar, Thuringia) became famous for complete specimens of *Chelocrinus carnalli* found on top of the Lower Schaumkalkbank. As described by Jaekel (1894) and Biese (1927), the Freyburg quarries yielded large slabs with bundles of up to 25 complete specimens (Fig. 185) that were current-oriented in one or two

Fig. 185. A bundle of *Chelocrinus carnalli* with a cluster of holdfasts probably encrusting larger shells (not preserved). Lower Muschelkalk, Schaumkalk; Freyburg/Unstrut. (Museum für Naturkunde, Berlin; from Biese 1927.) ×0.22.

directions (called *Blumennester*, flower nests, by the quarrymen). Many stems are preserved with their holdfasts, typically showing an attachment area perpendicular to the stem axis. This indicates anchorage on an unstable substrate because the stem would have broken above a tightly attached holdfast. The crinoids are on top of the oolitic Schaumkalk in an ochre-colored micritic limestone and were covered by greenish marl. The population, comprising individuals of different size, appears to have been rapidly killed and subsequently buried by mud. Very rarely, single specimens were found in skeletal layers inside the cross-stratified Schaumkalk Bed. In Gutendorf, the top of the Lower Schaumkalkbank is a hardground covered by a marly layer and by Wellenkalk. The hardground is extensively bored (*Trypanites*); it is encrusted by serpulids and individual *Chelocrinus* holdfasts (Fig. 186). These either are penetrated by *Trypanites* or fill up hollow *Trypanites* tubes with their stereom (Müller 1956b). Boreholes that are surrounded, but not covered, by stereom of holdfasts demonstrate that the boring organism was still alive. The stems of the Gutendorf specimens are broken either near the base or just below the cup. The crowns of the latter specimens are embedded with their 20 arms splayed out (Müller 1956a). The dominance of such individuals on the Gutendorf hardground indicates large areas favourable for settling. In Freyburg, on the other hand, groups of larvae had to anchor closely together on small solid patches on which they grew up in clusters. Such gregarious behaviour appears to have been the result of a lack of proper anchoring ground. With their purple to violet colour, the Freyburg *Chelocrinus carnalli* contrasts beautifully with the sediment.

DADOCRINUS: THE SMALL COUSIN OF *ENCRINUS*

At first glance, *Dadocrinus* looks like a juvenile *Encrinus*. Indeed, cladistic analyses indicate that *Dadocrinus* is more closely related to the encrinids than to the millericrinids, to which it had been assigned during previous decades. This Lower Anisian genus, with several species from the South Alpine and Germanic Muschelkalk, has uniserial arms. In the Germanic Basin, it is restricted to the eastern part, where dissociated ossicles built crinoidal limestones in the Lower Gogolin Beds. Their dispersal towards the west was prevented by a salinity barrier (Hagdorn & Gluchowski 1993). A hundred years ago, the classic site at Sakrau near Gogolin in Upper Silesia (now Poland) yielded beautiful slabs of a light yellowish grey limestone crowded with *Dadocrinus kunischi* (Fig. 187). Sedimentological and palaeoecological data indicate a muddy bottom with burrowing and mud-sticking bivalves (Hagdorn 1996). The crinoids grew in clusters, attached with their discoid holdfasts to the rear ends of the infaunal *Gervillella mytiloides* or to similar bivalves but also to the basal stem of other individuals. Holdfasts attached to hardgrounds have been observed at one locality. The size limitation of this species, with a maximum height of 20 cm, may be attributed to its preference for soft substrates; the bivalve byssus would not have been able to anchor larger crinoids on a muddy sea floor. In flat lenses, *Dadocrinus* specimens are concentrated on the upper and – less commonly – on the lower bedding plane with closely united holdfast clusters. In other localities, single specimens, stem fragments, cups and single, aborally recurved arms indicate stress-caused autotomy prior to burial. The *Dadocrinus* beds of Recoaro in the Vicentinian Alps (Italy) are very similar to those from Upper Silesia.

Its holdfast characterizes *Dadocrinus* as a primary hardground dweller. Living as an epibiont on mud-sticking mussels opened the possibility of secondary soft-ground colonization. This, however, limited the size of the crinoids and also prevented them from contributing to bioherm structures comparable to those of their larger encrinid relatives.

Fig. 186. Discoidal holdfasts of *Chelocrinus carnalli* on a hardground with *Trypanites* borings. Lower Muschelkalk, Schaumkalk; Gutendorf near Weimar. (Muschelkalkmuseum, Ingelfingen; photograph H. Hagdorn.) ×1.7.

Fig. 187. *Dadocrinus kunischi* on top of a bedding plane. Lower Muschelkalk, Gogolin Beds; Sakrau near Gogolin, Upper Silesia. (Muschelkalkmuseum, Ingelfingen; photograph H. Hagdorn.) ×1.6.

HOLOCRINUS: PRECURSOR OF ISOCRINIDS

Unlike encrinids and dadocrinids, holocrinids were not permanently fixed to the sea floor by a holdfast, but attached themselves actively with their cirri. Like their descendants, the isocrinids, they had stems with a preformed rupture point below the nodals so that after stress-induced autotomy the stem automatically ended with a cirrinodal. More or less smooth cryptosymplectial lower nodal facets, so typical of isocrinids, had not yet developed in holocrinids (Hagdorn 1983). However, taphonomic evidence from stem fragmentation patterns suggests that short ligament fibres connected nodals to infranodals. These were responsible for autotomy in holocrinids, and they must have been developed prior to the cryptosymplexies that are documented from Upper Ladinian onward (Baumiller & Hagdorn 1995). Complete *Holocrinus* skeletons remained extremely rare for a century, but recently several new sites have been discovered, yielding additional specimens. In the Upper Muschelkalk, isolated holocrinid ossicles occur in one single marker bed in southwestern Germany (Hagdorn 1983), whereas they are more diverse and more common in the Lower Muschelkalk. Their remains are found, commonly together with encrinids, on the tops of hardgrounds where the animals were living closely together. In the clastic marginal facies of the Lower Muschelkalk of Ralingen-Kersch near Trier (Rhineland-Palatine, Germany), isolated and worn ossicles of *Holocrinus* cf. *dubius* are concentrated in shelly tempestite beds, whereas the articulated specimens are embedded at the base of, and inside, gutters. The stems of these specimens may be intact or may have disintegrated into single noditaxes and cirri (Fig. 188), and the arms are shed at the radial facet. As judged by the fossil record, regeneration of the arms must have been very common among holocrinids. The typical accumulations of distal columnals can be explained by life-long stem autotomy. Obviously, holocrinids were well adapted to shallow, storm-dominated habitats where traumatic events caused arm-shedding and stem autotomy. The animals reattached themselves with their new terminal cirri and regenerated their arms.

Fig. 188. Stem fragments of *Holocrinus* cf. *dubius* demonstrating that breakage and disintegration of the stem preferentially occurred between nodals and infranodals (arrows), although this articulation was symplectial. Lower Muschelkalk, Muschelsandstein; Ralingen-Kersch near Trier. (Muschelkalkmuseum, Ingelfingen; photograph H. Hagdorn.) ×2.7.

THE MUSCHELKALK: A TRIASSIC CRINOID TREASURE-TROVE

The Germanic Muschelkalk provides the richest Triassic crinoid *Lagerstätten* with an excellent picture of crinoid diversification during Middle Triassic times after the end-Palaeozoic crisis. Unfortunately, such assemblages have still to be recovered from Lower and Upper Triassic strata. The famous Ladinian to Carnian St. Cassian Formation of the Dolomites (Italy) has yielded very diverse crinoid faunas. However, they consist mostly of disarticulated material that is far from being fully understood (Zardini 1976). One would hope that the Siberian or Chinese deposits will fill the Upper Triassic gaps of the fossil record in the future. The literature on Triassic crinoids has been reviewed by Hagdorn (1995).

IMPORTANT COLLECTIONS IN GERMANY

Muschelkalk crinoids, especially crowns of *Encrinus liliiformis*, are kept in most natural history museums. The less common species and large slabs with *Encrinus liliiformis*, giving an impression of its taphonomy and palaeoecology, are concentrated in the following museums in Germany:

Museum für Naturkunde an der Humboldt-Universität zu Berlin. With the types and originals of many classic crinoid papers, for example, those of von Schlotheim, von Buch, Quenstedt 1835, Beyrich, Biese and Picard. Large *Chelocrinus carnalli* slab on exhibit.

Bundesanstalt für Geowissenschaften und Rohstoffe, Aussenstelle Berlin (formerly Preussisches Geologisches Landesmuseum and Zentrales Geologisches Institut der DDR). Largest collection of Upper Silesian crinoids, with originals and types of von Meyer, Wagner and Assmann. At present no exhibits.

Geologisch-Paläontologisches Institut und Museum der Universität Göttingen. Excellent collection of North German Lower and Upper Muschelkalk crinoids, containing originals of von Koenen. Large slab with *Encrinus liliiformis* on display.

Institut und Museum für Geologie und Paläontologie der Universität Tübingen. Excellent collection of Crailsheim *Encrinus liliiformis* specimens, including Quenstedt's figured specimens and a few types.

Werksmuseum Lauffen der Heidelberger Zementwerke AG, Lauffen am Neckar. Large slabs from Linck's Neckarwestheim excavations.

Muschelkalkmuseum Hagdorn Ingelfingen. Representative collection of crinoids from the whole Muschelkalk with many figured specimens. Exhibits of most taxa.

22

Pentacrinites from the Lower Jurassic of the Dorset Coast of Southern England

MICHAEL J. SIMMS

The Lower Jurassic rocks exposed in the sea cliffs around Lyme Regis, on the Dorset coast of southern England, are justly renowned for the wealth of fossils they have yielded to collectors over nearly two centuries. Although it is the vertebrates for which the site is best known, including some of the earliest discoveries of ichthyosaurs, plesiosaurs and pterosaurs, it is the invertebrates that account for the great majority of fossils. Impressive and beautiful though some of the ammonites and other fossils are, none of these invertebrates, or indeed any of the vertebrates, can match the beauty of what is perhaps the best-known fossil species found here, the crinoid *Pentacrinites fossilis*. The preservation of some specimens is extraordinary. Not only do they occur as large tangled groups with virtually every ossicle intact, but commonly they are coated with a thin film of pyrite, giving the impression that the whole fossil has been cast in bronze or gold. Some specimens have been justly claimed to be among the most beautiful fossils ever found (Fig. 189).

How have they come to be preserved in this way? The answer is an equally remarkable story that has been a source of discussion for more than 150 years.

ANCIENT DORSET

Nearly all specimens of *Pentacrinites fossilis* from the Dorset coast have been recovered from a 2-m interval within the *obtusum* Zone of the Lower Jurassic (Sinemurian) Black Ven Marls (approximately 190 million years old). Early accounts considered the crinoids to be confined to a single impersistent band, the 'Pentacrinite Bed', but it is now known that the thin crinoid lenses may occur at any level within this 2-m thickness and occasionally outside it. This part of the Black Ven Marls is dominated by dark, well-laminated, oil-rich mudstones quite atypical of the environments in which crinoids are usually found. Indeed, the crinoids are the only large, apparently benthic organisms in these mudstones. The remainder of the fauna is dominated by pelagic or nektonic organisms, particularly ammonites. The presence of *Pentacrinites fossilis* here would appear, therefore, to be an enigma, because crinoids cannot survive where oxygen levels are low, yet both the lithology and the remainder of the fauna indicate that the environment was anoxic.

CRINOID LENSES AND FOSSIL WOOD

As early as 1836 William Buckland noted that groups of *Pentacrinites fossilis* were commonly associated with the thin lenses of coalified driftwood that are common in this part of the succession. More important, he noted that these pieces of fossil driftwood invariably lay above the crinoid lenses rather than beneath them. From this he concluded that they had lived attached to pieces of

Fig. 189. Intact crown of *Pentacrinites fossilis* on lower surface of 'Pentacrinites Bed', *obtusum* Subzone (Sinemurian), Charmouth. (The Natural History Museum, London; courtesy A. B. Smith.) ×1.

driftwood floating at the surface of the sea and that they had died when the driftwood and its crinoid cargo sank to the sea floor. This idea was not universally accepted, although its cause was championed elsewhere in relation to the similar occurrences of the giant crinoid *Seirocrinus* in the German Posidonienschiefer (see Chapter 23), and it remained a contentious issue, with *Pentacrinites fossilis* usually depicted in reconstructions as a benthic crinoid.

The relationship between *P. fossilis* and the associated driftwood is not, in itself, sufficient to convince everyone that these crinoids were pseudoplanktonic, living suspended beneath floating objects. Furthermore, the first *in situ* observations of living isocrinids in the early 1970s convinced some that the longer-stemmed pentacrinitids would easily have been capable of raising their crowns above the anoxic conditions close to the sea floor (Rasmussen 1977). However, with the development of the discipline of taphonomy, it was realized that if *P. fossilis* really had been pseudoplanktonic, then the way in which its remains became incorporated into the fossil record should be quite different from that of other fossil crinoids whose benthic mode of life was not in question. To establish whether the pentacrinitids really were different from other crinoids required detailed detective work into the morphology and preservation of these two groups of crinoids (Simms 1986).

DISARTICULATED BENTHIC CRINOIDS

Although fragmentary crinoids are not uncommon in the Lower Jurassic of the Dorset coast, intact specimens are decidedly rare. Other than *Pentacrinites fossilis* itself, the most common of these is a small isocrinid, *Balanocrinus gracilis*, which occurs in the *stokesi* Subzone of the Lower Pliensbachian in the area around Golden Cap, several kilometres to the east of Lyme Regis. This crinoid is not normally associated with driftwood, and because of its occurrence with a rich benthic fauna including bivalves, brachiopods and ophiuroids, there can be little doubt that it was truly benthic in habit. It provides a useful comparison with *P. fossilis*.

The rarity of intact crinoids reflects the speed with which the meagre soft tissues decay and the ossicles disarticulate following death. In most instances this disarticulation can be prevented only by rapid burial of the live crinoids by sediment thick enough to prevent scavengers and burrowers from reaching them subsequently. The specimens of *B. gracilis* from Golden Cap are preserved in lenses of fine sandstone or siltstone within a predominantly mudstone succession. The crinoids commonly appear to float within the sediment, individuals and parts of individuals being separated from others by a layer of sediment. Preservation is equally good whether the specimens are developed from the upper surface or the lower, although the stem and arms are commonly fractured at autotomy planes. Such preservation is just as one would expect in benthic crinoids that were enveloped by a sudden influx of sediment. In some instances this traumatic event appears to have triggered the autotomy of parts of the arms and stem.

The preservation of *P. fossilis* contrasts starkly with this. The crinoids occur as thin lenses composed almost entirely of crinoid ossicles cemented together with a syntaxial calcite overgrowth. Virtually no sediment occurs within these crinoid lenses, and there is no change in lithology above and below them. Furthermore, although the lower surface is commonly exquisitely preserved with every ossicle intact, the upper surface invariably shows some dissociation of the ossicles, with clear evidence of size-sorting by currents. It is clear from this that *P. fossilis* was not preserved by rapid burial in sediment and that, in fact, the animals' remains lay exposed on the sea floor for some time and were buried only slowly. It would appear that the softness of the sea floor muds, holding the ossicles in their original positions, and the absence of disruptive benthic organisms in this anoxic environment ensured the remarkable preservation of these crinoids.

PENTACRINITES FOSSILIS: A PSEUDOPLANKTONIC CRINOID

It might be argued that, rather than sinking from the surface to a more or less permanently anoxic sea floor, *Pentacrinites fossilis* inhabited the area during brief oxygenated spells and was then killed as anoxia returned. However, evidence from the associated pieces of driftwood establish quite unequivocally that *P. fossilis* must have been pseudoplanktonic. The position of the driftwood, above the crinoids, has already been mentioned and tends to support William Buckland's (1836) original suggestion that the crinoids were attached to floating driftwood. When the driftwood finally sank, waterlogged and overloaded by its cargo of crinoids and other organisms, the crinoids would reach the bottom first and the driftwood would come to rest on top. Benthic crinoids are known to use sunken pieces of driftwood as an anchorage (see Chapter 28), and it is conceivable that current activity rolled a piece of driftwood over so that it came to lie on top of the crinoids. However, a remarkable discovery in 1985 showed that *P. fossilis* must have colonized pieces of wood while floating at the surface and that this species was unable to survive in the anoxic conditions on the sea floor. The specimen in question is a fragment of highly compressed, coalified driftwood. The fragment is only 12 cm wide and 18 cm long, but it was clearly part of a much larger log at least 20 cm in diameter. Mature specimens of P. fossilis are preserved as a thin layer on the underside of the driftwood and extend a short way beyond its margin. The upper surface seemed devoid of any crinoid remains until examined closely. Along the one remaining original edge of the log, a band approximately 2 cm wide was covered with innumerable tiny white dots less than 1 mm across. Under the microscope, each was revealed to be the tiny attachment disc for the stem of a larval crinoid. A single, slightly larger attachment disc was present at one end of the driftwood fragment. The larval attachment discs were confined almost entirely to this marginal band, with only one small group further towards the centre. Their distribution coincides almost perfectly with a scenario in which P. fossilis colonized floating logs. The optimum position for the crinoids would have been on the underside of the log, a position that would have been occupied first (Fig. 190). As the log sank lower in the water, higher parts of it would have become accessible to and colonized by new larvae. However, the sinking of the log under its increasing burden would have been a sudden event and, as it entered the anoxic zone at the sea floor, would have resulted in the death of all the inhabitants of the log, from the earliest colonizers on the underside to the latest recruits along the upper edge. The virtual absence of attachment discs from the upper part of the log shows that, although it was now fully submerged as it lay on the sea floor and available for further colonization, none of the original inhabitants of the log survived to grow larger and no new recruits were able to occupy the vacant space.

Other aspects of the driftwood–crinoid relationship support the pseudoplanktonic model. The largest examples of *Pentacrinites* are confined to large pieces of driftwood, as might be expected, but in the case of benthic crinoids even large individuals may be found associated with quite small pieces of driftwood. Another feature of *P. fossilis*, and of its sister genus, *Seirocrinus*, is what has been termed the 'all or nothing rule' (Wignall & Simms 1990). Where these crinoids do occur on a piece of driftwood, they occur in great abundance and with a range of sizes, whereas other pieces are entirely devoid of crinoids. This suggests that a major barrier, in the form of vast stretches of open ocean, prevented all but a handful of crinoid larvae from ever reaching new, uncolonized pieces of driftwood. Instead, most larvae seem merely to have joined their parents on an increasingly overloaded ark. Such constant recruitment to the same piece of driftwood was clearly unsustainable, merely hastening the demise of the colony. In effect, each family group was committing a slow, inadvertent suicide. In contrast, benthic crinoids are not so dependent upon such specific substrates, but show a much more even distribution. Where they do occur associated with driftwood, it is commonly only in small numbers because many other niches are also available to them.

The stem length of *Pentacrinites fossilis* varies widely (Fig. 191). Short-stemmed individuals have closely spaced, long cirri. Such specimens resemble those of the invariably short-stemmed *Pentacrinites dichotomus* from the Toarcian of southern Germany (see Chapter 23). On the other hand, stems may reach 1 m in length; such stems commonly have highly cirriferous proximal regions and distal regions with very small, widely spaced cirri. However, one example of a distal-most stem with

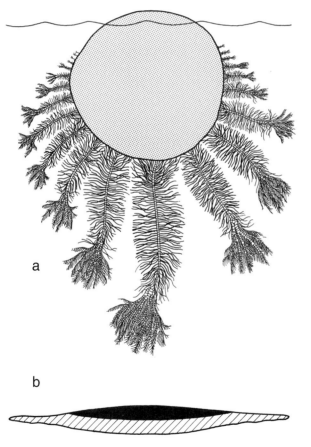

Fig. 190. (a) Reconstruction of distribution of short-stemmed *Pentacrinites fossilis* on floating driftwood reflecting colonization by successive generations of larvae on progressively higher parts of the log. (b) Cross-sectional appearance of coalified driftwood (black) after burial and compression; the larger crinoids are seen as a layer (hatched) extending from beneath the coalified driftwood. (After Simms 1986.) ×0.2.

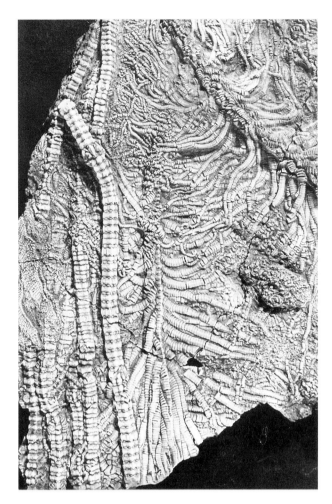

Fig. 191. Variation along the stem of *Pentacrinites fossilis* with cirriferous proximal stems in the right part of the figure and a bundle of stems from a more distal region with short, widely spaced cirri in the left part. Sinemurian, *obtusum* Subzone, Charmouth. (The Natural History Museum, London; also figured by Simms 1986.) ×1.

closely spaced nodals bearing long cirri is also known (Simms 1989). These differences may be related to the mode of attachment or to the size and shape of the supporting log.

Living attached to floating driftwood must have presented many hazards to these crinoids. Although the larval crinoids were attached by a cemented disc, this was abandoned when the crinoids were only 2–3 cm high and the cirri were used for attachment throughout the remainder of their life. Considering how crucial this was to their very existence, there does not seem to have been a standard strategy for ensuring firm attachment by the cirri. The development of cirri in *P. fossilis* shows interesting variation, with two distinct populations present even within single colonies. Some individuals have stems no more than 20 cm long along which the cirri are densely crowded and very long. In contrast, others have stems perhaps as much as 1 m in length along which the cirri are, except in the proximal region of the stem, widely spaced and very short. Commonly, these longer stems occur tangled together into large bundles, and occasionally at their distal end there is a second region of long, closely spaced cirri like those in the proximal region of the stem. Presumably these long cirri were used to grasp irregularities in the surface of the driftwood or, particularly in the case of the short-stemmed individuals, to attach to other individuals. Later species of *Pentacrinites*, such as the Toarcian *Pentacrinites dichotomus* and the Middle Jurassic *Pentacrinites dargniesi*, are only short-stemmed. A number of specimens of *P. dichotomus* are not associated with driftwood, suggesting that they may have exploited a different floating substrate, such as vesicular seaweed. As described in Chapter 25, *P. dargniesi* must have formed mats that drifted along the bottom. All in all, pentacrinitid colonies give the impression that their overall strategy for attachment was along the lines of: grab hold of anything and hope you don't fall off! This appears not always to have been entirely successful. Isolated specimens or groups of *Pentacrinites* are not uncommonly found in these anoxic mudstones and represent individuals that did become detached and sank to the sea floor.

Even for those that did not suffer this fate, there was no certainty that the driftwood would stay afloat for long under its increasing burden of crinoids. Yet *Pentacrinites* and *Seirocrinus* are among the largest crinoids known. How, and why, did they reach such huge sizes? First, the enormous obstacle that larvae faced in locating new pieces of driftwood in the vastness of the oceans may account for the large size of the adults, which need to produce a very large number of offspring. To ensure that they reached adulthood before the driftwood sank, they had an unusually fast growth rate, as indicated by growth lines on brachials (Simms 1986). This may have been achieved through a highly efficient, endotomous pattern of arm branching akin to the pattern of roads on an ideal banana plantation (Cowen 1981). The length of time that driftwood stays afloat may also greatly exceed the 1.5 years once thought to be the limit. Observations at Spirit Lake, near Mount St Helens, indicate that driftwood may remain afloat for more than 15 years, ample time for these crinoids to reach the spectacular size of some individuals.

The stratigraphic and geographic distribution of *P.*

fossilis is somewhat enigmatic. Although it is known to span several ammonite zones through the Sinemurian, it is common only in a 2-m interval that can be traced some distance. Rather than there having been a massive increase in the population of this crinoid, it seems more likely that the driftwood upon which these crinoids were dependent was caught in an oceanic gyre, rather akin to the Sargasso Sea, that became established during the deposition of this 2-m thickness of mudstones. The driftwood, and its cargo of crinoids, circled slowly in this gyre until sinking to the sea floor. Outside of such gyres this concentrating effect did not operate and hence *P. fossilis* is much more widely scattered.

IMPORTANT COLLECTION IN THE UNITED KINGDOM

The Natural History Museum, London

23

Lower Jurassic Posidonia Shale of Southern Germany

HANS HESS

ICHTHYOSAURS AND THE LARGEST SEA LILIES EVER FOUND

The Swabian Posidonienschiefer is named after the occurrence of the bivalves *Bositra buchi* and the very abundant *Steinmannia radiata*, both species being formerly assigned to *Posidonia*. The Posidonienschiefer have furnished to museums all over the world some of the most spectacular fossils, including ichthyosaurs, fishes, rare plesiosaurs, crocodiles, flying reptiles (pterosaurs) and the magnificent crinoid *Seirocrinus subangularis*. This sea lily, with its beautiful flower-like crowns and stems reaching more than 20 m in length, has been known for a long time. As early as 1742 Hiemer, a clergyman who wrote in Latin, described a slab from Ohmden with several individuals as *Caput Medusae* (head of medusa), comparing it to the ophiuroid with branched arms already known at that time. Hiemer was convinced that the animals were transported by the Flood from the Black Sea to the Stuttgart area, and this created considerable excitement at the time.

The Posidonienschiefer, which occur at the foot of the Swabian and Franconian Alb between the rivers Rhine and Main, belong to the Lower Jurassic (Lower Toarcian, Lias Epsilon, 185 million years before present) and are part of a facies widely distributed in Europe at that time (Fig. 192). They were laid down during three ammonite zones (*tenuicostatum*, but especially during the *falciferum* and *bifrons* Zones) (Fig. 193); a duration of about 0.5 million years has been estimated for the entire period of black shale deposition (Littke *et al.* 1991). The fossils from the communities of Holzmaden, Ohmden, Bad Boll and Dotternhausen, east and south of Stuttgart, are the best known. A few quarries are still in use and furnish highly priced fossils.

OIL SHALES, STILL QUARRIED

The finely layered Posidonia Shales were compressed from muddy sediments due to burial and the weight of the overlying strata. The black shales are rich in organic matter and in former times were burned for their oil. Some beds are more solid than others. For example, a single slab of 18-cm thickness (*Fleins*), which can be split into four parts, is still quarried, mainly for use in building tables. Directly underneath the *Fleins* is the thin *Hainzen* that overlies the *Koblenzer*, another solid bed with a thickness of 15 cm. The thickness of the Posidonienschiefer varies from about 1 m to about 40 m between the Rhine and the upper Main, and in the Holzmaden area it is between 5 and 14 m.

Similar to other bituminous black shales, the Posidonienschiefer contains many extremely well-preserved fossils, including soft parts. However, the majority of the fossils are not preserved intact. According to B. Hauff

Hans Hess, William I. Ausich, Carlton E. Brett, and Michael J. Simms, eds., *Fossil Crinoids.* © 1999 Cambridge University Press. All rights reserved. Printed in the United States of America.

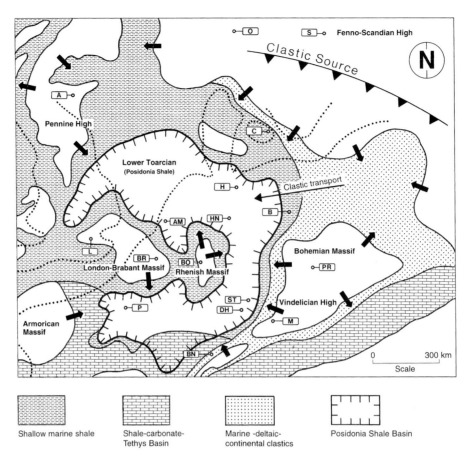

Fig. 192. Distribution of the Posidonia Shale facies in western Europe. Key: A, Aberdeen; O, Oslo; S, Stockholm; C, Copenhagen; H, Hamburg; B, Berlin; AM, Amsterdam, HN, Hanover; PR, Prague; BO, Bonn; BR, Brussels; P, Paris; ST, Stuttgart (Holzmaden area); DH, Dotternhausen; M, Munich; BN, Berne. (After Littke et al. 1991.)

(Hauff & Hauff 1981), about 30 quarries were exploited during 1939, and during that year only 8 complete ichthyosaurs were discovered, whereas about 100 ichthyosaurs were found disarticulated or incomplete.

These strata have become famous for their vertebrates, especially the attractive ichthyosaurs. The invertebrates include numerous, thin-shelled and compressed ammonites (*Dactylioceras, Harpoceras* and others), belemnites and squids, bivalves (*Steinmannia, Oxytoma* and *Pseudomytiloides* – formerly *Inoceramus*) and rare arthropods. Pieces of driftwood and remains of plants have also been found, but microfossils (foraminifera and ostracodes) are rare in the bituminous facies. Examples of all of these fossils are illustrated in Hauff and Hauff (1981) and Urlichs et al. (1994). The wood has been transformed into fossil hydrocarbon jet (*Gagat*), which can be carved and polished into jewellery; human iceage relics of jet have been discovered near Stuttgart, apparently made from material found in the Posidonia Shale.

FLOWERING LOGS

The Posidonienschiefer have furnished splendid specimens of *Seirocrinus subangularis*, mostly attached to logs (Fig. 194). These crinoids are concentrated in the lower (*tenuicostatum* Zone) part of the Posidonienschiefer, where they occur in the *Koblenzer* but mainly in the *Fleins* bituminous marly limestones composed partly of coccoliths (Fig. 193). Crinoids tend to be grouped around the ends of the logs; the food supply may have been more abundant in these regions when the logs were drifting. Almost all specimens have been prepared from the lower surface of the slabs, where they are better preserved and are partially pyritized. Large groups are restricted to logs and usually also contain small individuals, commonly attached to the stems of the adult specimens (Fig. 195). Isolated specimens appear to have been broken from driftwood, because their stems are rarely complete. The show-piece of the Hauff Museum in Holzmaden is a slab measuring 18 by 6 m. It contains

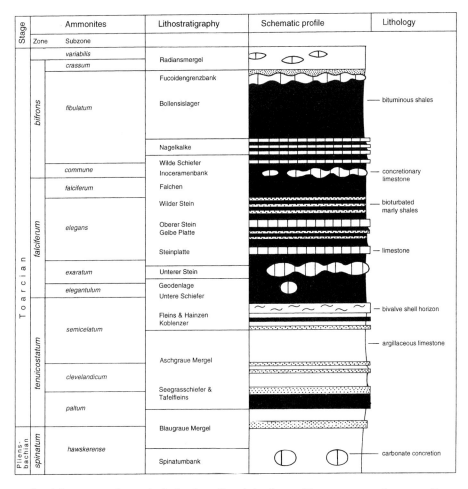

Fig. 193. Biostratigraphy, lithostratigraphy and idealized profile of the Lower Toarcian in southwestern Germany. (After Riegraf 1985.)

a 13-m-long piece of driftwood completely covered with bivalves (*Pseudomytiloides dubius*) and a giant group of about 280 large and small individuals of *Seirocrinus subangularis* with bundles of stems partly wound around the log (Fig. 196). This slab was discovered in 1908 in the *Fleins* at Holzmaden. Its preparation required 18 years of work and was not completed until 1970 for the opening of the museum. An even larger slab with a log close to 20 m in length and an area of about 500 m² covered with sea lilies has been discovered at Holzmaden and awaits preparation in the Stuttgart Museum. The largest slab exhibited in this museum measures 6 by 9 m. It was discovered in 1977 in the *Koblenzer* of Ohmden and contains 141 individuals on a partly exposed log. The stems are several metres long, and the crowns reach a diameter of 60 cm. The main point of interest in this colony is a number of crowns in oral view with their tegmen exposed. The sheer size of these logs with their attached crinoids and bivalves is overwhelming, and they are the largest and most spectacular invertebrate fossil remains of all time. Unfortunately, because of their enormous size it is impossible to reproduce them adequately in any book illustration (see Fig. 196).

DENSE CROWNS ON STRONG AND FLEXIBLE STEMS

Small individuals of *Seirocrinus* are superficially similar to *Pentacrinites dichotomus* (discussed in the next section), with a stem length of only a few centimetres. The columnals of large specimens are rounded subpentagonal to circular in cross section (hence the species name, *subangularis*) and typically vary in height (Fig. 197). The articular facets have large, strongly granulated radial areas, separating the crenulate, narrow pyriform petals. This structure gives the impression of considerable flex-

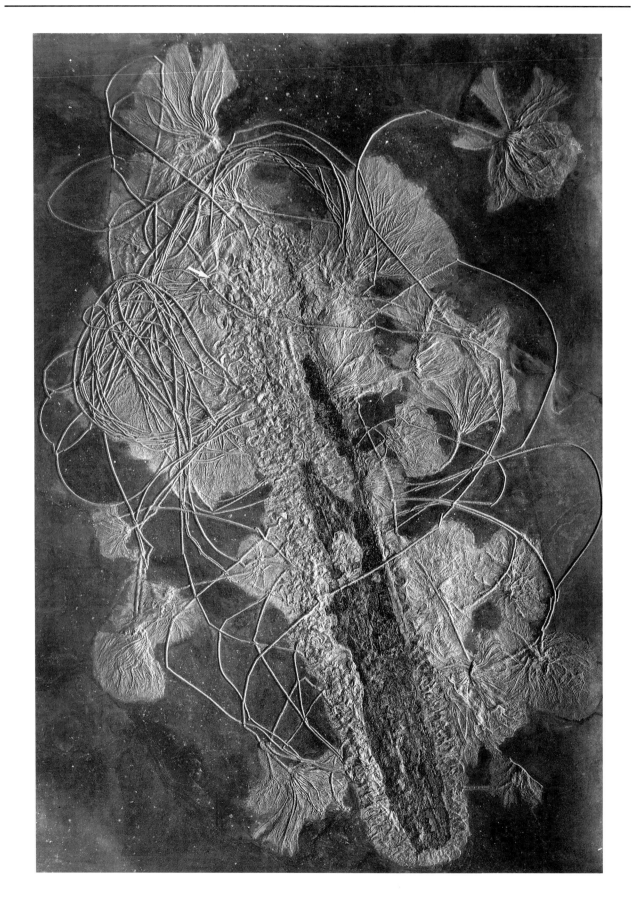

Fig. 195. *Seirocrinus subangularis*. Group with juvenile specimens attached to stem of adult individual; also contains *Pseudomytiloides dubius*. Posidonienschiefer, Ohmden. (Stuttgart Museum; photograph R. Harling; courtesy G. Dietl.) ×0.4.

Fig. 194. (*On facing page*) *Seirocrinus subangularis*. Colony with entangled stems on driftwood. Posidonienschiefer, Holzmaden. The slab was prepared from the lower side; one specimen, whose attachment is close to the upper end of the log (arrow), has a stem that crosses the log twice, proving that the log settled on the bottom on top of the crinoid. Size of slab 2.5 by 3.5 m. (Staatliches Museum für Naturkunde, Stuttgart; photograph R. Harling; courtesy G. Dietl.)

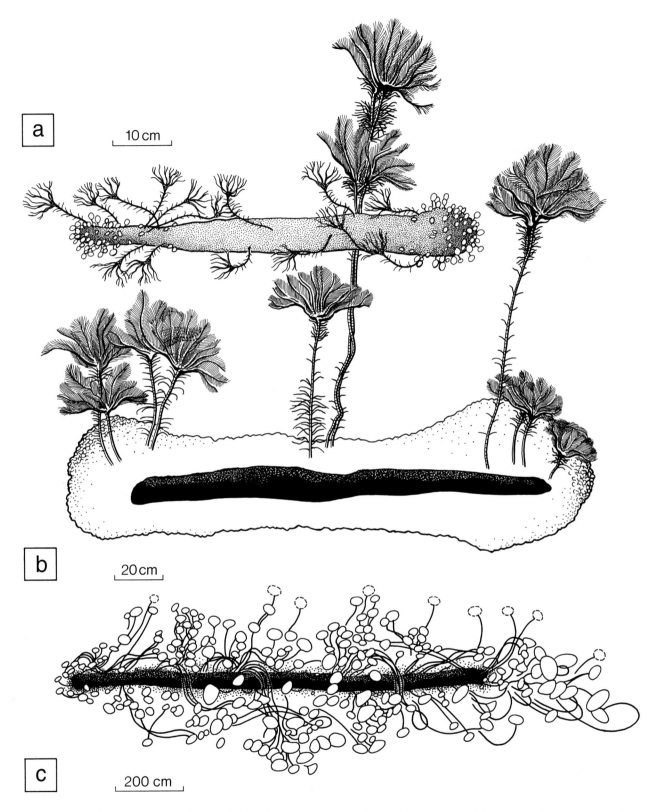

Fig. 196. Completely preserved colonies of *Seirocrinus subangularis* on logs of different size. (a) Colony of juvenile individuals (Göttingen Museum); (b) colony of subadult individuals (Senckenberg Museum; see also Fig. 199); (c) the giant colony of the Hauff Museum in Holzmaden. (Redrawn from Haude 1980.)

Fig. 197. Stem fragments, disarticulated cirrals and brachials of *Seirocrinus subangularis* on upper bedding plane of *Fleins*. Posidonienschiefer, Dotternhausen. (Courtesy Werkforum Dotternhausen.) ×1.

ibility combined with great strength. With increasing distance from the cup, the diameter of the stem is reduced, making it much more flexible. The length of the internodes gradually increases and may reach more than a thousand columnals near the end of a 20-m-long stem (Seilacher et al. 1968); the total length of the largest stem is more than 26 m (Simms 1989)! Stems of such extreme length suggest plankton-rich zones, and therefore a greater opportunity for food gathering, well below the surface (see also Chapter 11 for the similar case of the scyphocrinitids). In contrast to that in the isocrinids, the articulation between nodals and infranodals in *Seirocrinus* is symplectial. Synostosial autotomy planes appear to have evolved only among the isocrinids and, anyway, would be of no advantage to a pseudoplanktonic crinoid. The cirri are closely spaced in both the most proximal region and distal stem regions, with a dense tuft of cirri acting as a holdfast to the driftwood. The cirri vary from small to large individuals and from the proximal to the distal part of the stem. In small individuals, they are rhomboidal in outline and resemble those of *Pentacrinites*, whereas in large individuals the ossicles are rounded, short and slender. Divisions of the highly branched arms at primibrachials and secundibrachials are isotomous, but succeeding divisions are all endotomous (Fig. 198). Arm length may attain 50 cm in the largest individuals. Heterotomous arm branching provides a very dense filtration fan that would also be needed for rapid growth, as postulated for the pentacrinitids. Most brachial articulations are muscular. Syzygies are absent, allowing complete pinnulation of the arms. This improved the food-gathering process even more.

It is now widely accepted that *Seirocrinus* lived attached to floating logs. Such a pseudoplanktonic mode of life is supported by a number of facts (Haude 1980; Simms 1986). As observed in living crinoids, a filtration fan held into the current by crinoids anchored to the bottom would need stalks that were stiff near the bottom but flexible beneath the crown. In *Seirocrinus*, the reverse is true, and such a morphology is quite exceptional in fossil crinoids. (This genus derives its name from the long, rope-like stems.) As already noted by B. Hauff (Hauff & Hauff 1981), the size of the crinoids is proportional to that of the driftwood on which they are anchored (Fig. 196). The loading capacity of floating trunks has been discussed by Haude (1980). He calculated for the log in the Senckenberg Museum, measuring 190 cm by 8–12 cm (Fig. 199), a load capacity of about 12 kg. The weight in sea water of all *Seirocrinus* specimens on this log was estimated to be only around 0.5 kg, compared with a total weight of as much as 10 kg for the roughly 5,000 *Pseudomytiloides dubius* lamellibranchs. These figures are in agreement with a pseudoplanktonic life for *Seirocrinus*. Doubts have been raised as to whether the trunks could have floated long enough to allow *Seirocrinus* individuals to grow to such size. According to Haude, unloaded trunks of conifers (a type of tree common in the Lower Jurassic) remain buoyant for 2 years at the most. He therefore assumed that either the crinoids grew rapidly or the wood remained floating for longer periods, possibly because of its resin content or some other unknown feature that prevented decay and water-filling of the pores. Observations made after the eruption of Mount St Helens (Simms 1986) suggest that the trunks could have remained floating for much longer (see also Chapter 22). M. Jäger (pers. comm., 1993) has drawn attention to the absence of wood-boring lamellibranchs; these would have quickly destroyed the substrate. Whereas the first wood-boring, teredinid bivalves appeared toward the end of the Early Jurassic, these Jurassic forms appear to have colonized only wood lying on the bottom, not driftwood (Wignall & Simms 1990). On the large trunks floating near the surface, the large *Seirocrinus* individuals must have been hanging down (Fig. 199 gives an impression of this). Once the buoyancy fell below a certain point, large logs may have sunk quite rapidly, dragging the crowns along like parachutes. After the trunk settled, the crowns would have tilted over and sunk to the bottom. While

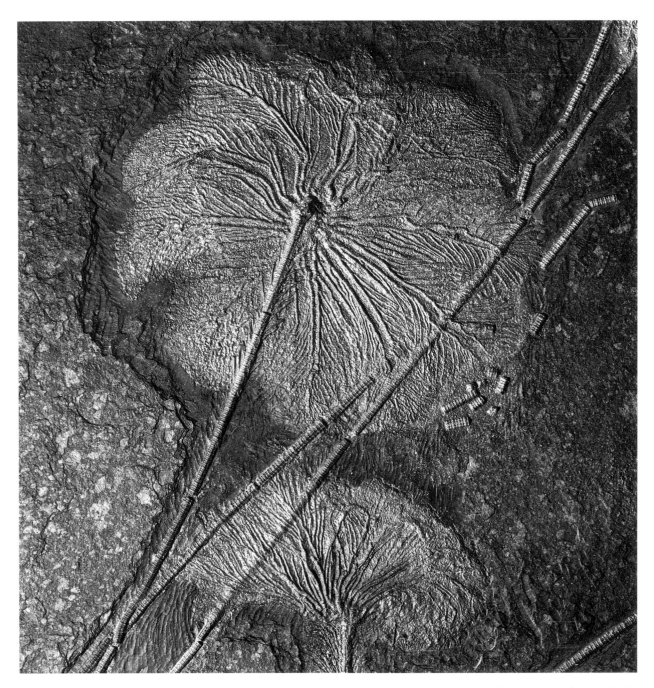

Fig. 198. *Seirocrinus subangularis*. Two crowns belonging to a large group, attached to driftwood on a slab of *Fleins*. Posidonienschiefer, Dotternhausen. Diameter of the larger crown approximately 40 cm. (Werkforum Dotternhausen; courtesy M. Jäger.)

smaller individuals and those with shorter stems came to lie on their sides, large ones with very long stems would have reached the bottom with their oral side downward (Haude 1980). Such a behaviour is shown by a very large group in the Tübingen Museum (Fig. 200) and the large colony exhibited in the Stuttgart Museum.

A trunk with a length of about 3 m in the Stuttgart Museum (Fig. 194) has a heavy overgrowth of *Pseudomytiloides dubius* and crinoids with twisted stems that lie partly over the log. Because this slab shows the lower surface, the log must have settled after some of the crinoids touched the bottom.

Fig. 199. *Seirocrinus subangularis*. 'Senckenberg log' with crowns hanging down to show presumed life position. This specimen from the Posidonienschiefer of Holzmaden is preserved in the Naturmuseum Senckenberg, Frankfurt, and has been reproduced repeatedly (e.g., by Breimer & Lane 1978), but only once in this position (Seilacher 1990b). Size of slab 250 ×260 cm. (Courtesy R. B. Hauff.)

PSEUDOPLANKTONIC *PENTACRINITES*

The short-stemmed *Pentacrinites dichotomus* McCoy (*Pentacrinites briareus württembergicus* Quenstedt and *Pentacrinites quenstedti* Oppel are synonyms; see Simms 1989) occurs in younger beds (*falciferum* and *bifrons* Zones) and, thus, was not contemporary with *Seirocrinus* (this species is restricted to the uppermost *tenuicostatum* Zone). *Pentacrinites dichotomus* is also much rarer. In contrast to the British occurrences of *Pentacrinites fossilis* in the Sinemurian, the species from the Posidonienschiefer is invariably very short-stemmed and, therefore, is comparable to *Pentacrinites dargniesi* described in Chapter 25. The largest specimens known, preserved in the Stuttgart collection, have stems about 20 cm long and an arm length of about 15 cm. Only the nodals are seen from the outside, being mostly covered by the closely spaced, compressed cirri with a length comparable to that of the arms (Fig. 201).

A superb slab with 153 individuals of *P. dichotomus*, with a diameter of 1.15 m, can be admired in the Stuttgart Museum (Urlichs et al, 1994). The slab was first figured by Beringer (1926), who gave detailed descriptions of the complete *Seirocrinus* and *Pentacrinites* specimens from the Posidonienschiefer known at the time. Most of the crinoids on this slab are of a similar size, but some small (i.e., juvenile ones) also occur. The crinoids appear to have grown radially from a central area, where the thickness suggests two layers of crinoids. Individuals on this slab are mostly presented on their side, but about a dozen specimens present their oral side. Another slab from Ohmden, preserved in the Stuttgart collection, contains a group measuring 55 by 40 cm with about a dozen individuals as well as two *Pseudomytiloides* (Fig. 202). Its preservation is quite similar to that of the large Stuttgart slab. These *Pentacrinites* were not found together with wood or with other remains of plants, and this is also true of some other groups. They seem to have been floating as a more or less spherical body with their long cirri intertwined for mutual support and grouped around some unfossilized material. It is questionable whether a material comparable to today's seaweeds (such as bladder-wrack) would be buoyant enough to support such a large group. It may well be that the crinoids were separated from their buoyant support before burial. In any case, a group of *P. dichotomus* about 3 m long, preserved in the Dotternhausen Werkforum, is attached to a log of similar length, proving a pseudoplanktonic lifestyle also for this species. The Dotternhausen slab was discovered in the Wilder Schiefer (*bifrons* Zone). It was badly damaged by blasting and had to be reassembled from about 300 fragments. According to F. Lörcher (pers. comm., 1996), who took 2 years to assemble and prepare this important show-piece, the group contains about 45 densely spaced individuals, which are inserted on one side (presumably the lower side) of a now-flattened log with a diameter of 10 cm. The crowns point away from the log, and their size is comparable to the specimens of the large slab at the Stuttgart Museum. The Dotternhausen collection also includes another group of *P. dichotomus* attached to wood.

Therefore, the Dotternhausen specimens are similar to the Dorset crinoids, which are regularly preserved underneath the lignitized driftwood to which they were attached (see Chapter 22). However, the Dorset *Pentacrinites fossilis* developed longer stems, similar to *Seirocrinus*. It has even been postulated by Hauff (1984) that the long and strongly flattened cirri (Fig. 201) of the *Pentacrinites* species could have been used for swimming, but the lack of muscle fibres between the segments contradicts this hypothesis. The stratigraphically younger *P.*

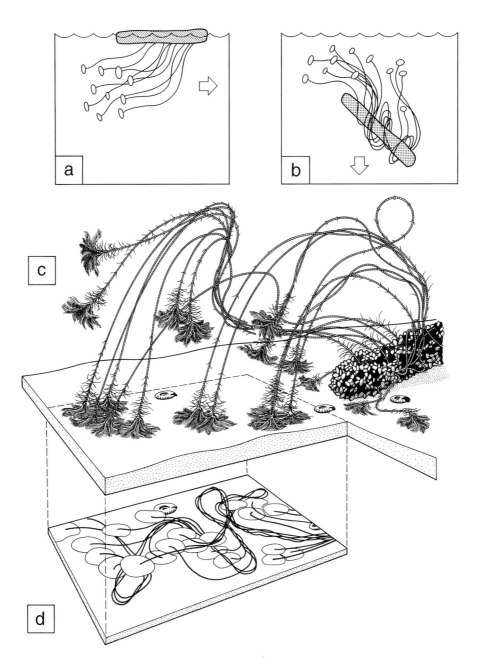

Fig. 200. Reconstruction of the sinking process of a large *Seirocrinus* colony, exhibited in the museum of the University of Tübingen. Length of slab approximately 10 m. (a) Floating log; (b) log at the beginning of the sinking process; (c) log reaching the bottom; (d) slab with the fossils. (The slab was figured by Seilacher *et al.* 1968; redrawn after Haude 1980.)

dichotomus may have evolved from *P. fossilis* by an increase in the number of brachials and the development of syzygies between secundibrachials 6 and 7, as well as by a shortening of the stems. These differences are presumably the result of a somewhat different lifestyle allowing mutual support of the animals.

A RARE COMATULID

A unique intact specimen of the free-moving comatulid *Procomaster pentadactylus* has been described (Simms 1988b). This exotic element in the crinoid fauna of the Posidonienschiefer is one of the oldest comatulids

Fig. 201. *Pentacrinites dichotomus*. Two individuals with the base of the crown and well-preserved cirri. Posidonienschiefer, Holzmaden. (Stuttgart Museum; courtesy G. Dietl.) ×0.5.

known. It is the only Lower Jurassic crinoid with only five unbranched arms; the cirri are robust, recurved and without a terminal claw (Fig. 203). The brachials are very low; their articulations are, like those of the other comatulids, either muscular or syzygial. The pinnules are well developed. This individual may have drifted into the basin attached to seaweed or some other floating substrate, which has now disappeared.

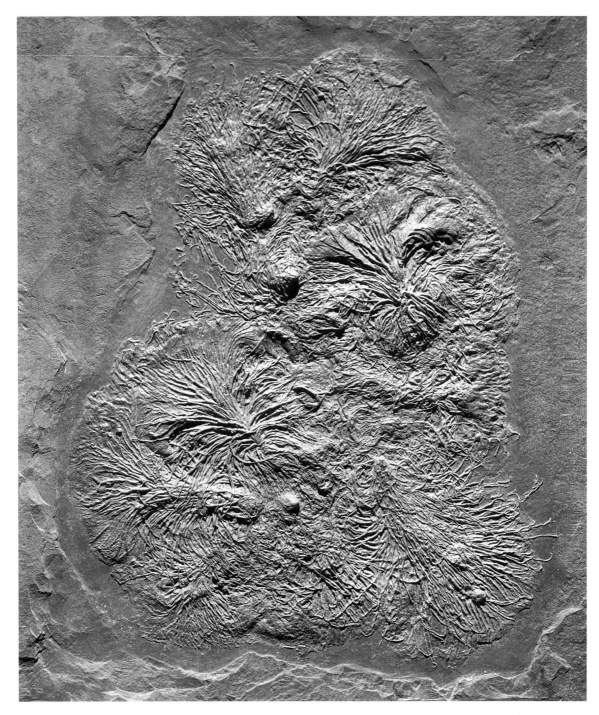

Fig. 202. *Pentacrinites dichotomus*. Posidonienschiefer, Ohmden (J. Fischer Quarry). Size of group 55 ×40 cm. (Stuttgart Museum; photograph R. Harling; courtesy G. Dietl.)

A LIFELESS BURIAL GROUND

The bituminous Posidonienschiefer are characterized by their high content of organic matter (kerogen), undisturbed fine laminations and almost total absence of benthic animals, as well as excellent preservation of many of the fossils. These facts point to stagnant, anoxic bottom water. The rich supply of organic material has been explained by upwelling of phytoplankton, caused by an opening of the Atlantic in the west and the Tethys

Fig. 203. *Procomaster pentadactylus*, holotype (Simms 1988b). Posidonienschiefer, Zell. (Stuttgart Museum; photograph H. Lumpe; courtesy G. Dietl.) ×1.5.

Ocean to the south (Riegraf 1985). As Littke *et al.* (1991) have pointed out, an inflow of nutrient-rich open waters from the Tethys over and through the entire shallow epicontinental Posidonia Sea during the long duration of Lower Toarcian time is difficult to conceive. Such a mechanism is also in conflict with the lack of evidence from much current activity. Therefore, these authors have proposed that there was a supply of nutrients from the many nearby emergent areas that were also the source of detrital clay (Fig. 192). Bioturbated horizons occur repeatedly, indicating short periods of higher oxygen content at the bottom (perhaps by increased water circulation), making life for benthic animals possible. The so-called Seegrasschiefer (containing the trace fossil *Chondrites*) is one of the earliest and best-known of the bioturbated horizons and appears to be the result of burrowing worms. Such horizons also contain benthic foraminifera, which become impoverished or disappeared under the anoxic conditions of the highly bituminous sediments (Riegraf 1985). Life on the bottom is also documented by the occurrence of the rare burrowing bivalves *Goniomya* and *Solemya* (Seilacher, 1990b). Recent *Solemya* carries chemosynthetic, anaerobic bacteria and may be gutless. The bivalve *Steinmannia radiata* (var. *parva*) is densely packed on certain bedding planes and was considered a hardy bottom dweller; a planktonic lifestyle with some kind of buoyancy device as postulated by Oschmann (1995) is questionable (Etter 1996). Encrusting algae (*Girvanella*) were quite common at times of reduced or interrupted sedimentation (Riegraf 1985) and also point to the presence of some oxygen at the bottom, with anoxic conditions underneath the algal mats. The absence of bioturbation and benthos in horizons with such mats may indicate increased salinity in the bottom layer. The beds with the *Seirocrinus* specimens, which are usually attached to driftwood as already discussed, contain few benthic foraminifera and lack benthic macrofossils. The same appears to be true of the horizons with *Pentacrinites*. It has been pointed out by M. Jäger (pers. comm., 1993) that the crinoids may have attached to bivalves (strongly anchored to the wood by their byssal threads) rather than to the naked wood. This view is supported by the fact that Posidonienschiefer crinoids almost always occur together with bivalves (*Pseudomytiloides dubius*), whereas wood occasionally contains bivalves but no crinoids, but rarely crinoids without bivalves. However, Simms (1986) described a specimen from the Lower Jurassic with larval attachment discs of *Pentacrinites fossilis* on a piece of coalified driftwood, and bivalves are lacking on this specimen (see Chapter 22).

It is now assumed that the Posidonienschiefer was deposited at a rate of about 0.1 mm per year in a shallow but extended marginal sea with a depth of about 20–50 m and warm, intermittently hypersaline water (Riegraf 1985). The accumulation and preservation of the organic matter were favoured by a density stratification of the water column caused by the influx of low-salinity nutrient-rich water from high latitudes (Prauss *et al.* 1991). Under conditions of reduced sedimentation, algal mats (cyanobacteria) produced carbonate crusts that stabilized the sediment. At times, weak water movement and/or bioturbation destroyed the algal mats. However, conditions were never favourable enough for bottom-living crinoids such as *Isocrinus* (*Chladocrinus*), a form otherwise common in Liassic strata. The spines of the tiny echinoid *Diademopsis crinifera* are found in large numbers in the *tenuicostatum* Zone before the appearance of the bituminous facies, indicating strongly increased algal growth on the bottom for a short period.

IMPORTANT COLLECTIONS IN GERMANY

Museum Hauff, Holzmaden. This splendid museum, devoted to the fossils from the Posidonienschiefer, was built near the famous fossil site (where collecting is still possible) and contains, among other specimens, a huge piece of driftwood covered with bivalves and crinoids.

Staatliches Museum für Naturkunde, Stuttgart (Museum am Löwentor). It has, together with the Hauff Museum in Holzmaden, the richest collection of Posidonienschiefer fossils, including a spectacular slab with *Pentacrinites dichotomus*.

University of Tübingen, Geological Institute. Exposed is the large slab described by Seilacher *et al.* (1968).

Werkforum Dotternhausen (Portlandzementwerk Rudolf Rohrbach, D-72359 Dotternhausen). This beautiful museum was opened in 1989 on the site of a Posidonienschiefer quarry and shows the fossils typical of these strata, including a unique slab with numerous individuals of *Pentacrinites dichotomus* attached to a log, as well as a large group of *Seirocrinus subangularis* from the *Fleins*, also attached to a piece of driftwood (Jäger 1993).

24

Middle Jurassic of Southern England

MICHAEL J. SIMMS

SHALLOW-WATER LIMESTONES

Throughout southern England the Middle Jurassic succession is dominated by shallow-water limestones, typically oolitic or bioclastic. These limestones are exposed along the coast and in a few crags inland, and they have been extensively quarried wherever they crop out. More massive beds have been the source of superb building stones, such as the famous Bath Freestone, whereas more thinly bedded units have been split for roofing slates, as in the case of the Stonesfield Slate of Oxfordshire.

Although commonly rich in marine invertebrates such as bivalves, brachiopods and gastropods, echinoderms are generally not as well preserved and, with the exception of the rich echinoid fauna, have largely been overlooked. Intact crinoids in the Middle Jurassic of southern England are rare, reflecting the predominantly high-energy environments in which the limestones were deposited. Any crinoids that lived in these environments disarticulated rapidly once they died, unless some exceptional event intervened.

Two of the most spectacular occurrences of intact Middle Jurassic crinoids in southern England present an interesting contrast in terms of the palaeoecology and taphonomy of the species involved and their history of discovery.

CRINOIDS FROM A CANAL

The earlier discovery, which is also the stratigraphically younger of the two, was made in the early part of the 19th century in a quarry alongside the Kennet and Avon Canal near Bradford-on-Avon, Wiltshire. The Forest Marble Formation, of Late Bathonian age (approximately 163 million years before present), is about 24 m thick in this area and is composed of mostly cross-bedded oolitic and bioclastic limestones. However, about 3 m above the base is a 3.5-m-thick band of clay with thin limestone partings, the Bradford Clay. This clay was exposed alongside the canal, and during excavation to puddle the canal, it was found that the lowest 10 cm, immediately above the underlying limestone, was packed with fossil debris. Among this fauna were numerous brachiopods, byssate bivalves, gastropods, regular echinoids and a host of other organisms encrusting or boring into the limestone beneath. However, without doubt the most spectacular and famous member of the Bradford Clay fauna was the millericrinid crinoid *Apiocrinites parkinsoni* (Fig. 204).

William Smith, the 'Father of Geology', was among the first to comment on the fauna of the Bradford Clay (Smith 1816, 1817) and even at that early stage realized that the fossils of this basal shelly layer of the clay had

Hans Hess, William I. Ausich, Carlton E. Brett, and Michael J. Simms, eds., *Fossil Crinoids*. © 1999 Cambridge University Press. All rights reserved. Printed in the United States of America.

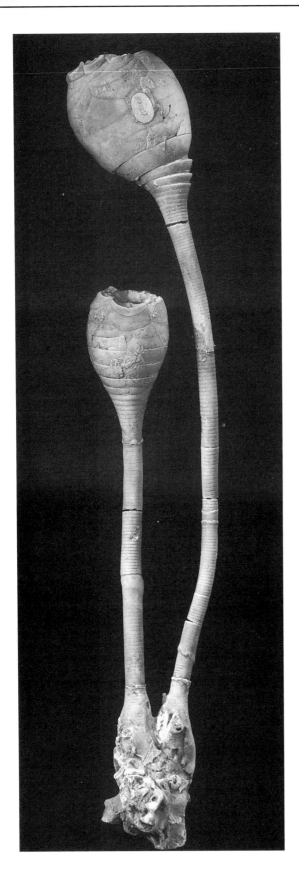

Fig. 204. *Apiocrinites parkinsoni*. Two specimens found isolated in the Bradford Clay, Forest Marble Formation (Late Bathonian), Bradford-on-Avon, Wiltshire. (The Natural History Museum, London; courtesy A. B. Smith.) ×0.8.

been associated with the top of the underlying limestone and had been buried by the clay. Where the clay had been stripped from the limestone beneath, the bare surface was commonly studded with the black holdfasts of *Apiocrinites* cemented firmly to it. Lying next to some of these holdfasts were intact specimens of the crinoid spread out flat on the limestone surface. Today, opportunities for collecting such specimens are rare, but many museum collections contain examples of the robust stems and the barrel-shaped cups of *Apiocrinites* from the original site. Far fewer complete specimens with the more delicate arms were collected, although fine examples can be found in the collections of the Bristol City Museum and the Natural History Museum in London.

THE BRADFORD HARDGROUND

The environment in which *Apiocrinites* and its associated fauna at Bradford-on-Avon lived is a classic example of a marine hardground (Palmer & Fürsich 1974). Early diagenetic lithification of carbonate sediments is commonplace in tropical marine environments today. Modern examples from the Persian Gulf bear a striking resemblance to the Bradford hardground. Lithification of the top few centimetres of the limestone occurred before deposition of the Bradford Clay, but appears to have been patchy. The hardground is highly irregular, with broad gullies separating and partly undercutting flat-topped hummocks. In places the limestone surface lacks the distinctive encrusting fauna, suggesting that it either remained unlithified in these areas or was lithified but covered by a layer of soft sediment preventing colonization, although evidence for periodic erosion of encrusting oysters does not support this latter scenario. The hardground is liberally encrusted and bored by a fauna of annelids, bivalves, bryozoans and other, more cryptic taxa such as acrothoracian barnacles, clionid sponges and phoronids. Palmer and Fürsich (1974) found a distinct polarization of the fauna between forms that occupied the top surface and those that occupied the undersurface of crevices and cavities within the hardground. The latter fauna was dominated by closely adpressed forms and contrasted strikingly with the fauna of the top surface, in which five distinct levels of ecological tiering could be recognized, from an endolithic layer through two distinct levels of encrusters to arborescent bryozoans and, finally, occupying the highest layer or tier, the crinoid *Apiocrinites* (Fig. 205). The crinoids themselves, particularly the robust holdfasts that survived long after the animal had died and the ossicles had been scattered by the currents, provided a substrate for yet more encrusters. Among these encrusters were the tiny holdfasts of juvenile *Apiocrinites*.

CATASTROPHIC BURIAL

The abundance and diversity of the encrusting and boring organisms of the Bradford hardground show that it remained exposed on the sea floor for a considerable time, perhaps several decades, yet the intact preservation of the crinoids beneath the Bradford Clay shows that burial was sudden and catastrophic. The underlying cause of this influx of mud is unknown. The limestone partings within the Bradford Clay indicate that it was deposited by several distinct influxes rather than a single event. This suggests a longer-term environmental change rather than an on–off event such as storm resuspension of sediment. In addition, the Bradford Clay is known to be laterally discontinuous, suggesting a relatively stable point source for the mud influxes. From the limited evidence available, it might be suggested that a medium-term relocation of a river mouth was the underlying cause of the mud influx responsible for the spectacular preservation of the *Apiocrinites* at Bradford-on-Avon.

A FORTUITOUS DISCOVERY

The second occurrence of intact Middle Jurassic crinoids considered here was discovered more than 150 years after that of the Bradford Clay *Apiocrinites* and in very different circumstances. The site was found by Paul Taylor, now a renowned authority on fossil bryozoa, while a student. In the summer of 1973 Paul was undertaking a geological mapping project in the Northleach area of the north Cotswolds. Although not an ideal area for mapping because of the limited exposures, Paul chose this area because of his interest in the Jurassic and also because of the opportunity for free accommodations offered by a relative. On one foray, coming across a low, partly vegetated bank, Paul broke off a small fragment of sandy-textured limestone to find it packed with echinoderm ossicles. A little excavation of the bank soon rewarded him with a small slab on which lay many intact crinoids with the arms splayed out. After this the site lay undisturbed until 1979, when, with a group of colleagues from Swansea University, Paul revisited the locality and excavated a substantial quantity of material. This formed the basis for a subsequent paper (Taylor 1983) in which the crinoids were described as a new species, *Ailsacrinus abbreviatus*, and its mode of life and taphonomy were interpreted. The great majority of this material is now housed in the Natural History Museum in London.

A CRINOID MONOCULTURE

Taylor's *Ailsacrinus* locality is a slightly lower stratigraphic level than that of the Bradford *Apiocrinites*. Paul identified the horizon within the Sharps Hill Formation, which lies close to the boundary between the Lower and Middle Bathonian. The *Ailsacrinus* Bed is approximately 25 cm of sandy and oolitic bioclastic limestone composed largely of shell debris with about 20% quartz sand. There is a gradational contact with the oolitic and bioclastic limestone below and a sharp contact with cross-bedded bioclastic limestone above. Several distinct bedding planes occur within the *Ailsacrinus* Bed where thin clay layers are present. The best-preserved crinoids are immediately above rather than below these clay layers and, hence, are on the undersurface of the bioclastic limestone units, the bulk of which contains only disarticulated crinoids. Although shell debris is abundant in the *Ailsacrinus* Bed, there are a few intact macrofossils other than the crinoids. Of the brachiopods, epifaunal bivalves and echinoids that have been found, all are abraded, suggesting that the *Ailsacrinus* Bed represents a virtual crinoid monoculture from which other macrofauna was excluded. The crinoids themselves occur crowded together on the bedding planes with population densities exceeding 200 per square metre in places (Fig. 206). More than half of them are preserved with the oral surface facing upward, suggesting life position, with most of the remainder lying on their side and only a few upside-down. Although the crinoids occur crowded together, their arms are not tangled but are spread out radially and are often flexed, sometimes ex-

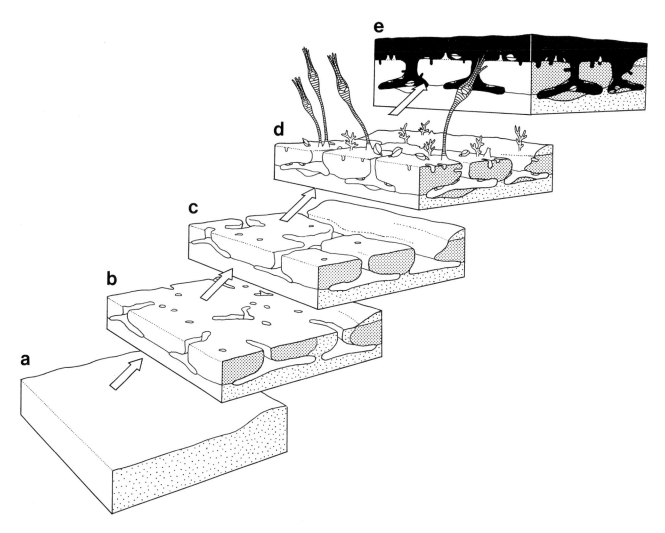

Fig. 205. History of the Bradford hardground. (a) Deposition of lime sand. (b) Burrowing by crustaceans and selective cementation. (c) Continued lithification and erosion of the uncemented material. (d) Colonization by boring, encrusting and other sessile animals occupying different tiers; periods of shell accumulation on the hardground alternated with periods of bioerosion, which removed some encrusters; floors of the crevices started to lithify. (e) Shell material derived from the hardground accumulated within crevices; clay deposition (black) buried the hardground and its fauna. (After Palmer & Fürsich 1974.)

tending for several centimetres through the overlying sediment. In some specimens traces of red and purple colours are preserved, as has been described for another Jurassic millericrinid (Fig. 62).

AILSACRINUS: A FREE-LIVING MILLERICRINID

Ailsacrinus, like *Apiocrinites*, is a representative of the order Millericrinida, yet the two taxa are, in many respects, very different morphologically. The most profound difference lies in the stem. Whereas *Apiocrinites* is cemented to the substrate by a holdfast at the base of the stem, the *Ailsacrinus* stem is greatly reduced, in some instances to a single ossicle, with a rounded distal end. At least in adult life, *Ailsacrinus* lacked any fixed attachment to a hard substrate. Whereas *Apiocrinites* was at the mercy of any adverse changes in its local environment, powerless to move away from its chosen spot, *Ailsacrinus* had a more flexible mode of life somewhat akin to that of the enormously successful comatulids that dominate most crinoid faunas today. Should things take a turn for the worst, *Ailsacrinus* could, in theory, pack its bags and move on. *Ailsacrinus abbreviatus* is

Fig. 206. Lower surface of bedding plane with *Ailsacrinus abbreviatus*. Sharps Hill Formation (Bathonian), Eastington, Gloucestershire. Most of the arms are broken at autotomy sites (syzygies); note rudimentary stems consisting of only one columnal (lacking in the specimen at the bottom). (Hess Collection; photograph S. Dahint.) ×1.

strikingly similar in appearance and population density to the Middle Jurassic comatulid *Paracomatula helvetica* (see following chapter), except that *Ailsacrinus* lacks cirri. Because comatulids can anchor temporarily to objects on and above the sea floor with cirri, they were able to occupy a much wider range of habitats than the less versatile *Ailsacrinus*. *Ailsacrinus* appears to have been a rather short-lived, although locally successful, experiment in millericrinid design, with only two species, *Ailsacrinus abbreviatus* and *Ailsacrinus prattii*, currently known.

DEATH AND RECOLONIZATION

As in the case of the Bradford Clay *Apiocrinites*, the intact preservation of *Ailsacrinus abbreviatus* points to rapid and catastrophic burial of a living assemblage of crinoids. However, unlike the Bradford Clay example, in which the fauna was completely wiped out, it is evident that in the case of the *Ailsacrinus* Bed there were repeated episodes of burial, each of which was followed by a period in which the crinoid population became re-established. The oral-up position of more than half of the crinoids and the lack of tangling of the arms indicate that they were buried *in situ* rather than swept into the area. The position of the clay layers *beneath* the well-preserved crinoids indicates that the crinoids were not killed by an influx of mud as in the Bradford Clay example. Instead, it seems that clay deposition was the normal background sedimentation for this environment and that the crinoids were instead buried by the oolitic bioclastic limestones that make up the bulk of each unit. Following each influx of sediment that buried the crinoids, slow mud deposition resumed. The thin mud layer that formed on the sea floor may have been stabilized by an algal film, which also may have inhibited scavenging and burrowing, thereby enhancing the preservation potential of the intact crinoids.

The underlying cause of each influx of coarse bioclastic sediment that buried the crinoids remains unclear. Storm resuspension appears unlikely because of the sorting and subparallel orientation of the grains in the bioclastic units. Symmetric wave ripples occur near the top of the *Ailsacrinus* Bed and indicate shallow water in which wave action may have moved material across the sea floor, periodically smothering any crinoids in its path. The fact that the area was repeatedly recolonized by an identical crinoid fauna suggests that the burial of crinoids at this particular site was a very localized phenomenon and that recruitment of the area from neighbouring populations occurred almost immediately following these burial events. However, the abrupt change to a cross-bedded shelly limestone suggests a broader environmental change, perhaps by the migration of a submarine dune field (sand waves; see following chapter), causing the final disappearance of *Ailsacrinus* from this area.

IMPORTANT COLLECTION IN THE UNITED KINGDOM

The Natural History Museum, London

25

Middle Jurassic of Northern Switzerland

HANS HESS

A SCHOOLBOY'S DELIGHT

In the surroundings of Basel, four different and unique crinoid beds occur. These fossils fascinated the author of this chapter when he was a young boy. The description of *Paracomatula helvetica* was also his first publication (Hess 1950).

The crinoid beds with four different species are exposed in the Jura Mountains of northwestern Switzerland (Fig. 207). One species also occurs in eastern France and ranges as far as England. Four different horizons, each characterized by its fauna, can be distinguished (Fig. 208):

1. Beds with well-preserved specimens of the isocrinid *Chariocrinus andreae* occur within an area of about 200 km² in the canton of Baselland. A small slab with *C. andreae* was mentioned and figured in Bruckner's monumental description of historical and natural sights (*Merkwürdigkeiten*, curiosities) of the Basel countryside, published between 1748 and 1763. At that time, the true nature of the *Chariocrinus* remains was unknown, and they were referred to as plants.
2. Beds with *Pentacrinites dargniesi* occur at several sites in the Swiss Jura and in eastern France; the species was first described in 1869 from the Moselle region. This species had a wider range than previously assumed, as documented by specimens from the Forest Marble of Wiltshire, now housed in the Natural History Museum in London, and from the Bajocian of Department Isère, France.
3. A lens with the comatulid *Paracomatula helvetica* was exposed in a trench dug in 1940 as a defensive measure by the Swiss Army near Hottwil, on a strategic hill not far from the Rhine. A few individuals of this species have also been found in the Schinznach Quarry mentioned in the next section.
4. A lens with the isocrinid *Hispidocrinus* (formerly assigned to *Chariocrinus*) *leuthardti* was discovered on a shooting range near Liestal and exploited between 1892 and 1903 by Franz Leuthardt (1904), a schoolteacher and renowned naturalist. This is the youngest of the four beds and the only one that the author has not seen in the field. So far this species has not been found elsewhere.

CELTIC PLATFORM AND SWABIAN REALM

All localities are of Middle Jurassic age, about 165 million years before present. At the beginning of this period, dark mudstones of Early Aalenian age were deposited in northern Switzerland under uniform, partly anoxic conditions. This Opalinum Clay is overlain by Late Aalenian and Early Bajocian shales, sandstones, sandy limestones and iron oolites. To the west, bioclastic (mainly crinoidal) limestones formed in the Late Bajocian. These belong to the western part of a shallow-

Hans Hess, William I. Ausich, Carlton E. Brett, and Michael J. Simms, eds., *Fossil Crinoids*. © 1999 Cambridge University Press. All rights reserved. Printed in the United States of America.

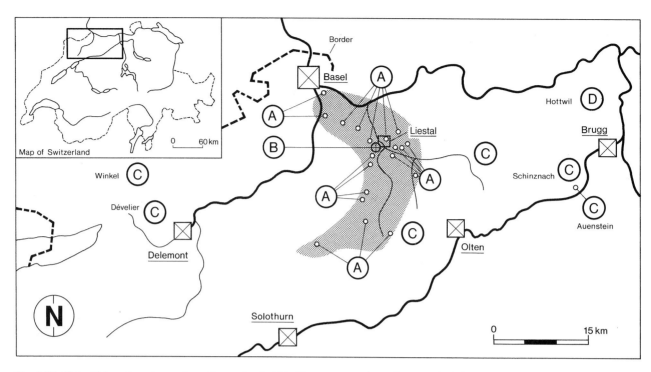

Fig. 207. Crinoid locations in northern Switzerland. (A) *Chariocrinus andreae* (Bajocian); (B) *Hispidocrinus leuthardti* (Bathonian); (C) *Pentacrinites dargniesi* (Bajocian); (D) *Paracomatula helvetica* (Bajocian).

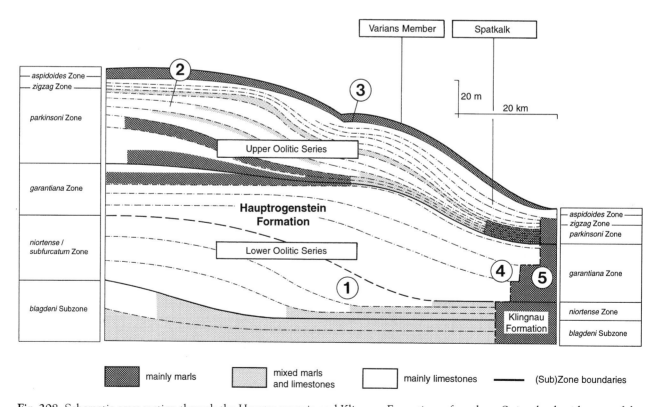

Fig. 208. Schematic cross section through the Hauptrogenstein and Klingnau Formations of northern Switzerland, with some of the main crinoid beds. (1) *Chariocrinus andreae*, Liestal; (2) *Pentacrinites dargniesi*, Develier and Winkel; (3) *Hispidocrinus leuthardti*, Liestal; (4) *Pentacrinites dargniesi* and *Isocrinus nicoleti*, Schinznach; (5) *Paracomatula helvetica*, Hottwil. Age of 4 and 5 is uncertain; see text. (After Gonzalez & Wetzel 1996.)

marine carbonate platform, the 'Burgundy Platform' (Celtic facies), covered by an epicontinental sea. In the central and eastern areas of the platform, a broad oolitic belt developed. To the south and east of the platform, tempestites and basinal calcareous mudstones (marls) of the Swabian facies accumulated under low-energy conditions; the boundary between the two facies types was just to the west of the Aare River (Gonzalez & Wetzel 1996, Fig. 1). The first two crinoid beds discussed in this chapter lie within the oolitic belt in the southeastern part of the Burgundy Platform, and the third bed (*Paracomatula helvetica*) was just east of the Celtic–Swabian facies boundary.

The eastern (Swiss) part of the Burgundy Platform is dominated by the shallow-marine oolitic barrier system that was controlled by tidal currents. The back-barrier system to the west includes micrites, oncolites and patch reefs; the off-barrier sediments to the east are characterized by tempestites and marls. In response to sea level changes, these facies belts migrated three times from west to east during Middle Bajocian to Middle Bathonian times, forming three shallowing-upward successions with a hardground on top (Gonzalez & Wetzel 1996).

The oolitic limestones of the platform belong to the Hauptrogenstein Formation ('main oolitic limestone' because of its wide distribution; *Rogen* = roe). The lower part of the Hauptrogenstein Formation, or 'Lower Oolitic Series', represents the first shallowing-upward succession. It reaches a thickness of 70 m in the western and 30 m in the eastern Jura Mountains. The accumulation of the rather thick sediments was made possible by a relatively slow rise in sea level combined with a steady subsidence (Gonzalez & Wetzel 1996). Oblique stratification (cross-bedding) indicates the presence of tidal sand waves on a shoal with a water depth estimated to be 1–10 m; the broad oolitic barrier was affected by north–south currents. At the same time, marly sediments of the Klingnau Formation accumulated in deeper water to the east. The second shallowing-upward succession started with sedimentation of calcareous mudstones and bioclastic limestones during a sea level highstand. A subsequent fall in sea level re-established ooid production, leading to the 'Upper Oolitic Series' with a thickness of up to 30 m in the western Jura. The facies change from the western oolitic limestones of the Hauptrogenstein Formation to the eastern marly basinal sediments of the Klingnau Formation occurs in the area west of the lower Aare River and is quite abrupt. The section of the Schinznach Quarry with beautiful sand waves is thus located in an area of significant changes in thickness, from 30 m to 6 m within a distance of a few kilometres. It has been placed by Hess (1972a) in the Upper Oolitic Series on the basis of the echinoderm fauna, which compares well with that of the western localities near Delémont clearly belonging to the upper part of the Upper Oolitic Series. According to Gonzalez and Wetzel (1996), the oolitic sediments of the western Upper Oolitic Series pass into marly deposits in this area, so that the Schinznach fauna appears to be in the Lower Oolitic Series. However, the exact age of the sediments at the Schinznach Quarry has not been determined; they may have been deposited later than indicated in Fig. 208.

The marly Klingnau Formation (formerly Parkinsoni Beds) locally includes crinoidal limestones such as the bed with *Paracomatula helvetica* near Hottwil (Fig. 209). *Chariocrinus andreae* is restricted to the Lower Oolitic Series. *Pentacrinites dargniesi* is characteristic of the similar facies of the Upper Oolitic Series in the western part of the Swiss Jura; but the comatulid *Paracomatula helvetica* and the stalked *Isocrinus nicoleti* (Fig. 214) cross the boundary from the basinal Klingnau Formation into the oolitic facies at the Schinznach Quarry, noted for its spectacularly diverse echinoderm fauna, which includes a lens with *Pentacrinites dargniesi*.

The third shallowing-upward succession was formed during the Bathonian, which follows stratigraphically above. On top of this sequence lie the marly limestone beds of the Varians Member deposited at a depth of 40–70 m (Meyer 1990); they have furnished the unique *Hispidocrinus leuthardti* lens.

BEDS WITH *CHARIOCRINUS ANDREAE*

Fragile Sea Lilies, Densely Packed

The beds with *Chariocrinus andreae* occur within the Celtic Hauptrogenstein facies, oolitic limestones that are partly cross-bedded but that also contain layers of mudstone. The crinoids appear near the base of the Lower Oolitic complex. More or less complete specimens have been found at about 30 sites in nearly the same stratigraphic position. *Chariocrinus andreae* commonly occurs in lenticular beds. In a number of sites, crinoids reach a density of 400 individuals per square metre and show excellent preservation, especially on the lower surface of the beds (Fig. 210). The crinoidal limestones are usually underlain by fine-grained sediments such as clay or marl (Fig. 211). As a rule, the upper surfaces contain only disarticulated fossils. Beds are

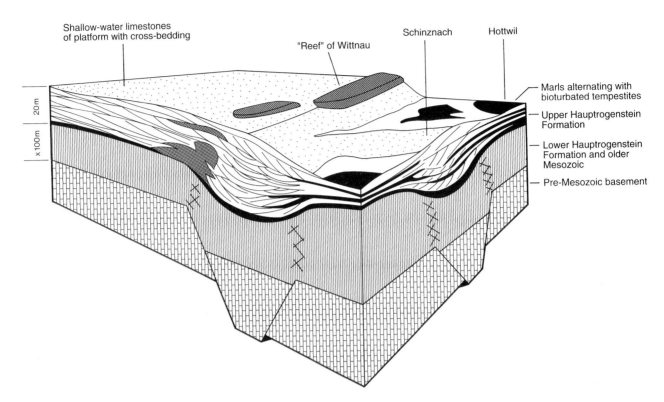

Fig. 209. Sedimentation and tectonics in the eastern Jura Mountains, showing the facies boundary between the western platform sediments and the eastern deeper water mudstones (marls) and tempestites. The facies at the eastern margin of the platform was probably influenced by a pre-Mesozoic trough. (Redrawn from Gonzalez 1993.)

composed almost exclusively of the dominant crinoid species (for a review of the fauna of this and the other beds see Hess & Holenweg 1985). A small bivalve, attached to the stems, occurs rarely, and asterozoans (mainly brittle stars) are present rarely in a few localities.

Chariocrinus andreae is a rather small species with a total height of 10–25 cm. The smallest individuals have a crown height of 3 cm; large crowns reach a height of 7 cm. Maximum stem length is difficult to determine because the stems are mostly broken between nodals and internodals. If unbroken, the stems taper somewhat distally and end with a whorl of cirri. One such stem, belonging to a nearly adult individual, reaches a length of 13 cm. The terminal cirri could have fixed the animals in the muddy bottom, but it appears more probable that, especially in environments affected by stronger currents, the distal part of the stem laid horizontally on the bottom and was held down by the terminal claws of several rows of cirri. Growth rings on a columnal indicate that the animals needed at least 4 years to reach maximum height (C. A. Meyer 1988). The size of the animals may vary considerably among localities, although larger (adult) animals are most common. In extended outcrops such as the Lausen Quarry, animals of different size occur in the different lenses up to 100 m along the quarry wall; however, in a given lens the superimposed beds (Fig. 211) usually contain individuals of similar size. Small individuals appear to be juvenile because their columnals and brachials are proportionally higher. Some of the beds have a comparatively large number of crowns on short stems, whereas others consist mainly of stems or their fragments. Because the preservation of crowns is identical, we assume that stem length varied between the different localities and that stems may have been autotomized repeatedly under less favourable conditions.

Rich Food at the Bottom of Sand Waves

Chariocrinus andreae is found in up to nine superimposed beds of 0.7–5 cm thickness in certain places, separated by thin layers of mudstone that may contain isolated ossicles. Thus, the colonies must have been repeatedly

Fig. 210. *Chariocrinus andreae*. Lower Hauptrogenstein (Lower Crinoid Beds, Lower Oolitic Series, Bajocian), Liestal. Lower surface of a very dense crinoid bed with complete specimens. (Hess Collection; photograph S. Dahint.) ×1.6.

Fig. 211. Unweathered section with well-developed *Chariocrinus andreae* Beds, Lower Hauptrogenstein, eastern part of quarry at the Lausen railway station. Exposed are two sets of crinoidal limestone beds interbedded with blue marl-clay and separated from oolitic limestone below and above by thicker layers of clay. In the lower set with three distinct beds (arrows), complete crinoids such as those of Fig. 210 occur on the lowest bedding plane, just above the hammer. The upper set passes into marly limestone in its upper part and into oolitic limestone to the right of the picture (not shown). The clay between the two sets has an elevated content of organic carbon (see text). The lower set of three crinoid beds thins to one bed, which wedges out at 3 m to the right of the section shown in the photograph. Clay near the top of the section is deformed through faulting. For detailed profiles in different parts of the quarry see C. A. Meyer (1988). (Photograph 1993 by H. Hess.) To view this figure in colour, see the colour plate section following page xv.

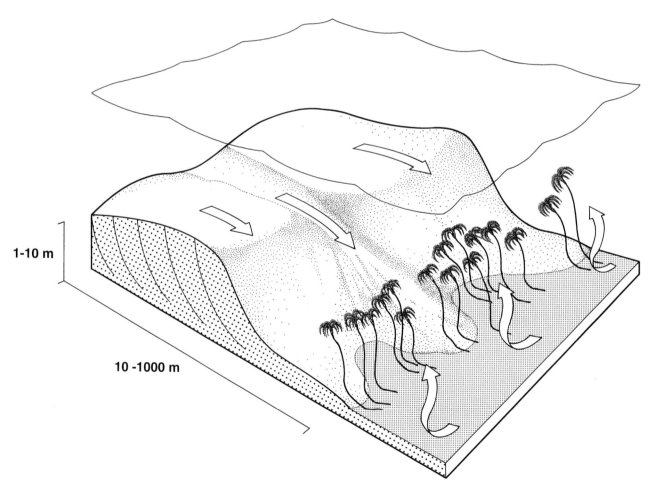

Fig. 212. Model of a sand wave with *Chariocrinus andreae* colonies on the leeward side. Arrows on top of the sand wave indicate sediment transport; arrows from bottom right represent weak currents resulting from flow separation (also depositing fine-grained material). Crinoid colonies will be displaced to similar new positions by movement of the sand wave. Size of crinoids exaggerated. (Modified from Gonzalez 1993.)

killed and replaced. The predominance, in some beds and at some localities, of small (presumably juvenile) individuals indicates that such colonies may have been rather short-lived; arrested or reduced growth from a lack of nutrients may be another explanation. Current-aligned individuals are sometimes present on bedding planes, and such orientation is much more pronounced for stem fragments on the upper bedding planes, where fossils are commonly disarticulated. Such preservation suggests that burial of the youngest colonies of the beds occurred some time after death. The main transport direction was from north to south (C. A. Meyer 1988; Gonzalez 1993), but it may be assumed that the crinoids established filtration fans using secondary flows in the opposite direction (Fig. 212). Well-preserved individuals mostly have intact crowns on stems of different length, as already explained. Specimens with regenerated arms are rare, indicating that predators were of minor importance in these dense crinoid populations. Traces from the action of scavengers as well as bioturbation are largely missing, with the exception of ooid-filled vertical burrows, which were probably produced by sea anemones.

The excellent preservation of crowns with attached stems at many localities and the high-energy, shallow-water facies seem to be mutually exclusive at first sight, but recent work may suggest how these beds were formed. The occurrence of clay and marl at the base of many of the beds indicates their formation at the margin of large fields of sand wave complexes (Gonzalez 1993), rather than in subtidal channels cut into oolitic sands as postulated by C. A. Meyer (1988). The dense colonies occurred in somewhat deeper water below fair-weather wave base (the water depth is estimated to be 10–15 m), where clay particles accumulated (Fig. 212). The sudden appearance of dense crinoid meadows indicates luxuriant growth based on a rich supply of planktonic organisms. The corresponding nutrients may have been brought into this epicontinental sea by rivers from land to the north; this was also the source of the detrital clay swept in various amounts. After a while the favorable conditions changed and the crinoids were killed, possibly through toxins resulting from red tides (blooms of phytoplankton) (C. A. Meyer 1988). In certain localities this process was repeated several times. This hypothesis is supported by the presence of remains of phytoplankton as well as the absence of larger amounts of clay minerals within the beds. In addition, the crinoid beds and the adjacent clays are darkly coloured in an unweathered section at the Lausen locality (Fig. 211), where an elevated content (0.5%) of organic carbon (hydrocarbons) was found in the clay overlying a set of crinoid beds (A. Wetzel, pers. comm., 1995). During further accumulation of the Hauptrogenstein sediments, oolitic sand became prominent. Crinoids disappeared from these, now very shallow, waters, and the bottom became inhospitable to crinoids due to shifting tidal sediments.

The occurrence of very dense isocrinid colonies on a soft bottom is without Recent analogue. Extant isocrinids prefer hardgrounds (Fig. 235, 236). Perhaps the greatest local concentrations of isocrinids occur in the Straits of Florida at 350–430 m, where *Endoxocrinus parrae* individuals, clustered along the upcurrent brow of isolated boulders, form groups of up to 39 in an area of 0.5 by 2 m, with a maximum density of 15 individuals in 0.25 m^2 (C. Messing, pers. comm., 1997). On unconsolidated sediment bottoms they attach to low outcrops, small bits of rubble and shell (Messing et al. 1990). One species (*Isocrinus blakei*) with delicate cirri has also been found on unconsolidated biogenic sand, but it is thought to be attached to hard surfaces buried underneath.

BEDS WITH *PENTACRINITES DARGNIESI*: STURDY MATS

This species occurs in the uppermost part of the Upper Oolitic Series of the Hauptrogenstein complex at a number of localities in northwestern Switzerland and across the border in France. It has also been found farther to the northwest near Toul and to the southeast near Tournus, in possibly contemporaneous strata. Bedding and preservation at the Swiss localities are comparable to the beds with *Chariocrinus andreae*, but *Pentacrinites dargniesi*, being a larger species than *Chariocrinus andreae*, achieves a density of only about 80 individuals per square metre. Near Delémont, where the beds are particularly well developed, one may count up to four complete individuals or generations lying on top each other (Fig. 213). Some of the beds contain echinoids (a species of *Acrosalenia* with very long spines). The Schinznach Quarry with its rich and diverse echinoderm fauna (Hess 1972a) included a lens of about 2 by 5 m with *P. dargniesi*; *Isocrinus nicoleti* (Fig. 214) and very rare specimens of mostly juvenile *Paracomatula helvetica* are not in this lens but occur in other parts of the quarry.

Pentacrinites dargniesi is a robust crinoid with highly branched, heterotomous arms (height of crown up to 10

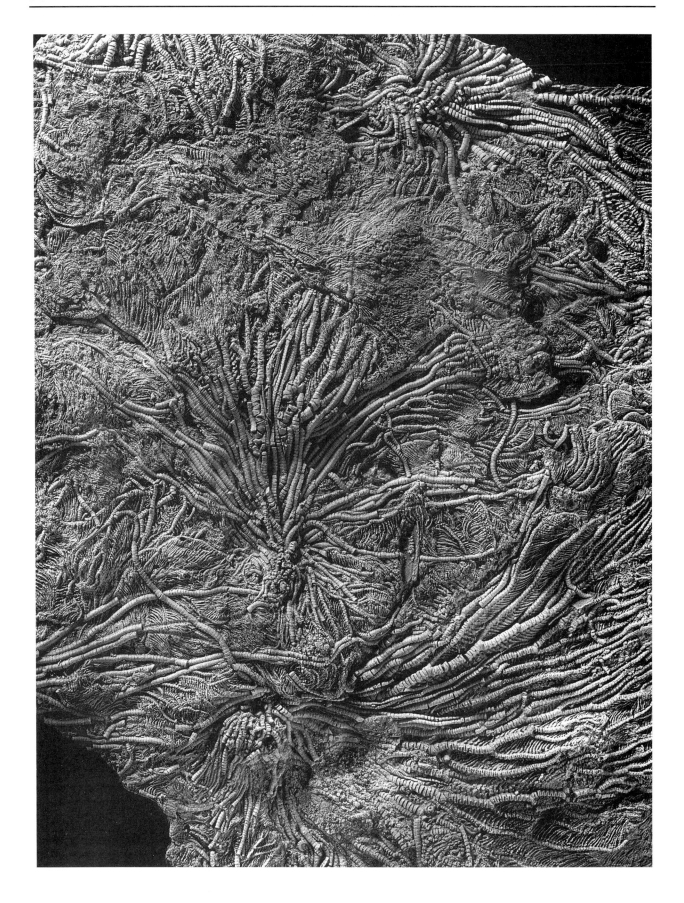

Fig. 213. (*On facing page*) *Pentacrinites dargniesi*. Upper Hauptrogenstein (Upper Oolitic Series, Bajocian), Develier. Lower bedding plane covered with complete specimens. The subadult individual in the centre has a complete stem, densely covered with long cirri. This oblique bedded slab with the crinoids in the lower part and oolitic limestone in the upper part rested on a layer of bluish clay. (Natural History Museum, Basel; photograph S. Dahint.) ×0.85.

Fig. 214. *Isocrinus nicoleti*. Hauptrogenstein (Bajocian), Schinznach. This isolated individual with its broken stem lies on a lens of marl embedded in clay and was buried by a sand wave. (Hess Collection; photograph S. Dahint.) ×1.5.

cm) and a very short, tapering stem (length 5 cm or less). Only stellate to pentagonal nodals are apparent from the outside, rudimentary internodals being hidden, so that the stems are densely covered with whorls of very long cirri (up to 12 cm) with a rhomboidal cross section and a terminal claw. Such cirri would have served well in aggregations stretching out horizontally, thus forming mats.

The individuals of *Pentacrinites dargniesi* must have lived interlinked with their long cirri forming mats, which may have drifted through the seas. In spite of the presence of rare pieces of wood in the beds, life on driftwood can be ruled out. The heavily branched crown formed a dense filtration device. The crowns would have stuck out to collect food, and the whole system could not have been rolled over easily. In view of the short, stiff stem with its dense cover of long cirri, a filtration fan could not have been formed. This view is supported by the wide distribution of such mats over the Burgundy Platform of the French and Swiss Jura. At the Schinznach Quarry, crinoids settled directly on carbonate sand. They are preserved in life position, commonly with their tegmen uncovered, on the upper side of an oblique stratified bed of oolite. An overlying layer of clay may have been the reason for the death of this colony. However, such conditions are exceptional and at other localities colonies grew on muddy bottoms. In the Winkel Quarry (Dept. Haut-Rhin, France), the crinoid beds are preserved *in situ* at the foot of a tidal delta (Gonzalez 1993), in a situation comparable to the beds with *Chariocrinus andreae* in the Lower Oolitic Formation. As in the *C. andreae* beds, individuals are mostly of similar adult size, but one bed at Villey-Saint-Étienne (Meurthe-et-Moselle, France) is composed entirely of very small, juvenile individuals with crown heights of only 1.5–2 cm (Hess 1972a), confirming the same size per bed rule.

PARACOMATULA HELVETICA: AN AGGREGATION OF FREE-MOVING CRINOIDS

The bed with *Paracomatula* belongs to the Klingnau Formation, an alternation of mudstones (marls) and bioturbated bioclastic marly limestones. These were laid down in deeper water (20 m or more) under a low-energy regime. The lens of Hottwil had a width of about 8 m, wedging out on one side and passing into a marly limestone with some isolated ossicles on the other side. The bed, which lies on a layer of bluish clay, has a maximum thickness of 8–9 cm and includes the remains of four to five generations of the comatulid without intermediate layers of marl. However, clay particles are dispersed throughout the bed, which occasionally contains pieces of wood. Crinoids lie in all directions with no sign of preferred orientation, with a density of close to 300 individuals per square metre (Fig. 215). The specimens are all of similar size. They are preserved intact on the lower surface of the bed (which is covered by a layer of marl hiding the fossils) but are more or less disarticulated on the upper surface. Some *Isocrinus nicoleti* also occur both in the bed and on the lower surface, where complete individuals lie on their side with half-opened crowns. Additional fossils include the irregular echinoid *Holectypus depressus* with preserved lantern and spines, as well as a few small asteroids.

Paracomatula helvetica is a species with only 10 slender arms (length up to 16 cm) and thin, circular cirri (length 5 cm) with a terminal claw. The cirri are inserted on a small pentagonal centrodorsal composed of 5 closely jointed, low columnals. Arm articulations are synarthrial, syzygial and, most frequently, muscular.

On the lower surface of the bed, the fossils are mostly lying on their side. The crowns are somewhat opened, as one would expect during life, and the cirri extend upward. Rare individuals are preserved with the aboral side down and extended arms. On the upper surface of the bed, some individuals are preserved in oral view, but most are disarticulated. Almost all individuals are adults. We think it unlikely that the assemblage was brought together by currents. It appears to be the only preserved dense assemblage of a comatulid in geological history. This must have been a local colony where several generations lived in a spot richly supplied with plankton; this place may have been close to an eddy, as indicated by pieces of wood contained in the bed. Disarticulated fossils on the upper surface suggest that burial of these individuals occurred some time after death. The few specimens of *Isocrinus nicoleti*, with stems exceeding 20 cm in length, were able to collect food (presumably smaller-sized plankton) at a higher level by raising their crowns above the comatulid colony. There is hardly any doubt that this colony was embedded where it lived. The presence of large amounts of phytoplankton in the bed (C. A. Meyer 1988) suggests that the animals may have died from toxins (caused by a red tide, as in the case of *Chariocrinus andreae*) or lack of oxygen.

With its exceptionally long arms, *Paracomatula helvetica* resembles the similarly gregarious *Ailsacrinus abbreviatus* from the Bathonian of southern England (see Chapter 24) and is also reminiscent of *Uintacrinus socialis* from the Niobrara Chalk (see Chapter 27). The development of long arms may have been a necessity for stemless crinoids living on muddy bottoms where currents were weak.

HISPIDOCRINUS LEUTHARDTI: SMALL AND SPINY SEA LILIES

A lens with *Hispidocrinus leuthardti* has been found embedded in clay. It was overlain by Bathonian marly limestones (Varians Member) with a rich fauna of

Fig. 215. (*On facing page*) *Paracomatula helvetica*. Klingnau Formation (Bajocian), Hottwil. Lower surface of the lens with complete specimens. Near the centre is a stem fragment of *Isocrinus nicoleti*. (Hess Collection; photograph S. Dahint.) ×1.2.

brachiopods, bivalves and irregular echinoids. The size of this lens, which was completely removed at the turn of the century, is estimated to be about 3 m², based on the material preserved in the Basel Natural History Museum. The bed has a maximum thickness of 12 cm and thins at the edges to a few millimetres. Approximately half of the bed is composed of crinoid ossicles. Both the lower and the upper surfaces of the lens are quite irregular. Finger-like patterns of infilled burrows, packed with small crinoid ossicles, occur on the lower surface. The size and shape of the burrows suggest the work of a callianassid crustacean.

This crinoid is smaller than *Chariocrinus andreae*, with a maximum stem length estimated to be 12 cm (the stems are usually broken) and a crown height reaching 5.5 cm. Otherwise it is similar, with a tapering stem ending in a whorl of cirri. Morphological characters are similar to those of *Chariocrinus andreae*. However, the axillaries bear large conical spines – hence the assignment by Simms (1989) to *Hispidocrinus*. Small individuals occur at the upper surface and close to the fringes of the bed; they seem to be juvenile because the axillary spines are very poorly developed. Even well-preserved specimens, which lie on their side, may be partly disarticulated. These individuals occur on the lower surface and also within the lens but are confined to the central part (Fig. 216). The edges and the upper surface are largely composed of disarticulated remains. No signs of preferred orientation by currents can be detected.

In contrast to the other beds, the *Hispidocrinus leuthardti* colony contains a considerable number of other fossils, namely well-preserved ophiuroids, bivalves ('*Avicula*'), brachiopods ('*Rhynchonella varians*'), serpulid worms and remains of two crustaceans (presumably responsible for the traces mentioned earlier). In such an environment, axillary spines may have been advantageous for defence. The similar *Chariocrinus andreae*, very rarely associated with other fossils, did not develop any spines.

It can be assumed that this occurrence was similar to that of *Paracomatula helvetica*, a strictly local colony where several generations were living in a spot well supplied with plankton. Again, there is no doubt that the colony was entombed where it lived. The relatively high percentage of disarticulated fossils in parts of the lens suggests that the colony was not rapidly buried.

IMPORTANT COLLECTION IN SWITZERLAND

Natural History Museum, Basel. This museum contains the majority of bedded crinoids found in Switzerland (collected by Leuthardt, Hess and others); some of them are shown in an exhibition gallery.

Fig. 216. (*On facing page*) *Hispidocrinus leuthardti*. Varians Beds (Bathonian), Liestal. Lower bedding plane of the lens with complete specimens and burrows. Note whorls of cirri at the end of some of the stems. (Natural History Museum, Basel; photograph S. Dahint.) ×2.2.

Upper Jurassic Solnhofen Plattenkalk of Bavaria, Germany

HANS HESS

A LIVELY TRADE

The extremely fine grained, even-layered lime mudstones north of Munich have been quarried since ancient times. The discovery of lithography by Alois Senefelder in 1796 has given this stone a worldwide reputation that was increased by the discovery of the remains of the early bird *Archaeopteryx*. Exploitation over many years has produced a plethora of marvellous fossils, even though they are rather rare. The only common fossil is the small crinoid *Saccocoma*, the focal point of this chapter. Quarrying is fortunately still going on because the slabs are valued today for floor and wall tiles. The best-known areas are (from west to east, named after neighbouring communities): Langenaltheim/Mörnsheim/Solnhofen, Eichstätt, Pfalzpaint/Gundolding, Schamhaupten/Zandt, Kelheim and Painten (Fig. 217).

PLATTENKALK: FLINZ, FÄULE AND KRUMME LAGEN

The Solnhofen Plattenkalk (Plattenkalke) belong to Early Tithonian and are about 150 million years old. The whole time of deposition of the Plattenkalk is within Weissjura or Malm Zeta 2 and belongs to the lower part of the Early Tithonian, which is at most half of the *Hybonoticeras hybonotum* Zone. This means that the whole series of the Solnhofen limestones represents 500,000 years or less (Barthel 1978). Only a minor part of this platy limestone has been used for lithographic purposes. The regularly bedded Plattenkalk reaches a total thickness of up to 100 m; they are interrupted by *Krumme Lagen*. These are irregularly folded beds, resulting from slumped soft sediment masses that collapsed downslope, possibly triggered by earthquakes. The very fine-grained, micritic limestone slabs of the Plattenkalk are called *Flinz* and occur in packets of sheets with an average thickness of 0.5–1 cm, reaching a maximum thickness of 30 cm in the Solnhofen area. The slabs frequently have some internal lamination but will not split along this due to an absence of clay minerals. They are separated by thinner, fissile shaly layers, called *Fäule*, which contain 10–20% clay. Stacks of *Flinz* can split along bedding planes, and it is along these that the majority of fossils are found. Both *Flinz* and *Fäule* contain the same microfossils (coccoliths together with rare foraminifera). According to most authors (Barthel *et al.* 1990) the Plattenkalk and *Krumme Lagen* were deposited in shallow basins with an average depth of 30–80 m between sponge–algal mounds. In certain outcrops, the transition of Solnhofen Plattenkalk into sponge–algal mound facies can be observed. Such beds are rich in fossils (brachipods, echinoids, bryozoans, serpulids) and contain abundant shell detritus. The sponge–algal mounds were progressively buried by carbonate sediments. In the eastern part and later also along the south-

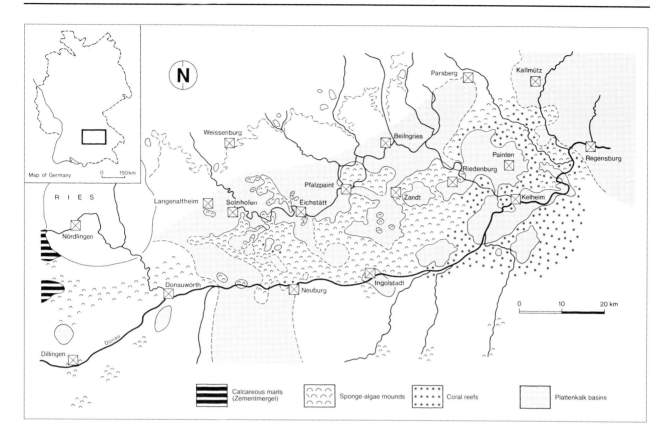

Fig. 217. Map of localities with outcrops and basins of Plattenkalke. (Redrawn from Viohl 1996.)

ern margin of the complex, sponges were replaced by corals, which produced more detritus (Fig. 217). There can be no doubt that deposition of the Plattenkalk occurred in relatively shallow water on the landward and most restricted part of a carbonate platform with a very irregular relief. To the north a landmass formed part of the Rhenish–Bohemian Massif. The origin of the Plattenkalk is discussed further in the section titled 'Plattenkalk Origin: Conflicting Theories and Open Questions'.

ANIMALS FROM LAND AND SEA

The celebrated Solnhofen Plattenkalk has furnished remains of at least 550 species of animals and plants (Frickhinger 1994), including land animals like insects, crocodiles, lizards, turtles, flying reptiles (pterosaurs) and the early bird *Archaeopteryx*. However, marine organisms, such as cephalopods, fish and crustaceans, are predominant. The largest number of fish are small or juvenile *Leptolepides sprattiformis*, which must have lived in shoals. There are considerable differences in both the content and number of fossils between the localities (Walther 1904). The Solnhofen area, which has given the formation its name, is rather poor in fossils compared with the Eichstätt area but is renowned for the thick, lithographic limestone beds occurring there. Impressions of jellyfish (*Rhizostomites*) are relatively common in the Pfalzpaint area, and, in contrast to the other fossils, these are preserved within the limestone beds. The famous horseshoe crab *Mesolimulus* has been found at many localities, but most frequently around Pfalzpaint. Mostly juvenile specimens of this animal have left numerous tracks in the Solnhofen–Eichstätt area, in most cases with the animal at the end of its trail (so-called death tracks). This leaves no doubt that these hardy animals remained alive on the bottom, at least for a limited period. Crustaceans, especially decapods (shrimps and the crayfish *Mecochirus*, which also left its tracks), are among the more common Plattenkalk fossils.

By far the most numerous macrofossil is the pelagic *Saccocoma*. The juvenile forms, in particular, are preserved in great numbers on the bedding planes near Solnhofen and especially near Eichstätt (Fig. 218), where *Saccocoma* occurs at seven different horizons. With the possible exception of the Triassic *Encrinus*

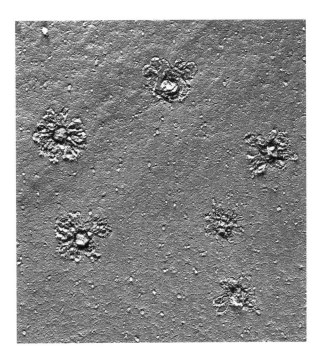

Fig. 218. *Saccocoma tenella*. Several curled-up specimens on the lower surface of a slab from Schernfeld near Eichstätt. (Natural History Museum, Basel; photograph S. Dahint.) ×0.7.

liliiformis, *Saccocoma* appears to have been the most numerous crinoid of all times. The cups of these crinoids split from the rock with the aboral side up, which makes them stand out as knobs. In many cases, the bedding planes split through the fossils, and parts of these adhere to both the underlying and overlying slab. The mode of preservation is discussed in more detail in the next section. Near Zandt a large number of the ophiuroid *Geocoma carinata* occur.[1] The comatulid *Pterocoma pennata* is found mainly near Solnhofen and Zandt, whereas echinoids are uncommon and asteroids (*Lithaster* and *Terminaster*) as well as holothurians are exceedingly rare. Microfossils include a diverse flora of coccoliths (Keupp 1977) and sparse benthic foraminifera.

SMALL CRINOIDS IN ENORMOUS SWARMS

Fixed crinoids are rare; only a few remains of a millericrinid have been reported. Conditions at the bottom must have been unsuitable for stalked crinoids. The abundant *Saccocoma* is a small, lightly built crinoid. These delicate crinoids were carefully described by Jaekel (1892) in one of his classic papers. We follow his characterization of the morphology (supplemented by our own observations on well-preserved specimens and from collecting in the field) to show how well these fascinating small animals adapted to a floating lifestyle. Two species have been mentioned: the smaller *S. tenella*, which differs from *S. pectinata*[2] by the presence of more pronounced lateral flanges or wings on the proximal brachials. In view of the commonly insufficient preservation (Goldfuss, the original author, mentioned that *S. pectinata* was badly preserved), the distinction would appear to need justification. We therefore accept only one species, *S. tenella*, the juvenile specimens being provided with comparatively larger wings. A third species, *S. schwertschlageri*, is a modification due to preservation (autotomy of arms) (Manni & Nicosia 1986).

These fossils occur on both the lower and upper surfaces of *Flinz*, but only on the lower side of *Fäule* beds (Janicke 1969). However, a number of small individuals have been discovered in a small quarry near Mühlheim on the upper surface of *Fäule* beds, some of them showing the oral side (C. Obrist, pers. comm., 1997). The majority of specimens occur on lower surfaces (Figs. 218, 219), showing the aboral side (the knob of the cup) or a corresponding impression, but never their oral side. In larger specimens the cup is always somewhat crushed, which shows that they were at most only partly infilled with sediment, similar to the preservation of the ammonite shells. Preservation of *Saccocoma* is best in the *Fäule* beds (Fig. 219); in *Flinz* beds, recrystallized calcite may obscure the details. The tips of the arms are usually coiled or lumped together (and then strongly recrystallized), or the animals are curled up as a whole on the bedding planes (Fig. 218).

Adult specimens reach an overall arm length of 25 mm (Fig. 219) with a cup height of 4 mm and a cup diameter of 5 mm. Juveniles have arms of 5 mm on a cup less than 1 mm high. The arms of these small specimens are branched only once at the first axillary. The cup, a hemisphere or open bowl without preserved tegmen, is formed by five very thin reticulate radials. They have a median ridge thickened on top with anastomosing ribs to carry a small articular facet for the first primibrachial, which is very short in aboral view (Fig. 220). This articulation is muscular, and the articulation between the first and second primibrachials (IBr_1–IBr_2) is weakly synarthrial (crypto-synarthrial) and outwardly sloping. The same type of articulation is between the first and second secundibrachials ($IIBr_1$–$IIBr_2$) and between the third and fourth secundibrachials ($IIBr_3$ and $IIBr_4$). Incidentally, IBr_1, $IIBr_1$ and $IIBr_3$ are wingless (see later in this section). The two distal articular facets

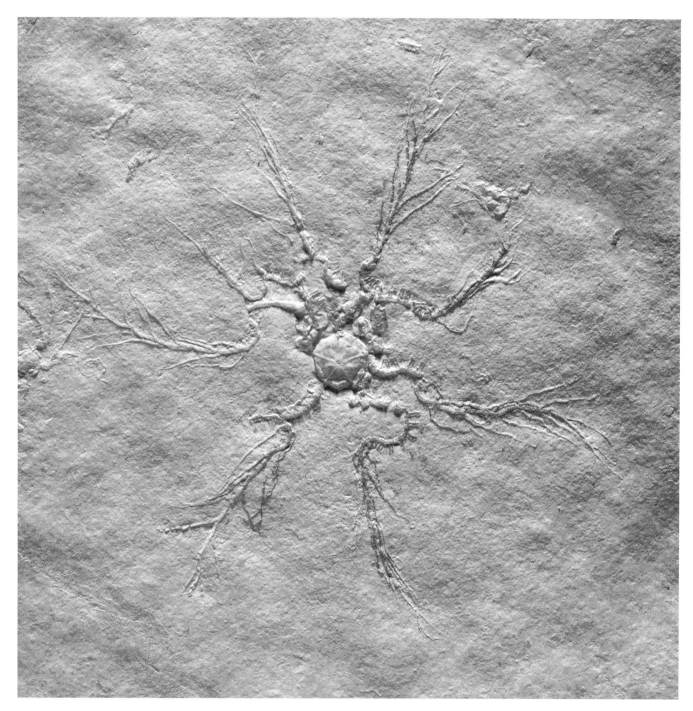

Fig. 219. *Saccocoma tenella* on a lower bedding plane, Langenaltheim. The specimen has unusually well-preserved *Schwimmplatten*, which also occur on the third secundibrachials; on one of the arms the second primibrachial is not axillary. (Hess Collection; photograph S. Dahint.) ×3. To view this figure in colour, see the colour plate section following page xv.

of the second primibrachial or first axillary are muscular. These facets (or at least their visible aboral ligament pits) slope outward at an angle of 45°. Due to the sloping facets of the primibrachials and the first secundibrachial, the arms are more or less horizontal at this point (Fig. 220b). Such a position is ideal for floating. Each second primibrachial divides in adult individuals into two, giving a total of 10 main arms with a number of side arms. These originate in some sort of pinnule socket, but their elements are essentially comparable to

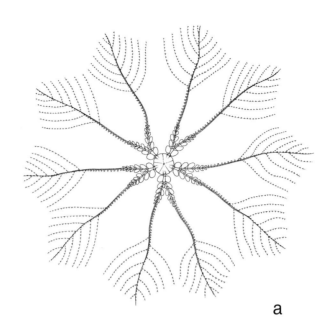

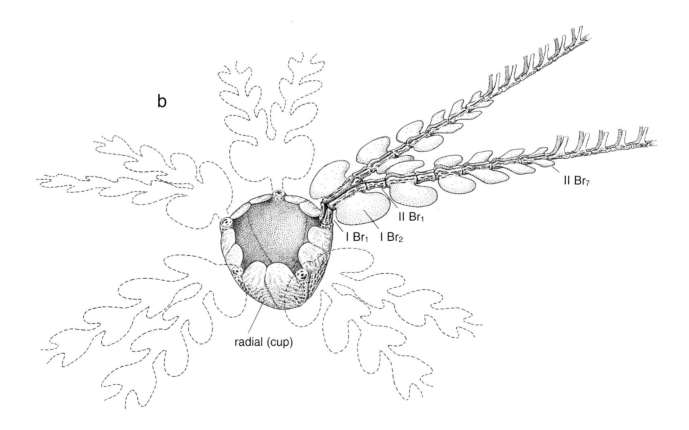

Fig. 220. Reconstruction of *Saccocoma tenella*. (a) Aboral view of a complete animal. (Redrawn from Jaekel 1918.) ×1.6. (b) Enlarged oblique oral view of cup and proximal arms, unknown soft parts in cup omitted. Key: IBr, primibrachials; IIBr, secundibrachials. ×6.

the elements of the main arms and cannot, therefore, be classified as true pinnules. As judged by the distinct aboral ligament pits, muscular articulations appear to be dominant in *Saccocoma*, and syzygies seem to be lacking altogether.

The axillary plate (primaxillary, IBr$_2$), as well as some of the following brachials (IIBr$_2$, IIBr$_{4-7}$), bears conspicuous, symmetric, dish-like lateral wings or flanges (Jaekel's *Schwimmplatten*) and weak oral processes (Figs. 219, 220). Interleaved between the winged brachials are the wingless IIBr$_1$ and IIBr$_3$, obviously to avoid interference of the winged brachials during movement of the arms. These wings are also on adult specimens (corresponding to *pectinata*). However, it appears that the wings are not any larger on adults than on smaller specimens (corresponding to *tenella*). Therefore, the lateral wings may not be so obvious in the adult individuals whose ability to float was supported by the finely branched arms. The dish-like lateral wings with their dense, porcelain-like, non-reticular structure certainly would have helped to keep the animal floating; such wings occur also in other Roveacrinida. A comparatively larger wing size in juveniles (whose arms were short and less branched) makes perfect sense for a pelagic lifestyle. Away from the cup, the lateral wings are progressively reduced and, in contrast, a pair of vertical projections or baffles is present on each brachial. The length of the vertical baffles (or, more accurately, their height) exceeds that of the corresponding arm ossicle. The baffles usually lie sideways on the bedding planes and were directed orally during life so that the food grooves were lying in a deep trough (Fig. 220b). On proximal elements the baffles are extensions of the arm ossicle with the same reticular structure, and their sides are stiffened by rather strong rods. The baffles occur to the very tips of the arms, where the brachials are reduced to needles and the lateral extensions are lacking altogether. In notes to her unpublished manuscripts, the late H. Sieverts-Doreck described the vertical projections as muscle fields, but such an interpretation is rejected here in favour of a baffle function.

By virtue of the porous nature of the plates as well as of the basal wings, *Saccocoma* must have been well adapted to a pelagic way of life. The branched arms provided a filtration network in which the captured plankton was passed along the food grooves towards the mouth, assisted in its progress presumably by ciliary action. The paired baffles would have improved the food-gathering process as well as protecting the soft parts in the furrow. The dominance of muscular articulations indicates considerable possibilities for displacement, such as moving the arms up and down to adjust the animal to different depths, a process helped by the *Schwimmplatten*, or moving them horizontally with the paired, wing-like baffles standing up, to create a current. A speed difference between the crinoids and the plankton it fed upon was essential; *Saccocoma* thus approached a nektonic lifestyle.

In two papers, a benthic lifestyle was proposed for *Saccocoma* (Milsom 1994; Milsom & Sharpe 1995), based on the specific gravity of the skeleton and the absence of a plausible buoyancy or swimming mechanism. However, none of the evidence from sedimentology, taphonomy and morphology supports this assumption.

At times *Saccocoma* flourished in the favourable conditions of the upper water layers. During such periods it must have been a key element in the food chain, converting and concentrating plankton (coccolithophorids and possibly other groups that were not fossilized) into food for larger animals such as small fish and small squid. The large number of *Saccocoma* feeding on coccolithophorids would have produced a huge amount of disaggregated coccoliths, contributing to the build-up of the limestone beds. The importance of *Saccocoma* in the food chain is proved by the common occurrence of coprolites (fossilized faeces), which are composed of their remains (Fig. 221). Müller (1969) thought that they originated from fishes, whereas Janicke (1970) claimed that they were excreted by teuthoid cephalopods. A specimen of the aspidoceratid *Physodoceras* from Nusplingen with remains of *Saccocoma* in the living chamber proves that certain ammonites fed on this crinoid (Lehmann & Weitschat 1973). Smaller, heap-like coprolites (Fig. 221) were thus probably excreted by cephalopods. The conspicuous, worm-shaped cylinders, named *Lumbricaria intestinum* and reaching a length of 170 cm (Barthel *et al.* 1990), may have been produced by a fish with a peristaltic gut. This is assumed from the constrictions of *Lumbricaria intestinum*. The coiled nature of *Lumbricaria intestinum* leaves little doubt that these coprolites descended directly to the sea floor from an animal living and feeding in the water above.

REEFS AS HABITAT OF BENTHIC CRINOIDS?

The comatulid *Pterocoma pennata* is fairly common in the Zandt quarries. It has also been found in the Solnhofen area (Fig. 222) but is rare around Eichstätt. This slender species has 10 long arms (up to 130 mm) with

Fig. 221. Coprolite, composed of *Saccocoma* remains. Plattenkalk of Schernfeld near Eichstätt, lower surface. (Natural History Museum, Basel; photograph S. Dahint.) ×3.

very long pinnules (up to 15 mm). The slender cirri are composed of very long cirrals and lack terminal hooks. Ligamentary articulations (syzygies) occur at intervals of four to five secundibrachials in the proximal part, where many fossils have arm breaks. The arms are typically coiled aborally and tangled in the proximal part, and the tips are straight with pinnules widely extended. This preservation is probably due to the more rapid decomposition of the oral (ventral) muscles than the aborally (dorsally) situated ligament, thus exerting a pull. Such arching is also present in many fish skeletons, and this may also be due to the shrinkage of ligaments after death.

Another comatulid, the sturdier '*Pachyantedon*' *gracilis*, is exceedingly rare (Walther 1886; later referred to *Solanocrinites*). It has been found near Kelheim and must have come from neighbouring sponge–algal mounds or from patch reefs. The same is true of an almost complete specimen of *Millericrinus* (now *Pomatocrinus*) *nobilis* described by Walther (1886) and other millericrinid remains (incorrectly named *Millericrinus mespiliformis* by Frickhinger 1994). *Pterocoma pennata*, on the other hand, with its slender build and its elongated cirri without a terminal hook, seems well adapted to life on a muddy bottom. The extremely long pinnules may have been an adaptation to an environment where currents were weak and, therefore, the supply of food was rather limited. The absence of autochthonous benthic forms associated with *Pterocoma* seems to exclude *in situ* burial at the place of living, possibly higher, sufficiently oxygenated soft grounds (Fig. 223).

PLATTENKALK ORIGIN: CONFLICTING THEORIES AND OPEN QUESTIONS

The formation of this sediment has long been and still is a puzzle, because no exact Recent analogue is available for comparison. Barthel, who devoted much of his all-too-short life to these sediments, as well as other scientists thought that the Solnhofen Limestone was deposited in partly closed basins (*Wannen*) with re-

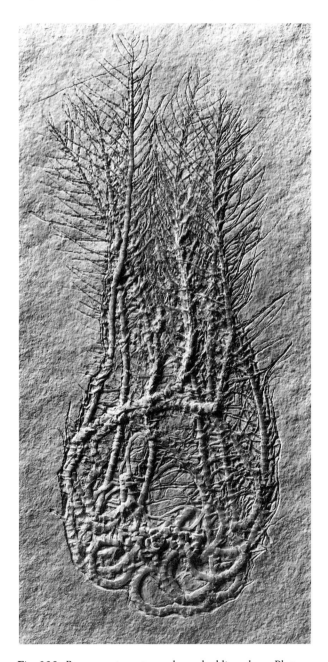

Fig. 222. *Pterocoma pennata*, on lower bedding plane, Plattenkalk, Solnhofen. (Jura Museum, Eichstätt; photograph left by H. Doreck.) ×1.6.

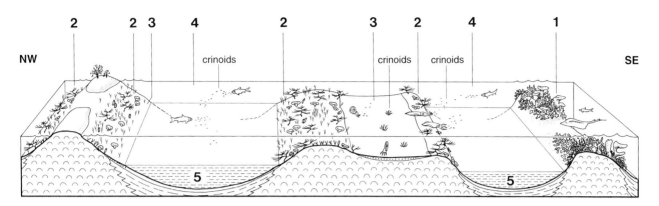

Fig. 223. Deposition areas of Plattenkalke. (1) Coral patch reef; (2) hardgrounds on top of ancient sponge mounds, penetrating into surface waters with normal salinity and colonized by seaweeds, sponges and octocorals; (3) soft bottoms with various invertebrates, including *Pterocoma pennata*; (4) surface waters with plankton (including *Saccocoma*) and nekton; (5) hypersaline, stagnant bottom waters. (Redrawn from Viohl 1996.)

stricted water flow. A tropical climate could have caused rapid evaporation. The bottom environment was very calm, as demonstrated by settling marks next to fossils and scratch marks of decaying fish (Viohl 1994). Many animals, such as the shoals of young fish and also large fish with prey still in their mouth, must have been rapidly killed, presumably due to a lack of oxygen rather than by an increase in salinity, which would not have killed them so quickly. The scarcity of burrows in the sediment together with the lack of pyrite and normal concentrations of iron and manganese (high concentrations would indicate anoxic conditions as in the Posidonia Shale) appears paradoxical. According to Barthel *et al.* (1990), the salinity of the lower water layers must have been sufficiently high (at least 4–5%) that burrowing animals could not exist in the basins.

The sudden death of marine vertebrates, so obvious in these sediments, is paralleled by the occurrence of *Saccocoma*, deposited mainly articulated and in life position (i.e., aboral side of cup downward). The reason for the coiling of the arm tips or the complete curling up of the arms so frequently observed is not known, but it seems unlikely that this preservation was due to posthumous changes such as dehydration. The cause of death of huge swarms of *Saccocoma* is unknown. Theories include an influx of fresh water from rain or poisoning by blooms of phytoplankton, a rather unlikely explanation in view of the low organic productivity of the basins (Barthel *et al.* 1990). Reproduction of *Saccocoma* may not have been uniform throughout the basins, but dependent on temporary favourable local conditions of water movement and food supply. Once these conditions failed, the crinoids would have died off quickly.[3]

One of the keys to the reconstruction of the environment is the origin of the extremely fine-grained lime mud. To what extent coccolithophorids (planktonic algae), possibly the main food of *Saccocoma*, contributed to the lime mud making up the beds is uncertain. Coccoliths are found mainly on the bedding planes. Barthel (1978) concluded from the scarcity of fossils in the *Flinz* Beds that these were rapidly deposited, with the bulk of sediments washed in during storms. A possible Recent analogue may be on the Bahama Bank, where during the present highstand of sea level, fine-grained carbonates on the bank tops are stirred up during storms and swept away in suspension (Heath & Mullins 1984). These bank-derived muds then settle in adjacent basins which can be tens of kilometres away. A geochemical analysis of beds from Eichstätt suggests that carbonate and clay particles may have been transported into the area from an outside source as a ready mixture (Bausch *et al.* 1994). The occurrence of *Krumme Lagen* points to some relief in the Plattenkalk area; and as a consequence, some beds may also have been the result of gravity flows or turbidites of uniform, fine-grained lime mud. This carbonate ooze is the fine debris originating from calcareous algae, but may also contain carbonate resulting from the precipitation of tiny aragonite needles in shallow water. (Such whiting of the water occurs regularly in the Bahamas and the Persian Gulf, where precipitation is caused by the removal of CO_2 by assimilating algae.)

On the basis of fish taphonomy, Viohl (1994) inferred very rapid deposition of the *Flinz*, with several laminae representing less than a year. One problem in the *Flinz* layers is that appreciable diagenesis has altered the original nature of the fine-grained carbonate material to a micrite, the origin of which is now difficult to determine. *Flinz* periods were followed by quiet, marly *Fäule* episodes. During such times cyanobacteria, whose spherical remains occur in some of the *Fäule* beds, could have formed mats on the lime mud, reflecting the presence of some oxygen in the lower water layers. The scarcity of benthic organisms points to rather hostile conditions on the bottom, but rich life must have been possible at times in the upper water layers, as documented by *Saccocoma* and the coprolites containing them (Fig. 223).

THE WIDE DISTRIBUTION OF SACCOCOMIDS IN THE MESOZOIC

Saccocoma tenella is a wide-ranging Tithonian species that has been reported from Russia to Cuba (Nicosia & Parisi 1979). Remains of saccocomids are frequently encountered in thin sections of pelagic Jurassic and Cretaceous limestones of the Mediterranean zone and the Atlantic area. Elements isolated from marly sediments include an Upper Oxfordian species with lateral wings (*Schwimmplatten*) on brachials and branched, antler-like spines on the cup, called *Saccocoma quenstedti* (Verniory 1961; see also Pisera & Dzik 1979). Secundibrachials, described by Verniory (1962) from the Tithonian of the French Alps, have lateral wings (Verniory's *expansions natatoires*) of amazing size, with a length up to six times that of the corresponding brachials! A number of ossicles assigned to several Upper Jurassic species were recovered during Leg 11 of the Deep Sea Drilling Project off the Florida coast (Hess 1972b).

IMPORTANT COLLECTIONS IN GERMANY

Many museums all over the world contain fossils from the Solnhofen Plattenkalk. Fine specimens are, for example, preserved in the Bayerische Staatssammlung für Paläontologie und historische Geologie in Munich. We would, however, recommend a visit to the local museums, such as the very interesting Jura Museum in Eichstätt, beautifully set in a castle (Willibaldsburg), and one of the private museums – for example, the Maxberg Museum near Solnhofen, which also has an interesting exhibition of the technique to produce lithographic plates. Such a trip could be combined with a visit to one of the quarries – for example, in the Eichstätt area – which abound with *Saccocoma*.

NOTES

1. The ophiuroids *Sinosura kelheimensis* and *Ophiopetra lithographica* are from the somewhat younger Mörnsheim Beds (Malm Zeta 3) of the Kelheim/Eining/Weltenburg area. They appear to have lived on the bottom at or near the place of burial.
2. The paired vertical, orally directed projections that give the arms their pectinate appearance on the bedding planes are common to both species. The corresponding brachials are usually preserved sideways, in contrast to the proximal brachials with their lateral wings, which present their aboral side in accordance with the cup.
3. Very fine sediment brought in by storms would have impaired the crinoids' feeding abilities, preventing further movement. The immobilized animals then sank to the bottom in a cloud of settling sediment; by virtue of their higher density and their center of gravity near the cup, the crinoids would have reached the bottom in life position before the bulk of the sediment. This new theory, reached by the author only at the page proof stage, gives a plausible explanation for the taphonomy observed. It may also resolve the mystery of the curled-up specimens with their orally enrolled arms (Fig. 218). This unusual behaviour must have resulted from muscular contractions of the brachials, presumably to protect the food grooves against invasion of mud. Posthumous decomposition of the muscles can be ruled out because this would have caused the arms to become aborally flexed, as in the case of *Pterocoma pennata* (Fig. 222).

27

Uintacrinus Beds of the Upper Cretaceous Niobrara Formation, Kansas, USA

HANS HESS

A LARGE, SHALLOW INLAND SEA

During most of the Cretaceous, the western interior of the North American continent was covered by an epicontinental sea. During the Late Cretaceous, at the peak of the transgression, this shallow sea extended from the Rocky Mountains to close to the Mississippi Valley and from northwestern Canada into the Gulf of Mexico, flooding one-third of North America and covering an area of about 1,500 by 6,000 km (Fig. 224). The Western Interior Seaway was eventually filled with sediments from the rising western highlands, ancestors of the Rocky Mountains. From west to east, the sediments include non-marine lowland sands and silts, littoral marine sands and silts and offshore marine muds. The Cretaceous transgression reached western Kansas by the Late Cretaceous, about 85 million years before present. The sequence, with a total thickness of 450 m, starts here with deltaic sediments, represented by the sands and silts of the Dakota Formation. This interfingers laterally with the finer-grained Graneros Shale, deposited seaward of the delta. With continuing transgression, the eastern margin of the sea moved farther to the east. In western Kansas, shallow-water limestones began to accumulate, followed by laminated chalk (Lower and Upper Greenhorn Limestone). Partial regression of the sea is reflected in the deposition of the Carlile Shale; renewed transgression led to the deposition of the Niobrara Chalk, a fine-grained, rather pure limestone derived from shells, planktonic foraminifera and coccoliths. During this period, the bottom waters became stagnant at times, limiting life on the sea floor. In the lower part of the Niobrara Formation, an occasional increase in salinity led to the precipitation of gypsum. However, at other times, an abundant fauna existed. Finally, the Pierre Shale, a succession of muds resulting from erosion of lands far to the west and to the east, concludes the sequence with the final regression of the Cretaceous Interior Sea (Fig. 225).

THE PROVINCIAL FAUNA OF THE WESTERN INTERIOR SEAWAY

Because of the great distance from the open oceans, the shallow Western Interior Seaway lacked periodic interchange with the Cretaceous oceans of the world. Even at their best, marine assemblages had a low diversity, and many strata lack shelly invertebrates altogether, suggesting hostile conditions at the bottom. The fauna is usually dominated by a few species of planktonic foraminifera, crustaceans, swimming and bottom-dwelling cephalopods (*Baculites* and *Scaphites*) and huge, thin-shelled benthic inoceramid clams. Oyster beds (*Pseudoperna congesta*) are very common in places. Remains of marine vertebrates, such as fishes, reptiles and turtles,

Hans Hess, William I. Ausich, Carlton E. Brett, and Michael J. Simms, eds., *Fossil Crinoids*. © 1999 Cambridge University Press. All rights reserved. Printed in the United States of America.

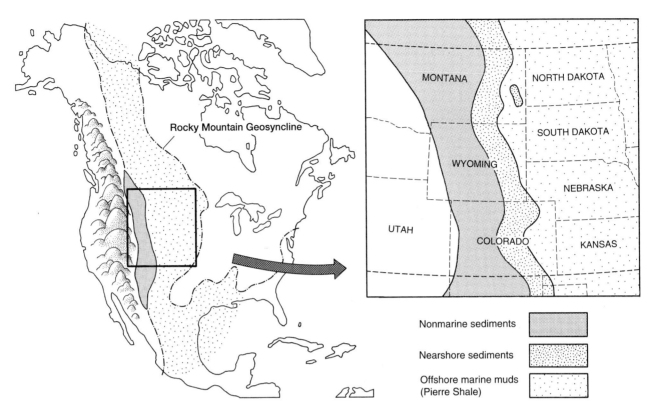

Fig. 224. Palaeogeography of the Western Interior in the Late Cretaceous. (Redrawn from Laporte 1968.)

weather out of the soft strata that lie exposed to summer heat and winter snows on the barren plains. *Uintacrinus*, but even more so reptiles and the diving bird *Hesperornis regalis*, have contributed to the fame of these sediments. The reptiles include the powerful mosasaurs, the long-necked plesiosaurs and the flying *Pteranodon*. Small and large fishes, as well as ammonites, provided the base for a substantial reptilian food chain. Some animals were of extraordinary size, such as the fishes *Portheus* and *Xiphactinus*, which reached lengths of more than 5 m. On land, emerging from the receding Late Cretaceous seas, an abundant fauna and flora developed, dominated by the widely publicized dinosaurs. However, remains of land animals and plants are rare in the Niobrara Chalk, and the same is true of stalked crinoids and echinoids.

THE LARGEST STEMLESS CRINOIDS

The first specimens of a large, stemless crinoid were discovered by Professor Marsh of dinosaur fame in 1870 in the Santonian of the Uinta Mountains, Utah – hence the name of this crinoid. Large slabs containing complete specimens were later collected from the Niobrara Chalk of western Kansas over a considerable geographic range. Well-known localities are near Martin and Elkader in Kansas, and in recent years complete specimens have been uncovered from the same strata in Colorado. Isolated ossicles have been recorded from the Santonian of England, France and Germany, as well as from Italy and even from Australia (Rasmussen 1961).

The Niobrara Chalk specimens are the best known and are exhibited in many museums. More or less complete individuals occur on the lower surface of thin lenticular limestone slabs that are interbedded with chalk. The localities near Elkader (Logan County, Kansas) furnished the large slabs so suitable for museum exhibits. As described by Springer (1901), several lenses were found within a radius of 10–15 km. At the most productive locality (Springer's No. 1), crinoidal remains formed a bed of about 15 by 6 m, with a thickness of 1.5 cm at the centre and thinning on all sides 'to the thinness of cardboard'. Springer stressed the fact that the crinoids could not be separated easily from the underlying chalk, so that in many cases only moulds of calyces were collected. The crinoids in this large colony were mostly adults with an uncrushed cup diameter of 2–5 cm. They are preserved in all positions, but most

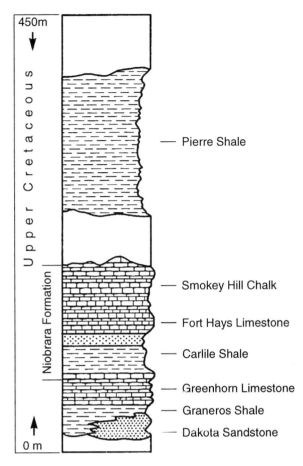

Fig. 225. Stratigraphy of the Upper Cretaceous in western Kansas. (Redrawn from Laporte 1968.)

are lying on their sides with the arms opened out. Many were embedded with the curvature of the cup downward, which presents an aboral view. In such cases the arms can rarely be observed because they penetrate into the slab and lie in another plane. Only a few animals were buried with their oral side down, showing the tegmen with the anal tube. A small colony (Springer's locality No. 2 and situated only about 800 m from locality 1), covering about 1.5 m², consisted of only very small crinoids. At another locality (Springer's No. 3, about 20 km from No. 1) the crinoids were preserved with arms closely folded together, resembling bundles of parallel rods. The lenses described by Springer from this area did not all belong to exactly the same stratigraphic horizon. The crinoids were never found as single, detached specimens, but were always more or less entangled with others (Figs. 226, 227). Densities of 25–50 animals per square metre were common. Quite generally, these crinoids are almost completely flattened, with only the cups standing out in some relief.

FLEXIBLE CALYX AND LONG ARMS WITH THIN BRACHIALS

The calyx of *Uintacrinus socialis* is globose and flexible. It may reach a diameter of 7.5 cm and is composed of numerous smooth, thin, closely fitting plates. The elements making up the calyx belong to three categories (Bather 1896): (1) the base with a central piece, five infrabasals, five interradial basals and five radials; (2) the fixed brachial elements with two primibrachials (the second one axillary) and about eight secundibrachials, which become gradually modified from the flat ossicles to the low, free brachials; and (3) the supplementary, interbrachial interradial plates. In adult individuals, up to four pairs of pinnules are fixed into the calyx. The transition from the plates of the calyx to those of the arms is gradual. The large body size, obviously connected to the lifestyle of this crinoid, is thus the result of the incorporation of many brachials, the proximal pinnules and a number of interradials into the calyx (Fig. 24).

The mouth lies close to one of the margins, and a large, conical anal tube is located in the centre of the tegmen (Fig. 228). The tegmen is commonly preserved as a jet-black, carbonized membrane with small, irregular granules. The ambulacral grooves are not covered by any plates. There is no trace of an attachment structure such as stem, cirri or holdfast. The 10 arms, resulting from a single division at the second primibrachials, may reach a length of 125 cm (Fig. 229). Springer (1901, p. 15) mentioned an individual with a cup width of 6.2 cm and an incomplete arm of 100 cm. Muscular articulations alternate with scattered syzygial articulations. In adult individuals, the syzygies occur at regular intervals of three to six brachials and, more distally, the intervals become much more irregular and commonly longer. The syzygies with their prominent ridges resemble those of comatulids. In spite of the rather large number of syzygies, broken or regenerated arms are quite exceptional. The brachials are thin or low; their height near the calyx is about 1.5 mm in large specimens, with a width of 8 mm. Distally the height is nearly the same and the width diminishes to 1.5 mm. The oral muscular fossae on brachials are rather small. The pinnules, which reach a length of 20 mm near the cup, alternate from one side to the other, leading to a wedge-shaped appearance of the brachials, especially the fixed and more distal brachials. Distally the pinnules become extremely fine, almost thread-like. The low and, therefore, very numerous brachials with pinnules constitute a very dense food-

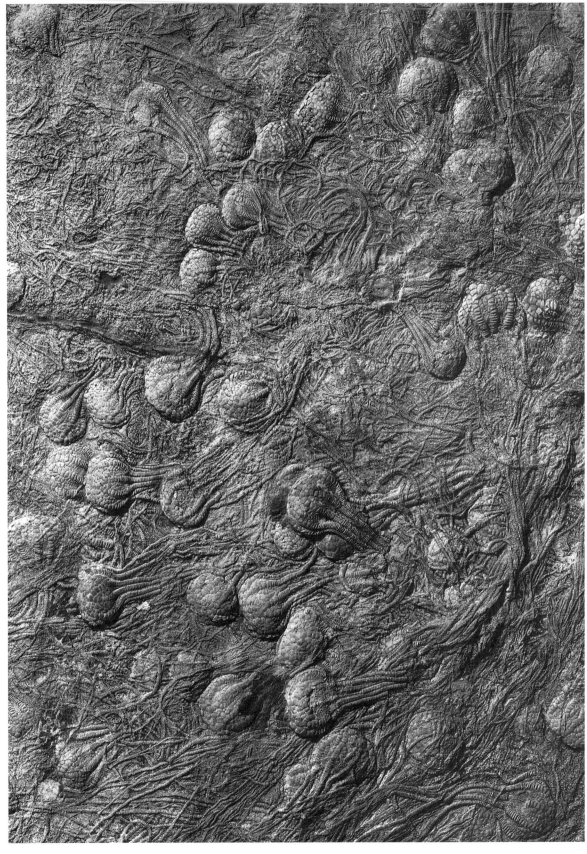

Fig. 226. Large slab of *Uintacrinus socialis*. Niobrara Chalk, Elkader (Logan County), Kansas. As judged from its preservation, this slab comes from Springer's original locality; it shows the lower surface and is exhibited in the Museum für Geologie und Paläontologie, University of Tübingen. (Photograph W. Gerber; courtesy H. P. Luterbacher.) ×0.3.

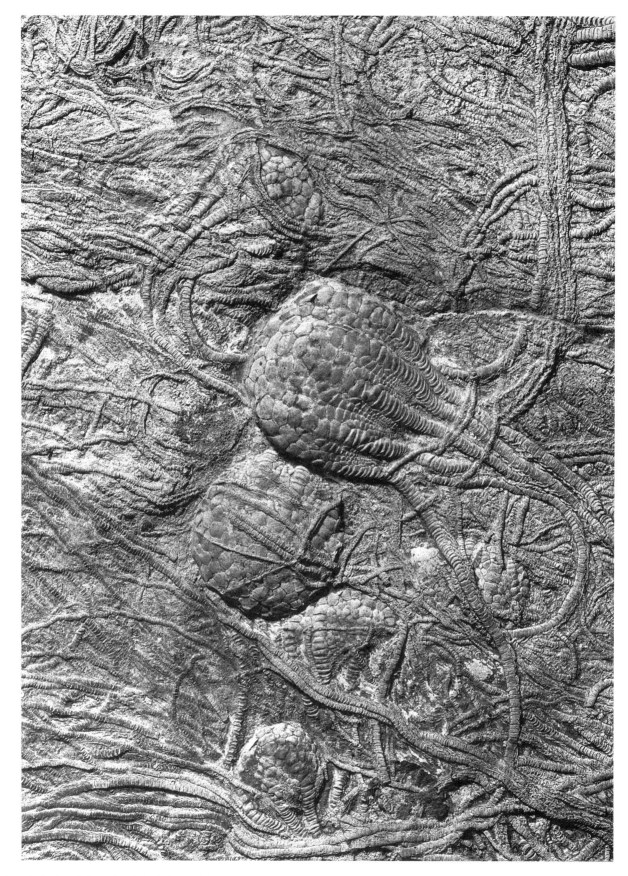

Fig. 227. Detail from the slab of Fig. 226. ×1.

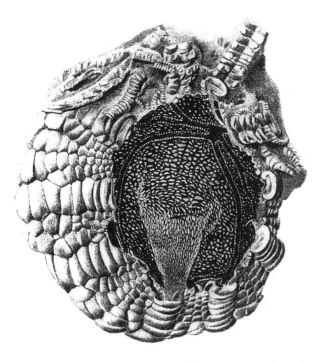

Fig. 228. Close-up of a specimen of *Uintacrinus socialis* with tegmen (note eccentric mouth with food grooves) and central anal cone. Niobrara Chalk, Elkader (Logan County), Kansas. (From Springer 1901.) ×2.

collecting apparatus. The muscular articulations suggest that *Uintacrinus* could move its arms actively to collect food and possibly also to disentangle them from neighboring animals.

UINTACRINUS: A PELAGIC CRINOID?

Most authors dealing with this peculiar crinoid have described it as having floated. Springer (1901) explained the excellent preservation of the fossils by 'the presence of the soft, muddy bottom of a quiet Mediterranean sea, or lagoon. Those crinoids that were at the time at the lowest part of the floating mass rested directly upon the soft mud, and settled into it, in the position in which they happened to be. . . . The others piled on top of them, and not having any such soft or plastic bed to receive and preserve them, were crushed out of shape, disarticulated, and their cup plates and brachials were indiscriminately mixed up.' Bather (1896, Pl. 55) figured *Uintacrinus socialis* 'as swimming, five alternate arms raised, and five in the act of depression.' Abel (1927) assumed that *Uintacrinus socialis* had a pelagic lifestyle on the high seas and was swept into the shallow Niobrara Sea, becoming stranded near the shore. In view of the excellent preservation of the Niobrara specimens, Kirk (1911) suggested that the large swarms of *Uintacrinus* aggregated in the shallow waters near the shore for spawning. Hyman (1955) thought that these crinoids 'lived in floating swarms, in which the long arms were inextricably intermingled'; she illustrated a specimen with the cup uppermost, the long arms hanging down.

As discussed by Struve (1957), a few authors assumed that *Uintacrinus* lived on the bottom, and one even considered that this crinoid could creep starfish-like along the bottom, oral face downward. Others thought it could have used its arms for locomotion, similar to today's comatulids (Milsom *et al.*, 1994). Struve favoured a life on the firm bottom for *Uintacrinus*. He rejected a life in dense colonies and burial at the place of living. He thought that each animal must have occupied a space corresponding to its diameter with outstretched arms, that is, 2 m across. For Jaekel (1918) the skeleton was too heavy for a free-swimming lifestyle, and he considered that the animals lived in groups on the bottom, with outstretched arms.

It is interesting to compare here the two supposedly pelagic crinoids, *Uintacrinus* and *Saccocoma* (see Chapter 26). Both crinoids have a wide geographic distribution, which supports the notion of a pelagic life for these animals. *Uintacrinus socialis*, with its exceedingly long arms, is very different from the tiny *Saccocoma*. The very long, pinnulate and flexible arms of *Uintacrinus*, with their muscular and syzygial articulations, are not very different from those of benthic crinoids, and contrast with the highly specialized and branched arms of *Saccocoma*. The conspicuous, central anal cone and the eccentric mouth are reminiscent of modern comatulids, such as *Comatula*, a fact already recognized by Springer. One gets the impression that the large *Uintacrinus* could have had a pelagic lifestyle only by virtue of a buoyant calyx with entrapped gas. Breimer and Lane (1978) have suggested that *Uintacrinus* might have modified the body cavity by developing special oil or gas compartments in its expanded, thin-walled calyx, but such structures have not been found. In the case of a pelagic, floating lifestyle, the buoyant calyx would have been uppermost, with the arms more or less hanging down (as figured by Hyman and contrary to the restoration of Bather), similar to a jellyfish, but unique for a crinoid.

Following this reconstruction one would expect stranded communities to have touched the bottom with their arms first, followed by the calyx. This view is not supported by the position of the fossils, with the calyx normally standing out in some relief from the lower surface of the slabs. The arms usually lie sideways and extend into the slabs. Milsom *et al.* (1994) proposed that *Uintacrinus* lay on the sea floor with the calyx embedded in the sediment. In their reconstruction, the proximal sections of the arms were on the sediment surface, stabilizing the crinoid, and the distal ends curved upward to form a feeding bowl. This view is not supported by the high density and the taphonomy of these crinoids with their arms stretched out to one side (Fig. 226). Moreover, a collecting bowl as proposed by these authors would appear to have been of limited use in the soupy substrate, which would also have been unsuitable for moving across the bottom. In our view, *Uintacrinus socialis* lived gregariously on the soft bottom and was buried in life position; its colonies may have resembled dense patches of tall eel grass.

Undoubtedly, the bulbous, hollow calyx prevented *Uintacrinus* from sinking into the soupy ground of the chalk, at the same time anchoring the animal. As pointed out by Jablonski and Bottjer (1983), the inoceramids, dominant clams of the Niobrara Chalk, adopted two strategies for living on the soupy substrate. The huge, flattened, thin-shelled *Inoceramus (Platyceramus) platinus* (reaching the incredible length of 2 m) floated snowshoe-like on the soft muds; and *Volviceramus grandis* (reaching 30 cm in size) and similar forms chose an iceberg strategy, with their highly inflated bowl-shaped left valve buried in the bottom and their flattened right valve exposed. *Uintacrinus*, with its large, bowl-shaped calyx, was well suited to an iceberg strategy. It is true that inoceramids and *Uintacrinus* do not occur together, but the large size and numbers of the inoceramids and of *Uintacrinus* indicate excellent conditions for benthic animals that could adapt to the muddy chalk. The rich supply of nannoplankton in the water and making up the chalk helped *Uintacrinus* to reach such a large size. Current-oriented specimens are uncommon, indicating weak currents before burial. Death may have been caused by events similar to those that killed other dense colonies, discussed in Chapter 25 (planktonic blooms, with corresponding poisoning, even though there are no indications of anoxic conditions in the *Uintacrinus* beds; perhaps also a sharply increased rate of sedimentation, leading to suffocation). The example of the Jurassic co-

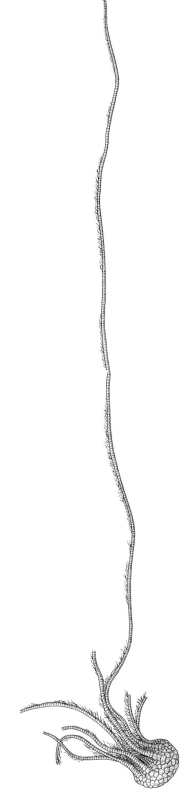

Fig. 229. Arm length of *Uintacrinus socialis* in relation to cup size. (Redrawn from Springer 1901.) ×0.2.

matulid *Paracomatula helvetica* (see Chapter 25) demonstrates that unattached forms could easily have formed very dense aggregations. This gregarious form also has unusually long arms, and the position of the individuals (which lie mostly on their sides) is comparable to the present one.

The related *Marsupites testudinarius* from the English Chalk has a cup of similar size, but with fewer and larger plates, the arms being shorter. This species of worldwide distribution (Sieverts 1927) has also been assumed to be probably pelagic (Breimer & Lane 1978), but we consider it a benthic form, too, populating soft bottoms. As pointed out by Milsom et al. (1994), who came to the same conclusion, the widespread distribution of *Marsupites* and *Uintacrinus* may reflect a planktonic juvenile stage of unusual duration among crinoids.

IMPORTANT COLLECTIONS IN THE UNITED STATES

Museums with Slabs on Display

Denver Museum of Natural History, Denver, Colorado
Museum of Comparative Zoology, Harvard University, Cambridge, Massachusetts
Kansas Museum of Natural History, Lawrence
Sternberg Museum at Ft. Hayes State Museum, Hayes, Kansas
Fick Fossil Museum, Oakley, Kansas

Significant Collections

National Museum of Natural History, Smithsonian Institution, Washington, D.C.
Yale Peabody Museum, New Haven, Connecticut

28

Tertiary

HANS HESS

THE POOR RECORD OF INTACT TERTIARY CRINOIDS

As a result of the distribution of sea and land during the Tertiary, which was similar to that of today, most marine sediments were deposited in shallow water close to present coastlines. Therefore, there are only a few chances to collect from deposits in which crinoids might be well preserved. The predominance among echinoderms of echinoids, and the extreme rarity of well-preserved crinoids in such sediments, is therefore no surprise. The presence of isocrinids (*Metacrinus*) in shallow-water sediments of Palaeocene to Oligocene age in New Zealand and the Antarctic Peninsula, as described later in this section, reveal the shift of these stalked crinoids to a deeper environment during the later Palaeogene (Stilwell et al. 1994). It must be assumed that, starting from a low at the Cretaceous–Tertiary boundary, the number of crinoid species has steadily increased during the Tertiary to arrive at the present diversity (see Fig. 3). The poor fossil record of Tertiary crinoids is, therefore, most probably due to non-preservation rather than to a lack of species. In fact, a remarkable number of crinoid remains have been found in the Danian (lowermost Palaeocene), and these include species of *Cainocrinus, Calamocrinus, Isselicrinus, Nielsenicrinus* and *Bourgueticrinus* and also comatulids (Rasmussen 1972).

Tertiary crinoids are commonly known only as stem fragments and cups; these remains belong to bourgueticrinids (*Bourgueticrinus, Bathycrinus, Conocrinus, Democrinus*) and isocrinids (*Cainocrinus, Isselicrinus, Nielsenicrinus*). In addition, dorsal cups (centrodorsals with radials) of comatulids are found. Complete specimens are very rare. Well-known examples are crowns of *Isselicrinus subbasaltiformis* from the Lower Eocene of Denmark and England. In Denmark such crowns are known from the Rösnäs Formation, a red clay with pyritized fossils (Rasmussen 1972), and in southeastern England they occur in the contemporaneous London Clay, deposited in deeper waters on the continental shelf. These crinoids are associated with commonly occurring driftwood from tropical trees, such as palms. In contrast to the Lower Jurassic assemblages (Chapters 22 and 23), the wood appears to have been colonized only after reaching the bottom (Wignall & Simms 1990). Numerous remains of the comatulid *Amphorometra inornata*, found within 1 m of a fossil log in the London Clay, indicate that the animals were attached to the log with their strong, curved cirri while feeding (Paul 1992).

In the New World, the Keasey Formation (probably Lower Oligocene) of northwestern Oregon has furnished complete crowns of *Raymondicrinus oregonensis* (Moore & Vokes 1953). From the Miocene of Japan some rather well-preserved specimens of the recent species *Teliocrinus springeri* (now confined to the Indian Ocean) have been described by Oji (1990). These were found near the base of turbidite beds and appear to have lived in

Hans Hess, William I. Ausich, Carlton E. Brett, and Michael J. Simms, eds., *Fossil Crinoids*. © 1999 Cambridge University Press. All rights reserved. Printed in the United States of America.

an offshore environment similar to that of extant isocrinids, their preservation resulting from rapid burial by turbidity flow (Fig. 230).

The Eocene, shallow-water La Meseta Formation of Seymour Island, Antarctic Peninsula, has furnished numerous well-preserved specimens of the isocrinid *Metacrinus fossils* and some specimens of the comatulid *Notocrinus rasmusseni* (Meyer & Oji 1993). More recently, Baumiller and Gazdzicki (1996) described a new isocrinid, *Eometacrinus australis* (Fig. 231), as well as the comatulid *Notocrinus seymourensis* and the cyrtocrinid *Cyathidium holopus* from the lower units of this formation. *Eometacrinus australis* is characterized by five primibrachials, with synarthrial articulation between IBr_{1+2} and a muscular $IIBr_{1+2}$ articulation. The extant genera *Metacrinus* and *Saracrinus* also have more than two primibrachials and a muscular articulation between $IIBr_{1+2}$, but IBr_{1+2} is syzygial. Surface water temperatures in this region during the Late Eocene were estimated to be 10–15°C and were thus well within the range of temperatures in the habitat of modern *Metacrinus* species in Japan. The separation of Australia from Antarctica at the end of the Eocene and the opening of the Drake

Fig. 231. Crown of *Eometacrinus australis*. Eocene La Meseta Formation, Seymour Island, Antarctic Peninsula. (Courtesy A. Gazdzicki; also figured by Baumiller & Gazdzicki 1996.) ×1.

Fig. 230. *Teliocrinus springeri*. Proximal stem with parts of cirri and proximal part of crown. Morozaki Group (Miocene), Utsumi, Japan (figured by Oji 1990). Note the relatively large basals and the cryptosyzygial articulation between IBr_1 and IBr_2. IBr_2 and $IIBr_2$ are axillary. (Courtesy T. Oji.) ×2.4.

Passage in the Oligocene, enabling the development of the cold circum–Antarctic current, resulted in a sharp decline of sea temperature that made life for *Metacrinus* and other isocrinids in these waters impossible.

THE OREGON SEA LILIES

For Tertiary sediments, the Oregon crinoids are quite exceptional and merit a brief description, which is based on the detailed work of Moore and Vokes (1953). The crinoids were discovered in a relatively unstratified to massive tuffaceous siltstone that is associated with some hard calcareous beds and a few layers of ashy tuff. The calcite of the ossicles was generally dissolved to some extent, and the fossils preserved partly as moulds. The difficulties of preparing and exposing the specimens from

Fig. 232. *Raymondicrinus oregonensis*. Three crowns with attached proximal stems. Note three regenerated arms on specimen in the middle. Keasey Formation (probably Lower Oligocene), Mist, Columbia County, Oregon. (Hess Collection; photograph S. Dahint). ×0.85.

the hard matrix have been vividly described by the authors; and some good specimens were obtained by dissolving the calcite of the crinoids in acid, which resulted in sharp external moulds. Material recently collected at the Mist outcrop on the Nehalem River includes perfectly preserved specimens (Fig. 232). From the presence of ash pebbles and well-preserved land plants, it was concluded that the tuffaceous materials were produced by volcanic activity and thus derived from land not far away behind a shoreline. The associated invertebrate fauna consists mostly of molluscs that appear to have lived and to have been buried in relatively deep water. Associated with crinoids were a nearly complete ophiuroid with its spines, and also well-preserved leaves of an oak and a bayberry species, suggesting proximity of land.

The Oregon crinoid differs from most other isocrinids (including *Isocrinus*) in having a muscular articulation between $IIBr_1$ and $IIBr_2$, which led Klikushin (1982) to propose for this species the genus *Raymondicrinus*. Such an articulation occurs also in the Recent genera *Metacrinus* and *Saracrinus*, but in these the articulation between IBr_1 and IBr_2 is syzygial (synarthrial in the Oregon specimens), the basals are large and the primibrachials number more than two (there are generally seven primibrachials in *Metacrinus* and four in *Saracrinus*). Syzygies occur sporadically, the first one normally between $IIBr_3$ and $IIBr_4$. The rays of *Raymondicrinus oregonensis* bifurcate isotomously three times, making well-preserved crowns of this large crinoid objects of great beauty. The stem was rather long, and the nodals with their small cirrus sockets are hardly different from the six to seven internodals; the cirri have a terminal claw (Fig. 232). A second species, *Isocrinus nehalemensis*, was described by Moore and Vokes (1953) from the same area, but this may be just the juvenile form of *R. oregonensis*.

The lack of disarticulation of the crinoids, as well as the presence of a well-preserved ophiuroid on crinoid arm fragments and of weakly hinged bivalves with their valves still associated, suggests rapid burial in deep water below wave base, possibly as a result of volcanic activity on the nearby land. Thick sediment layers must have accumulated rather rapidly to prevent action by scavengers. The crinoids and the other members of the fauna appear to have lived in a zone marked by an upwelling of cool waters, such as occur along the present California coast. Therefore, their mode of life is comparable to that of existing stalked crinoids.

PLEISTOCENE CRINOIDS

A moderately rich crinoid locality in the Early Pleistocene of eastern Jamaica includes three isocrinids and a bourgueticrinid; they all belong to extant taxa and suggest a minimum depositional depth of about 180 m for these strata (Donovan 1994).

29

Recent

HANS HESS

About 600 extant species of crinoids are known, mostly free-moving feather stars living in shallow tropical seas and described in Clark's monumental monograph (Clark 1915–1950; Clark & Clark 1967). Comatulids exhibit an extraordinary morphological plasticity that generates major taxonomic problems and suggests that their diversification continues (Messing 1997). The approximately 80 species of stalked sea lilies are restricted to deeper water. All living forms belong to the Articulata.

DOMINANCE OF COMATULIDS OR FEATHER STARS IN SHALLOW-WATER REEF ASSEMBLAGES

The most diverse crinoid assemblages occur in the extensive coral reefs of the tropical Indo-Pacific (Meyer & Macurda 1980), whereas the West Indian faunas are not nearly as rich. The largest number of species at a single locality was reported from Lizard Island on the Great Barrier Reef with a total of more than 30 shallow-water comatulids (Macurda & Meyer 1983). As Meyer & Macurda (1977) pointed out, this cannot match some fossil assemblages, like the one from the famous Lower Mississippian beds of Crawfordsville, Indiana, where more than 60 species lived in a non-reefal environment (see Chapter 18). On the north coast of Papua New Guinea, densities and richness reach 115 specimens and 12 species per square metre (Messing 1997). In the western Caribbean coastal area, Macurda and Meyer (1977) reported densities exceeding 20 individuals per square metre for comatulids, living at depths of 1–3 m. The highest densities are observed in tidal passes, whereas reef areas are characterized by populations with a high species diversity rather than high densities of one species. In any case, the densities reported for living comatulids are a far cry from that of the fossil *Paracomatula helvetica* with up to 300 individuals per square metre preserved in a single lens (see Chapter 25). Recent forms may also be useful for recognizing variability within one species. In favourable environments of the Lesser Antilles, one species of *Nemaster* was reported with 40 arms up to 30 cm long, but in the less favourable Bahamas it had only 20 arms with a maximum length of 20 cm. The implications of such differences in a single species should be kept in mind when fossil communities are evaluated. A good example of variability within a species is the isocrinid *Chariocrinus andreae* from the Middle Jurassic of Switzerland. This species has a variable number of brachials and large size differences from one site to another, presumably the result of differences in food supply.

Comatulids occupy different niches within the reefs. Some flourish in areas of high current flow; these climb to prominent locations such as coral heads or sponges (Figs. 233, 234). Others avoid stronger current and hide their disc in crevices while extending their arms into spaces among corals. There is also a strong bathymetric preference between 1 and 70 m. On coral heads or other sites exposed to currents, crinoids may aggregate in small

Hans Hess, William I. Ausich, Carlton E. Brett, and Michael J. Simms, eds., *Fossil Crinoids*. © 1999 Cambridge University Press. All rights reserved. Printed in the United States of America.

Fig. 233. *Oxycomanthus bennetti*, attached to a sponge, with a trumpet fish hovering among the arms. The fish is not preying on the crinoid but rather is using the crinoid for shelter, waiting to ambush small fish that approach within range (D. L. Meyer, pers. comm., 1996). Off Manado, northern tip of Sulawesi, depth around 20 m. (Photograph O. C. Honegger.) To view this figure in colour, see the colour plate section following page xv.

clusters. Some multi-armed comatulids with as many as 100 arms lack cirri altogether and cling to the substrate by means of some of the arms (Meyer & Macurda 1980). Arms are also deployed as different types of fans for feeding, as beautifully illustrated by Meyer and Macurda (1980) among crinoids of the Palau and Guam reefs. The majority of these crinoids are active during both the day and night, but crinoids living in very shallow water tend to be active at night only. In a classic paper, Magnus (1963) described the behavior of *Heterometra savignii* in the Red Sea, where the animals hide during the day with enrolled arms (possibly to avoid predation by fishes) and appear in large numbers at night for feed-

ing on the plankton rising up from deeper water. The arms are extended into a filtration fan, mouth downcurrent. This species avoids soft bottoms, preferring eel grass or rocky outcrops, including crevices in piers.

Comatulids rarely live on soft bottoms; those that do have long, slender cirri that prevent the animals from sinking into the mud. Live observations of dense aggregations of deep-sea comatulids on soft bottoms, similar to those of the fossil *Paracomatula helvetica* (see Chapter 25), have not been made so far. Together with some stalked forms, comatulids are surprisingly common in colder northern and Antarctic waters. This is particularly true of members of the large family Antedonidae, which are widely distributed from the seaweed-rich littoral zone of western Europe and the Mediterranean to the deep, cold waters of the North Atlantic.

STALKED ISOCRINID SEA LILIES

The shallowest recorded locations of living stalked crinoids are from about 100 m in the western Pacific and about 170 m in the western Atlantic. Their decline and eventual disappearance from shallow water in today's oceans were explained by increased predation by fishes since the Late Mesozoic (Meyer & Macurda 1977; Meyer 1985). This hypothesis was substantiated by Oji (1996), who counted significantly more regenerated arms, the result of sublethal predation, in specimens of Caribbean *Endoxocrinus parrae* from shallower water than in those from deeper water. This is in line with an extremely low frequency of regenerated arms in fossil isocrinids, in contrast to the usually high frequency of regenerated arms in Recent isocrinids.

Because stalked crinoids are out of reach of scuba divers, very little has been known until recently about their life in their natural habitat. Their preference for hardgrounds such as rocks and mounds makes them relatively inaccessible to trawling. We are fortunate today in having an increasing number of photographs of these picturesque animals taken from manned submersibles or by unmanned cameras lowered from research vessels (Figs. 1, 235, 236).

Macurda and Meyer (1974) were among the first to figure stalked isocrinids, living at depths of 200–300 m. The most prominent of these crinoids was *Cenocrinus asterius* with its nearly 50 arms extended into a parabolic filtration fan. The stem, almost 1 m long, was anchored to the loose substrate, with four or five whorls of cirri from the distal part of the stem running along the bottom. These pictures provided the basis for the new reconstructions of fossil isocrinids. The two individuals figured in Fig. 235 are anchored to a hard substrate, very similar to the specimen of *Endoxocrinus parrae* shown in Fig. 236.

Today, stalked crinoids prefer areas of moderate currents. Populations of crinoids are much denser on topographic elevations exposed to increased current flow. This is demonstrated by a recent survey undertaken in the Straits of Florida, where crinoids were abundant along the flanks of lithoherms and surrounding hardgrounds at 500–700 m depth. However, the current-swept crests and upcurrent ends of the mounds were the domain of scleractinian corals and zoanthids (Messing *et al.* 1990). The most abundant crinoids on the flanks were the isocrinids. *Diplocrinus* (?) *maclareanus* with densities of 15–20 individuals per square metre and the large *Endoxocrinus parrae*. They were attached to the hard bottom with whorls of cirri from the distal part of the stem, and the crowns formed the typical parabolic filtration fan, oral side downcurrent (Fig. 236). The bourgueticrinid *Democrinus* was also quite common in this area, and the tiny, paedomorphic comatulid *Comatilia iridometriformis* reached densities of 65 individuals per square metre. Intermound areas were populated by the same crinoids as well as by another isocrinid, *Isocrinus blakei*. Crinoids were attached to hardgrounds, but where the sediment bottom was unconsolidated they clung to low outcrops or small pieces of rubble and shell. Unconsolidated sediments without possibilities for attachment were virtually barren. In somewhat shallower water the greatest local concentrations of isocrinids occur on isolated boulders or larger mounds with a vertical relief of up to 30 m.

Fig. 234. *Oxycomanthus bennetti* showing cirri attached to a sponge; the terminal claws of several cirri are embedded in the surface of the sponge. Some of the cirri that are down have epibionts attached to the distal ends; such cirri might be raised for periodic cleaning by the comb-bearing oral pinnules (see also Fig. 6) (D. L. Meyer, pers. comm., 1996). Note ophiuroid hiding under the cirri. Off Manado, northern tip of Sulawesi at 20 m. (Photograph O. C. Honegger.) To view this figure in colour, see the colour plate section following page xv.

Fig. 235. Two individuals of the living insocrinid *Cenocrinus asterius* with 40–50 arms, attached to hardground. Crowns are in the wilted-flower posture (Baumiller *et al.* 1991) with the oral side directed upward, indicative of a slack current. Hogsty Reef, Bahamas at 310 m. (Courtesy Harbor Branch Oceanographic Institution, Inc., through D. L. Pawson.) To view this figure in colour, see the colour plate section following page xv.

On the continental slope of the Bay of Biscay, at a depth of 1,246 m, a rather high population density of the small stalked isocrinid *Annacrinus wyvillethomsoni* was observed during an unmanned underwater survey (Conan *et al.* 1981). In patches, these aggregations reached 8 or 10 specimens per square metre on a sea floor strewn with pebbles and rocks. The distal group of cirri grasped pebbles or the upper edges of rocks, and again the crown was deployed as a parabolic filtration fan, with the arms strongly recurved into a current with a velocity of approximately 20 cm per second.

These observations may indicate that the fossil soft-bottom communities of isocrinids with their extremely high densities (such as *Chariocrinus andreae*; see Chapter 25) have no Recent analogue.

SMALL STALKED CRINOIDS WITHOUT CIRRI AT MODERATE TO ABYSSAL DEPTHS

Two groups of small, stalked crinoids with normally unbranched arms have been dredged from deeper waters or have been observed during deep-sea explorations. Both groups originated in the Late Cretaceous and occur throughout the Tertiary. Despite their small size, they can usually be identified to the level of family or even genus in underwater photographs (Macurda & Meyer 1976a). These small forms deploy their crowns as filtration fans, similar to the more conspicuous isocrinids, bending the stem in its proximal part near the cup (Roux 1980).

The first group, the family Hyocrinidae (order Miller-

Fig. 236. Living isocrinid *Endoxocrinus parrae* on coral. Crown with 32 arms in oral view. In this parabolic filtration fan posture (Macurda & Meyer 1974), the stem is bent proximally at a right angle (behind the crown), the oral side of the crown is directed downcurrent and the distal part of the arms recurve aborally into the current. San Salvador, Bahamas, at 692 m. (Courtesy Harbor Branch Oceanographic Institution, Inc., through D. L. Pawson.) To view this figure in colour, see the colour plate section following page xv.

icrinida, suborder Hyocrinina), includes crinoids with a simple column of disc-shaped columnals having symplectial articulations with radiating ridges (commonly known from other Millericrinida). The column lacks cirri and is fixed by a disc to the substrate. The cup is composed of thin plates, and the five arms are narrow with respect to the radials but are densely pinnulate. They range from approximately 700 to nearly 5,000 m, but occur mainly at depths below 2,000 m.

The second group includes crinoids with compact cups and high columnals united by synarthries (Fig. 16). They belong to the Bourgueticrinida. Most forms have branching, dendroid roots for life on soft substrates. The best-known species are the Bourgueticrinidae, which are the most common crinoids encountered in photographs of the bottom. They have been observed at depths of more than 6,000 m (Macurda & Meyer 1976b). The same authors (1976a) figured a group of 20 individuals of *Democrinus* on a muddy skeletal sand bottom in the Straits of Florida at 600–700 m. The Phrynocrinidae and Porphyrocrinidae are two rare families that, along with the Bourgueticrinidae, have synarthrial articulations in the stem and are attached by a terminal disc. They occur in depths ranging from 400 to almost 4,000 m.

CYRTOCRINIDS: LIVING FOSSILS

The stemless, barnacle-like *Cyathidium* was thought to be extinct by Late Cretaceous times, but in 1967 and

1971 specimens of the extant species *Cyathidium foresti* were found by German (Fechter 1973) and French expeditions (Cherbonnier & Guille 1972) in the mid-Atlantic (see Heinzeller & Fechter 1995 for a comprehensive study). The animals were attached to boulders at depths of 380–900 m. In 1994 a new species, *Cyathidium plantei*, was discovered by H. Fricke during dives with the submersible *Jago* on the submarine slopes off Grande Comore at a depth of 200 m (Heinzeller *et al.* 1996). The almost black specimens were cemented to cave ceilings by their cup. *Cyathidium foresti* and *C plantei* are very similar to *Cyathidium depressum* from the Cenomanian, the presumed ancestor of these living fossils. *Cyathidium* hardly resembles a crinoid at first sight (Fig. 237). The cup is composed of the fused radials; the large radial cavity may be tightly sealed by an outer circlet of primibrachials and an inner circlet of first secundibrachials, completely hiding the small curled arms. Heinzeller and Fechter (1995) postulated that these animals had the possibility of raptorial feeding, in addition to the usual suspension feeding.

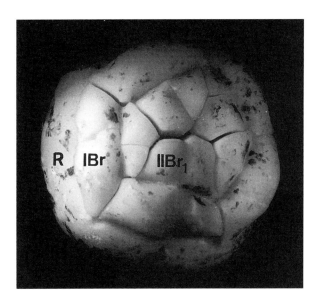

Fig. 237. Oral view of *Cyathidium foresti* (formerly *C. meteorensis*). Tegmen and curled-up arms are hidden beneath an outer circle of 5 axillary primibrachials (IBr) and an inner circle of 10 somewhat irregular first secundibrachials (IIBr$_1$). The base is formed by the fused radials (R). The originally black specimen was dredged in 1967 off the Great Meteor Bank from a depth of about 650 m and is preserved in the Zoologische Staatssammlung, Munich. (From Heinzeller & Fechter 1995, with permission.) ×3.5.

The likewise stemless *Holopus rangii* occurs commonly at depths of 300–400 m in the Caribbean, cemented to rock walls or even to roofs of submarine caves (Grimmer & Holland 1990). It has long been thought that this peculiar crinoid was restricted to the Caribbean, but another species, *Holopus alidis*, has been discovered by a French oceanographic expedition near New Caledonia (Bourseau *et al.* 1991) in depths of approximately 400 m (Fig. 31). The same expedition also made the sensational discovery of a living representative of the Jurassic *Gymnocrinus* (Figs. 32, 238). The two cyrtocrinids, *Holopus alidis* and *Gymnocrinus richeri*, appear to be quite numerous on seamounts at depths between 300 and 500 m. As discussed in Chapter 1, the discovery of these extant forms confirms that most cyrtocrinids, living and fossil, preferred hard substrates with low sedimentation rates and bottom waters well ventilated. Similar cyrtocrinids, with a short stem and a crown with arms that could be enrolled fist-like or enclosed within well-developed radials, were widely distributed in the Jurassic of Europe (Figs. 33, 34). Many individuals of the Recent *Gymnocrinus richeri* were in the course of arm regeneration. Signs of regenerations and pathological changes, possibly the result of predation, are also common in fossil cyrtocrinids. The discovery of a living *Gymnocrinus* confirms the brilliant reconstructions of these animals by Jaekel (1918).

Gymnocrinus richeri was accompanied by *Guillecrinus neocaledonicus*, a curious form with 5 weakly pinnulate arms and brachials united by syzygies and muscular articulations. The disc and the food grooves are gelatinous and not covered by ossicles. The stem of *Guillecrinus*, with a preserved length of nearly 40 cm, has cylindrical columnals with radiating ridges that are reduced, in the central part of the column, to 5 broad, crenulate ridges, separated by deep ligamentary pits. The animal was attached by means of a terminal disc. The relationships of this and a second species, *Guillecrinus reunionensis*, found near Réunion, are uncertain. The fauna off New Caledonia also furnished *Proisocrinus ruberrimus*, a large stalked crinoid with 20 weakly pinnulate arms, widely distributed from the Philippines to New Caledonia at depths of 800–1,900 m. This form has a proximal stem with isocrinid characters, such as pentalobate columnals and nodals with small cirri (but without cryptosymplexy on the distal face), but the median and distal parts of the stem have columnals reminiscent of millericrinids. The mode of attachment for *Proisocrinus* is unfortunately unknown. Rasmussen (1978) proposed a family of its

Fig. 238. *Gymnocrinus richeri*. Oral view of a complete specimen with two short columnals and outstretched arms. This individual was dredged from the top of a seamount off New Caledonia at about 450 m. Closely fitting extensions of the first primibrachials (which may be axillary and then carry the secundibrachials) seal the oral side of the cup with the visceral mass. In this living cyrtocrinid, the first primibrachials lack the ring commissure typical of radials (T. Heinzeller, pers. comm., 1995) and thus appear to be in the position that radials occupied in the similar fossil forms. In the extant species, radials are rudimentary or fused with the proximal columnal. Note the asymmetry of the cup, leading to pronounced asymmetry of the arms. Only a few of the pinnules at the lower arms are extended. In view of its weakly developed muscles, this crinoid was probably not adapted to raptorial feeding (Heinzeller et al. 1994). (Courtesy J.-P. Bourseau; from Bourseau et al. 1991.) ×1.

own for this crinoid, the Proisocrinidae, which he assigned to the Isocrinida. Simms (1988a) noted a pattern of stem development reminiscent of Encrinidae, but Bourseau et al. (1991) considered it to be the sole survivor of the fossil family Millericrinidae. It certainly appears that the rudimentary cirri were not functional in the sense of an isocrinid stem. Still another curious species collected off New Caledonia is the 10-armed *Caledonicrinus vaubani* with radials forming a prominent circlet. The stem has rounded columnals lacking cirri and is attached by a disc to the bottom. Given its neuroanatomical features, *Caledonicrinus* may form a group of its own (Heinzeller et al. 1995). The western Pacific and especially the New Caledonian crinoids thus appear to represent the most archaic crinoid fauna in Recent oceans, being closely related to the fossil fauna of the Mesozoic Tethys Ocean.

EPILOGUE: PRESENT AND PAST

The success of the comatulids in today's seas may be attributed to their ability to move around freely; to some extent this is also true of the isocrinids, which, after becoming detached, can anchor themselves again with the terminal cirri. The small stalked forms belonging to the Bourgueticrinida and the Cyrtocrinida either are firmly attached by a disc to a hard substrate or are anchored with branching roots in a muddy bottom. These mostly deep-sea and abyssal forms appear to be remnants of a host of fossil crinoids, using the same mechanisms to fix them in the currents.

The present-day crinoid faunas are rich in shallow-water tropical seas, but are also prominent at moderate depths on continental shelves, which appear to offer a number of good sites for the stalked isocrinids. Most of today's crinoids prefer hard substrates, a lifestyle that was not as common in many ancient seas. Prodeltaic environments, with the sites richest in fossil species diversity, and soft bottoms at the base of sand waves or tidal deltas, with the highest fossil densities, seem to lack crinoids in today's oceans. Some modern crinoids are able to swim, presumably to find better feeding sites. However, planktonic or pelagic crinoids, so apparent in some pre-Tertiary strata, are also absent from Recent seas.

Appendix I
Geological Time Table with Crinoid Assemblages

HANS HESS AND WILLIAM I. AUSICH

Age (million years)		System		Series/Stage (selection)	
Cenozoic	23	Quaternary			Pleistocene
		Tertiary	Neogene	28	Pliocene / Miocene
			Palaeogene		Oligocene / Eocene / Paleocene
	65				✷ Danian
Mesozoic		Cretaceous		27	✷ Maastrichtian / Santonian
					✷ Valanginian
					✷ Berriasian
	146	Jurassic		26	✷ Tithonian
					✷ Oxfordian
				24	✷ Bathonian
				25	✷ Bajocian
				23	Toarcian
					✷ Pliensbachian
				22	✷ Sinemurian
	208	Triassic			✷ Norian
				21	✷ Carnian / Ladinian
	245				✷ Anisian

Note: Asterisks refer to the overview in Chapter 3, numbers to chapters in which assemblages are discussed.

Hans Hess, William I. Ausich, Carlton E. Brett, and Michael J. Simms, eds., *Fossil Crinoids.* © 1999 Cambridge University Press.
All rights reserved. Printed in the United States of America.

Age (million years)	System		Series/Stage (selection)	
245		Permian	20	Guadalupian / Artinskian / Sakmarian
290	Carboniferous	Pennsylvanian	19	Stephanian/ Missourian
323	Carboniferous	Mississippian	17,18 / 16	✲ Chesterian / ✲ Meramecian / ✲ Osagean/Visean / ✲ Kinderhookian/ Tournaisian
363		Devonian	14,15 / 13 / 12	Fammenian / Frasnian / ✲ Givetian / ✲ Eifelian / Emsian / Siegenian / Lochkovian
409		Silurian	11 / 9,10	Pridolian / ✲ Ludlovian / ✲ Wenlockian / ✲ Llandoverian
439		Ordovician	8 / 6,7	✲ Ashgillian / Cincinnatian / ✲ Mohawkian/ Caradocian / Tremadocian
510		Cambrian		

Palaeozoic

Appendix II
Glossary of Rocks

HANS HESS

Sediments can be classified according to *chemistry* [e.g., silicates (clay minerals, quartz), carbonates], *grain size* (clay, silt, sand, pebbles, boulders) or *components* (mud content, fossil fragments, ooids).

Clastic ('fragmented') sediments are derived from the erosion of rocks.

SILICICLASTIC SEDIMENTS

These sediments are erosional products of igneous and metamorphic rocks in the first instance. They are composed of quartz or clay minerals. They are classified according to grain size:

Clay (< 0.002 mm or 2 μm) → hardened into *claystone*. Fissile claystones are called *shales*; fissility results from horizontal alignment of the clay minerals during deposition, but mainly during compaction (example: Posidonia Shale).
Silt (4–63 μm) → hardened into *siltstone*.
Sand (0.063–2 mm) → hardened into *sandstone*, frequently with cement of calcium carbonate.
Pebbles (2–64 mm).
Boulders (>64 mm).

Some of these sediment terms have been incorporated into modern carbonate rock nomenclature.

Clay and silt are the components of mud; siltstones, claystones and shales are therefore varieties of *mudstones*, also called argillaceous rocks. Most sedimentary grains of silt and sand resulting from the weathering of terrigenous rocks are composed of quartz. *Slate* is a finer-grained siliciclastic rock in which aligned platy minerals produce fissility like that of shale. The mineral alignment of slate results, however, from deformation pressures through tectonic movements rather than from orientation of the clay particles during sedimentation (example: Hunsrück Slate).

CARBONATE SEDIMENTS

These sediments are composed of calcium carbonate. (Dolomite, a mixture of calcium and magnesium carbonate, is uncommon in crinoid assemblages.) The main carbonate rock is *limestone* with at least 95% calcium carbonate; it is of biological or chemical origin. Most limestones are now composed of calcite, but many were originally made up of the unstable modification aragonite, precipitated as needles in warm, shallow seas. Limestones with fine-grained siliciclastic components are called *marls* (terminology used mainly in Europe). Marls contain up to 75% carbonate or up to 65% clay; according to the percentage of the component such terms as *marly clay* or *marl-clay*, *calcareous marl* and *marly limestone* are commonly used.

Hans Hess, William I. Ausich, Carlton E. Brett, and Michael J. Simms, eds., *Fossil Crinoids*. © 1999 Cambridge University Press. All rights reserved. Printed in the United States of America.

Like the siliciclasts, carbonate sediments can be classified according to the size of the particles. Carbonate mud (<20 μm, also called *micrite*) and carbonate silt (20–63 μm) are cemented into *lime mudstone*. Carbonate sand (63 μm to 2 mm), or *calcarenite*, usually made up of skeletal fragments or of ooid grains, is cemented into *bioclastic* or *oolitic limestone*. Depending on the amount of larger components and therefore of their packing in the microcrystalline (*micritic*) or more coarsely crystalline (*sparitic*) matrix, the following terms have been used: *mudstone* or *micritic limestone* (<10% components); *skeletal limestone* or *packstone*; or *calcarenite* or *grainstone* (in these, the components are in contact with each other). Many limestones are compound materials containing ooids, bioclasts, faecal pellets or oncoids (irregular grains of 2 mm or more, coated by algae) cemented with sparry (crystalline) or microcrystalline (micritic) calcite – hence the terms *oosparite*, *oomicrite*, *biosparite* and *biomicrite*. Oncolitic limestones were usually formed in shallow subtidal or intertidal zones and are therefore not connected with crinoid assemblages.

Lime mud is formed by chemical precipitation, biologically or mechanically by the degradation of larger components or by the accumulation of tiny shells of calcareous plankton (coccoliths) and foraminifera. *Chalk* consists largely of almost ultramicroscopic blades and spines from coccoliths (tests of minute algae). *Biomicritic limestones* have at least 10% recognizable fossil remains. It is commonly difficult to determine the origin of mud-sized particles in micritic limestones (see Chapter 26 on Solnhofen Limestone), but most sand-sized or larger carbonate particles are so-called bioclasts, fragmental particles derived from the breakdown of any sort of calcareous shell or skeleton or even of an earlier limestone. Bioclastic limestones include *crinoidal limestones*, which are made up to a large degree of remains of crinoids, mostly columnals and stem fragments. Oolitic limestones, associated with crinoid assemblages on carbonate platforms, consist of sand-sized particles (ooids) formed in very shallow, agitated tropical sea water by precipitation of calcium carbonate around a shelly or some other nucleus.

Limestones are usually bedded as a result of interbedded layers of clay, silt or marl. These result in 'shale-like' fissility (e.g., the thinner '*Fäule*' beds of the Solnhofen Limestone). The term *calcareous shale* is used here in connection with mudstones that contain carbonates.

Chert nodules may occur around crinoid assemblages; they are the result of chemical precipitation of extremely small quartz crystals from pore water.

In shales, with a calcium carbonate content of less than about 10%, the calcite is normally concentrated as layers of ellipsoidal *calcareous nodules* or concretions; these are of diagenetic origin (precipitated from pore water during compaction).

Interrupted sedimentation (*diastem*) in shallow-water carbonates led to the formation of *hardgrounds*, usually encrusted by oysters. These are preferred settling sites for many crinoids. *Firmgrounds* have a coherent, 'tough' surface and may also be settled by crinoids.

Molasse sediments or *sequences* were formed in foreland basins by the weathering of mountain chains. They include non-marine settings, such as alluvial fans, rivers and floodplains, and can accumulate to great thicknesses. Shales and turbidites, deposited in foreland basins during mountain building, are called *flysch*.

TRANSPORT MECHANISMS OF SEDIMENTS

Sediments may have been transported by gravity or turbidity currents into deeper water, forming *turbidites* (many of these show graded bedding, with the fine material on top), or they may have been deposited through the action of storms in more shallow water and are then called *tempestites*. Both events buried many of the crinoid assemblages described in this book. In certain basinal sediments, such as the Solnhofen Limestone, the mechanism of deposition (turbidite or tempestite) is difficult to decipher.

Winnowed beds are caused by (periodic) storms or bottom currents that selectively erode fine sediments on the sea floor well below the normal wave base. Erosion of the soft mud leaves behind discontinuous, matrix-poor deposits, containing much shelly and other fragmentary material.

Bibliography

HANS HESS

Abel, O. (1927): *Lebensbilder aus der Tierwelt der Vorzeit* (2d ed.). Gustav Fischer. 714 pp.
  (1939): *Vorzeitliche Tierreste im deutschen Mythus, Brauchtum und Volksglauben*. Gustav Fischer. 304 pp.
Aigner, T. (1985): Storm depositional systems: Dynamic stratigraphy in modern and ancient shallow-marine sequences. *Lecture Notes in Earth Sci.* 3, 1–174. Berlin.
Aigner, T. & Bachmann, G. H. (1992): Sequence-stratigraphic framework of the German Triassic. *Sediment. Geol.* 80, 115–135. Amsterdam.
Allison, P. A. & Briggs, D. E. G. (1991): The taphonomy of soft-bodies animals, in: S. K. Donovan (ed.), *The Processes of Fossilization*, 120–140. Belhaven Press.
Amemiya, S. & Oji, T. (1992): Regeneration in sea lilies. *Nature* 357, 546–547.
Anderson, W. I. (1969): Lower Mississippian conodonts from northern Iowa. *J. Paleont.* 43, 916–928.
Archer, A. W. & Feldman, H. R. (1986): Microbioherms of the Waldron Shale (Silurian, Indiana): Implications for organic framework in Silurian reefs of the Great Lakes area. *Palaios* 1, 133–140.
Arendt, Y. A. (1968): Novye dannye o rannemelovykh krinoidejakh Kryma (New information about Early Cretaceous crinoids of the Crimea). *Bjull. Mosk. Obsk. Ispyt. Prir. Otd Geol.* 43, 156–157 (in Russian).
  (1974): *Morskije liliji cirtokrinidy* (The Cyrtocrinid Sea Lilies). Trudy Paleont. Inst. (Akad Nauk SSR) 144, Moscow (in Russian). 251 pp.
  (1976): Ordovikskie iglokozhie Hemistreptocrinoidei (Ordovician echinoderms Hemistreptocrinoidea). Moskovskoe Obshchestvo Ispytatelei Prirody, Byulletin, *Otdel. Geol.* 51, 63–84 (in Russian).
Arendt, Y. A. & Rozhnov, S. V. (1995): Concerning hemistreptocrinids. *Paleont. Zhurnal*, 1995, 1, 119–121.
Ausich, W. I. (1980): A model for niche differentiation in Lower Mississippian crinoid communities. *J. Paleont.* 54, 273–288.
  (1988): Evolutionary convergence and parallelism in crinoid calyx design. *J. Paleont.* 62, 906–916.
  (1996a): Crinoid plate circlet homologies. *J. Paleont.* 70, 955–964.
  (1996b): Phylum Echinodermata, in: R. M. Feldmann (ed.), *Fossils of Ohio*, 242–261. Ohio Geol. Survey Bull. 70.
  (1997): Regional encrinites: A vanished lithofacies, in: C. E. Brett (ed.), *Paleontological Events: Stratigraphic, Ecologic, and Evolutionary Implications*, 509–519. Columbia University Press.
  (1998): Early phylogeny and subclass division of the Crinoidea (phylum Echinodermata). *J. Paleont.* 72, 499–510.
Ausich, W. I. & Babcock, L. E. (1998): The phylogenetic position of *Echmatocrinus brachiatus*, a probable octocoral from the Burgess Shale. *Palaeontology* 41, 193–202.
Ausich, W. I. & Baumiller, T. K. (1993): Taphonomic method for determining muscular articulations in fossil crinoids. *Palaios* 8, 477–484.
Ausich, W. I. & Bottjer, D. J. (1982): Tiering in suspension-feeding communities on soft substrata throughout the Phanerozoic. *Science* 216, 173–174.
  (1985): Echinoderm role in the history of Phanerozoic tiering in suspension-feeding communities, in: B. F. Keegan

& B. D. S. O'Connor (eds.), *Proceedings of the Fifth International Echinoderm Conference, Galway*, 3–11. Balkema.

Ausich, W. I. & Kammer, T. W. (1990): Systematics and phylogeny of the late Osagean and Meramecian crinoids *Platycrinites* and *Eucladocrinus* from the Mississippian stratotype region. *J. Paleont.* **64**, 759–778.

Ausich, W. I. Kammer, T. W. & Baumiller, T. K. (1994): Demise of the Middle Paleozoic crinoid fauna: A single extinction event or rapid faunal turnover. *Paleobiology* **20**, 345–361.

Ausich, W. I., Kammer, T. W. & Lane, N. G. (1979): Fossil communities of the Borden (Mississippian) delta in Indiana and northern Kentucky. *J. Paleont.* **53**, 1182–1196.

Ausich, W. I. & Sevastopulo, G. D. (1994): Taphonomy of Lower Carboniferous crinoids from the Hook Head Formation, Co. Wexford, Ireland. *Lethaia* **27**, 245–256.

Austin, T., Sr. & Austin, T., Jr. (1843–1849): *A Monograph on Recent and Fossil Crinoidea*. Bristol. 122 pp.

Bartels, C. (1995): Die unterdevonischen Dachschiefer von Bundenbach, in: W. K. Weidert (ed.), *Klassische Fundstellen der Paläontologie* 3: 38–55. Goldschneck-Verlag.

Bartels, C. & Blind, W. (1995): Röntgenuntersuchung pyritisch vererzter Fossilien aus dem Hunsrückschiefer (Unter-Devon, Rheinisches Schiefergebirge). *Metalla* **2.2**, 79–100. Bochum.

Bartels, C. & Brassel, G. (1990): *Fossilien im Hunsrückschiefer*. Museum Idar-Oberstein 7. 232 pp.

Bartels, C., Briggs, D. E. G. & Brassel, G. (1998): *The fossils of the Hunsrück Slate*. Cambridge University Press. 309 pp.

Barthel, K. W. (1978): *Solnhofen. Ein Blick in die Erdgeschichte*. Ott Verlag, Thun. 393 pp.

Barthel, K. W., Swinburne, N. H. M. & Conway Morris, S. (1990): *Solnhofen: A study in Mesozoic paleontology*. Cambridge University Press. 236 pp.

Bassett, M. G., Bluck, B. J., Cave, R., Holland, C. H. & Lawson, J. D. (1992): Silurian, in: J. C. W. Cope, J. K. Ingham & P. F. Rawson (eds.), *Atlas of Palaeogeography and Lithofacies*. *Geol. Soc. Mem.* **13**, 37–56.

Bassler, R. S. & Moodey, M. W. (1943): Bibliographic and Faunal Index of Paleozoic Pelmatozoan Echinoderms. *Geol. Soc. Amer. Spec. Pap.* **45**. 734 pp.

Bather, F. A. (1893): The Crinoidea of Gotland. Pt. 1. The Crinoidea Inadunata. *K. Svensk. Vetenskaps-Akad. Handl.* **25**, 1–200.

(1896): On *Uintacrinus*: A morphological study. *Proc. Zool. Soc. London* **1895**, 974–1004.

(1900): The Echinodermata, the Pelmatozoa, in: E. R. Lankester (ed.), *A Treatise on Zoology*. Pt. 3. The Crinoidea, 94–204. Adam & Charles Black.

(1907): The discovery in West Cornwall of a Silurian crinoid characteristic of Bohemia. *Trans. Royal Geol. Soc. Cornwall* **13**, 191–197.

(1924): *Saccocoma cretacea* n. sp.: A Senonian Crinoid. *Geol. Ass. London, Proc.* **35**, 111–121.

Baumiller, T. K. (1990a): Physical modeling of the batocrinid anal tube: Functional analysis and multiple hypothesis testing. *Lethaia* **23**, 399–408.

(1990b): Non-predatory drilling of Mississippian crinoids by platyceratid gastropods. *Palaeontology* **33**, 743–748.

(1993): Boreholes in Devonian blastoids and their implications for boring by platyceratids. *Lethaia* **26**, 41–47.

Baumiller, T. K. & Ausich, W. I. (1992): The "broken stick" model as a null hypothesis for crinoid stalk taphonomy and as a guide to the distribution of connective tissues in fossils. *Paleobiology* **18**, 288–298.

(1996): Crinoid stalk flexibility: Theoretical predictions and fossil stalk postures. *Lethaia* **29**, 47–59.

Baumiller, T. K. & Gazdzicki, A. (1996): New crinoids from the La Meseta Formation of Seymour Island, Antarctic Peninsula, in: A. Gazdzicki (ed.), *Palaeontological Results of the Polish Antarctic Expedition*. Pt. 2. *Palaeontologia Polonica* **55**, 101–116.

Baumiller, T. K. & Hagdorn, H. (1995): Taphonomy as a guide to functional morphology of *Holocrinus*, the first post-Palaeozoic crinoid. *Lethaia* **28**, 221–228.

Baumiller, T. K., LaBarbera, M. & Woodley J. D. (1991): Ecology and functional morphology of the isocrinid *Cenocrinus asterius* (Linnaeus) (Echinodermata: Crinoidea): In situ and laboratory experiments and observations. *Bull. Marine Sci.* **48**, 731–748.

Baumiller, T. K. & Plotnick, R. E. (1989): Rotational stability in stalked crinoids and the function of wing plates in *Pterotocrinus depressus*. *Lethaia* **22**, 317–326.

Bausch, W. M., Viohl, G., Bernier, P., Barale, G., Bourseau, J.-P., Buffetaut, E., Gaillard, C., Gall, J.-C. & Wenz, S. (1994): Eichstätt and Cerin: Geochemical comparison and definition of two different Plattenkalk types. *Géobios* M. S. No. 16, 107–125.

Beede, J. W. (1906): Fauna of the Salem Limestone of Indiana, Echinodermata. *Indiana Department of Geology and Natural Resources Annual Report* **30**, 1243–1270.

Beringer, K. C. (1926): Die Pentacriniten des schwäbischen Posidonienschiefers. *Jh. Ver. Vaterl. Naturkunde Württemberg* **82**, 1–49. Stuttgart.

Biese, W. (1927): *Ueber die Encriniten des unteren Muschelkalkes von Mitteldeutschland*. Abh. Preuss. Geol. L.-Anst. N.F. **103**. Berlin. 119 pp.

Bigot, A. P. D. (1938): Crinoïdes du Bathonien du Calvados. *Ann. Paléont.* **27**, 1–38.

Blake, D. B. (1984): The Benthopectinidae (Asteroidea: Echinodermata) of the Jurassic of Switzerland. *Eclogae Geol. Helv.* **77**, 631–647.

Blumer, M. (1960): Pigments of a fossil Echinoderm. *Nature* **188**, 1100–1101.

Bottjer, D. J. & Ausich, W. I. (1986): Phanerozoic development of tiering in soft substrata suspension-feeding communities. *Paleobiology* **12**, 400–420.

Bourseau, J.-P., Améziane-Cominardi, N., Avocat, R. & Roux,

M. (1991): Echinodermata: Les Crinoïdes pédonculés de Nouvelle-Calédonie, in: A. Crosnier (ed.), *Résultats des Campagnes Musorstom* 8. Mém. Mus. Natn. Hist. Nat. (A) **151**, 229–233. Paris.

Bourseau, J.-P., David, J., Roux, M., Bertrand, D., & Clochard, V. (1998): *Balanocrinus maritimus* nov. sp., crinoïde pédonculé nouveau du Kimméridgien inférieur de La Rochelle (Charente-Maritime, France). *Géobios* **31**, 215–227.

Bowsher, A. L. (1955): *Origin and Adaptation of Platyceratid Gastropods*. University of Kansas Paleont. Contrib., Mollusca, Art. 5. 11 pp.

Breimer, A. (1978): General morphology, recent crinoids, in: R. C. Moore & C. Teichert (eds.), *Treatise on Invertebrate Paleontology.*, Pt. T. *Echinodermata* 2, Vol. 1, T9–T58. Geological Society of America and University of Kansas Press.

Breimer, A. & Lane, N. G. (1978): Ecology and paleoecology, in: R. C. Moore & C. Teichert (eds.), *Treatise on Invertebrate Paleontology*. Pt. T. *Echinodermata* 2, Vol. 1, T316–T347. Geological Society of America and University of Kansas Press.

Brenchley, P. J. (1989): The late Ordovician extinction, in: S. K. Donovan (ed.), *Mass Extinctions: Processes and Evidence*, 104–132. Belhaven Press.

Brett, C. E. (1978): Host-specific pit-forming epizoans on Silurian crinoids. *Lethaia* **11**, 217–232.

— (1981): Terminology and functional morphology of attachment structures in pelmatozoan echinoderms. *Lethaia* **14**, 343–370.

— (1983): Sedimentology, facies and depositional environments of the Rochester Shale (Silurian: Wenlockian) in western New York and Ontario. *J. Sediment. Petrol.* **53**, 947–971.

— (1984): Autecology of Silurian pelmatozoan echinoderms, in: M. G. Bassett & J. D. Lawson (eds.), *Autecology of Silurian Organisms*, 87–120. Spec. Pap. Paleont. **32.**

Brett, C. E. & Baird, G. C. (1994): Depositional sequences, cycles and foreland basin dynamics in the Late Middle Devonian (Givetian) of the Genesee Valley and western Finger Lakes region. *N.Y. State Geol. Ass. 66th Annual Meeting Guidebook*, 505–586.

Brett, C. E. & Baird, G. C. (in press): Revised stratigraphy of the upper Trenton Group in its type area, central New York State: Sedimentology and tectonics of a Middle Ordovician shelf-to-basin succession, in: C. E. Mitchell & R. D. Jacobi (eds.), *Taconic Convergence: Orogen, Foredeep, and Craton Interactions*. Geol. Soc. America Spec. Pap.

Brett, C. E., Boucot, A. J. & Jones, B. (1993): Absolute depths of Silurian benthic assemblages. *Lethaia* **26**, 25–40.

Brett, C. E. & Eckert, J. D. (1982): Paleoecology of a well-preserved crinoid colony from the Silurian Rochester Shale in Ontario. *Royal Ontario Museum, Life Sci. Contrib.* **131**, 1–20.

Brett, C. E. & Liddell, W. D. (1978): Preservation and paleoecology of a Middle Ordovician hardground community. *Paleobiology* **4**, 329–348.

Brett, C. E. & Seilacher, A. (1991): Fossil Lagerstätten: A taphonomic consequence of event sedimentation, in: G. Einsele, W. Ricken & A. Seilacher (eds.), *Cycles and Events in Stratigraphy*, 283–297. Springer-Verlag.

Brett, C. E. & Taylor, W. L. (1997): The *Homocrinus* beds: Silurian crinoid *Lagerstätten* of western New York and southern Ontario, in: C. E. Brett & G. C. Baird (eds.), *Paleontologic Events: Stratigraphic, Geologic, and Evolutionary Implications*, 181–223. Columbia University Press.

Broadhead T. W. & Waters, J. A. (1980): *Echinoderms: Notes for a short course*. University of Tennessee, Department of Geological Sciences, Studies in Geology 3. 235 pp.

Brookfield, M. E. (1988): A mid-Ordovician temperate carbonate shelf: The Black River and Trenton Limestones Groups of southern Ontario, Canada. *Sediment. Geol.* **60**, 137–153.

Brookfield, M. E. & Brett, C. E. (1988): Paleoenvironments of the Mid-Ordovician (Upper Caradocian) Trenton Limestones of southern Ontario, Canada: Storm sedimentation on a shoal-basin-shelf model. *Sediment. Geol.* **57**, 75–105.

Brower, J. C. (1974): Upper Ordovician xenocrinids (Crinoidea, Camerata) from Scotland. *Univ. Kansas Paleont. Contr. Pap.* **67**, 1–25.

— (1975): Silurian crinoids from the Pentland Hills, Scotland. *Palaeontology* **18**, 631–656.

— (1992a): Cupulocrinid crinoids from the Middle Ordovician (Galena Group, Dunleith Formation) of northern Iowa and southern Minnesota. *J. Paleont.* **66**, 99–128.

— (1992b): Hybocrinid and disparid crinoids from the Middle Ordovician (Galena Group, Dunleith Formation) of northern Iowa and southern Minnesota. *J. Paleont.* **66**, 973–993.

— (1994): Camerate crinoids from the Middle Ordovician (Galena Group, Dunleith Formation) of northern Iowa and southern Minnesota. *J. Paleont.* **68**, 570–599.

— (1995a): Eoparisocrinid crinoids from the Middle Ordovician (Galena Group) of northern Iowa and southern Minnesota. *J. Paleont.* **69**, 351–366.

— (1995b): Dendrocrinid crinoids from the Ordovician of northern Iowa and southern Minnesota. *J. Paleont.* **69**, 939–960.

Brower, J. C. & Strimple, H. L. (1983): Ordovician calceocrinids from northern Iowa and southern Minnesota. *J. Paleont.* **57**, 1261–1281.

Bruckner, D. (1748–1763): *Versuch einer Beschreibung historischer und natürlicher Merkwürdigkeiten der Landschaft Basel*. 23 Vols. 5. Emanuel Thurnecysen.

Buckland, W. (1836): *Geology and Mineralogy, Considered with Respect to Natural Theology*. Vol. 1. Bridgewater Treatises **6**. W. Pickering. 618 pp.

Burdick, D. W. & Strimple H. L. (1982): Genevievian and

Chesterian crinoids of Alabama. *Geol. Survey Alabama Bull.* **121**. 277 pp.

Cain, J. D. B. (1968): Aspects of the depositional environment and paleoecology of crinoidal limestones. *Scottish J. Geol.* **4**, 191–208.

Carpenter, P. H. (1884): Report upon the Crinoidea collected during the Voyage of H.M.S. Challenger during the years 1873–76. *Rept. Sci. Results Explor. Voyage H.M.S Challenger, Zoology.* Pt. 1. *General Morphology with Description of the Stalked Crinoids.* HMSO. 442 pp.

Caster, K. E. (1981): Cincinnatian contributions to knowledge of the Lophophora, in: J. T. Dutro, Jr. & R. S. Boardman (eds.), *Lophophorates: Notes for a Short Course*, 237–251. University of Tennessee, Department of Geological Sciences, Studies in Geology **5**.

Cherbonnier, G. & Guille, A. (1972): Sur une espèce actuelle de crinoïde crétacique de la famille Holopodidae: *Cyathidium foresti* nov. sp. *Comptes Rendus Acad. Sci. Paris, Sér. D*, **274**, 2193–2196.

Chesnut, D. R., Jr. & Ettensohn, F. R. (1988): Hombergian (Chesterian) echinoderm paleontology and paleoecology, south-central Kentucky. *Bull. Amer. Paleont.* **95**. 102 pp.

Clark, A. H. (1915–1950): *A monograph of the existing crinoids.* U.S. Natl. Museum, Bull. **82**, Vol. 1: *The comatulids*, Pt. 1 (1915a) 406 pp.; Pt. 2 (1921) 795 pp.; Pt. 3 (1931) 816 pp.; Pt. 4a (1941) 603 pp.; Pt. 4b (1947) 473 pp.; Pt. 4c (1950) 383 pp.

Clark, A. H. & Clark A. M. (1967): *A monograph of the existing crinoids.* U.S. Natl. Museum, Bull. **82**, Vol. 1: *The comatulids*, Pt. 5. 860 pp.

Conan, G., Roux, M. & Sibuet, M. (1981): A photographic survey of a population of the stalked crinoid *Diplocrinus (Annacrinus) wyvillethomsoni* (Echinodermata) from the bathyal slope of the Bay of Biscay. *Deep-Sea Research* **28A**, 441–453.

Conway Morris, S. (1993): The fossil record and the early evolution of the Metazoa. *Nature* **361**, 219–225.

Cowen, R. (1981): Crinoid arms and banana plantations: An economic harvesting analogy. *Paleobiology* **7**, 332–343.

Davis, R. A., & Cuffey, R. J. (eds.) (in press): *Sampling the Layer Cake That Isn't: The Stratigraphy and Paleontology of the Type Cincinnatian.* Ohio Division of Geological Survey Guidebook **13**.

DeRiccardis, F., Iorizzi, M., Minale, L., Riccio, R., Richer de Froges, B. & Debitus, C. (1991): The Gymnochromes: Novel marine brominated phenanthroperylenequinone pigments from the stalked crinoid *Gymnocrinus richeri*. *J. Org. Chem.* **56**, 6781–6787.

Destombes, J., Hollard, J. & Willefert, S. (1985): Lower Palaeozoic rocks of Morocco, in: C. H. Holland (ed.), *Lower Palaeozoic of North-western and West-central Africa*, 91–336. Wiley.

Donovan, S. K. (1988): The early evolution of the Crinoidea, in: C. R. C. Paul & A. B. Smith (eds.), *Echinoderm Phylogeny and Evolutionary Biology*, 235–244. Clarendon Press.

— (1989a): The improbability of a muscular crinoid column. *Lethaia* **22**, 307–315.

— (1989b): The significance of the British Ordovician crinoid fauna. *Modern Geol.* **13**, 243–255.

— (1991): The taphonomy of echinoderms: Calcareous multi-element skeletons in the marine environment, in: S. K. Donovan (ed.), *The Processes of Fossilization*, 241–269. Columbia University Press.

— (1992a): A field guide to the fossil echinoderms of Colpow, Bellman and Salthill Quarries, Clitheroe, Lancashire. *North West Geologist* **2**, 33–54.

— (1992b): New cladid crinoids from the late Ordovician of Girvan, Scotland. *Palaeontology* **35**, 149–158.

— (1992c): A new crinoid from the Ashgill Starfish Bed, Threave Glen. *Scott. J. Geol.* **28**, 123–126.

— (1993): Contractile tissues in the cirri of ancient crinoids: Criteria for recognition. *Lethaia* **26**, 163–169.

— (1994): Isocrinid crinoids from the late Cenozoic of Jamaica. *Atlantic Geol.* **30**, 195–203.

— (1995): Pelmatozoan columnals from the Ordovician of the British Isles. Pt. 3. *Palaeontogr. Soc. Monogr.* **149** (597), 115–193.

Donovan, S. K. & Franzén-Bengtson, C. (1988): Myelodactylid crinoid columnals from the Lower Visby Beds (Llandoverian) of Gotland. *Geol. Fören. Stockholm Förhand.* **110**, 69–79.

Donovan, S. K., Kallmeyer, J. W. & Veltkamp, C. J. (1995): Functional morphologies of the columns of Upper Ordovician *Xenocrinus* and *Dendrocrinus*. *Lethaia* **28**, 309–315.

Donovan, S. K., Paul, C. R. C. & Lewis, D. N. (1996): Echinoderms, in: D. A. T. Harpe & A. W. Owen (eds.), *Fossils of the Upper Ordovician.* Palaeont. Ass. Field Guides to Fossils **7**, 202–267, 277–294.

Donovan, S. K. & Pawson, D. L. (1998): Proximal growth of the column of bathycrinid crinoids (Echinodermata) following decapitation. *Bull. Marine Sci.* **61**, 571–579.

Donovan, S. K. & Sevastopulo, G. D. (1989): Myelodactylid crinoids from the Silurian of the British Isles. *Palaeontology* **32**, 689–710.

Donovan, S. K. & Veltkamp, C. J. (1990): *Barycrinus* (Crinoidea) from the Lower Carboniferous of England. *J. Paleont.* **64**, 988–992.

Dott, R. H., Jr. & Batten, R. L. (1976): *Evolution of the Earth* (2d ed.). McGraw-Hill. 504 pp.

Eckert, J. D. (1988): Late Ordovician extinction of North American and British crinoids. *Lethaia* **21**, 147–167.

Eltysheva, R. S., see Yeltisheva, R. S.

Erwin, D. H. (1993): *The Great Palaeozoic Crisis: Life and Death in the Permian.* Columbia University Press. 327 pp.

Etter, W. (1996): Pseudoplanktonic and benthic invertebrates in the Middle Jurassic Opalinum Clay, northern Switzerland. *Palaeogeography, Palaeoclimatology, Palaeoecology* **126**, 325–341.

Falk, H., & Mayr, E. (1997): Concerning bay salt and peri chelate formation of hydroxyphenanthroperylene quinones (Fringelites). *Chemical Monthly* **128**, 353–360.

Falk, H., Mayr, E. & Richter, A. E. (1994): Simple diffusive reflectance UV–Vis spectroscopic determination of organic pigments (Fringelites) in fossils. *Microchim. Acta* **117**, 1–5.

Fechter, H. (1973): *Cyathidium meteorensis* spec. nov., ein neuer Crinoide aus der Familie Holopodidae. *Helgoländer wiss. Meeresunters.* **25**, 162–169. Hamburg.

Foote, M. (1995): Morphological diversification of Paleozoic crinoids. *Paleobiology* **21**, 273–299.

Fox, W. T. (1962): Stratigraphy and paleoecology of the Richmond Group in southeastern Indiana. *Geol. Soc. Amer. Bull.* **73**, 621–642.

Franzén, C. (1974): Epizoans on Silurian – Devonian crinoids. *Lethaia* **7**, 287–301.

— (1977): Crinoid holdfasts from the Silurian of Gotland. *Lethaia* **10**, 219–234.

— (1982): A Silurian crinoid thanatotope from Gotland. *Geol. Fören. Stockholm Förhand.* **103**, 469–490.

— (1983): *Ecology and Taxonomy of Silurian Crinoids from Gotland*. Acta Universitatis Upsaliensis, Abstracts of Uppsala Dissertations from the Faculty of Science **665**. 31 pp.

Frickhinger, K. A. (1994): *The Fossils of Solnhofen*. Goldschneck-Verlag. 333 pp.

Goldring, W. (1993): *The Devonian Crinoids of New York*. New York State Mus. Bull. Mem. **16**. 670 pp.

Gonzalez, R. (1993): *Die Hauptrogenstein-Formation der Nordschweiz (mittleres Bajocien bis unteres Bathonien)*. Ph.D. Thesis Geol. Paläont. Inst., University of Basel, No. 2.

Gonzalez, R. & Wetzel, A. (1996): Stratigraphy and paleogeography of the Hauptrogenstein and Klingnau Formations (middle Bajocian to late Bathonian), northern Switzerland. *Eclogae Geol. Helv.* **89**, 695–720.

Grimmer, J. C. & Holland, N. D. (1990): The structure of a sessile, stalkless crinoid (*Holopus rangii*). *Acta Zool.* **71**, 61–67. Stockholm.

Grimmer, J. C., Holland, N. D. & Hayami, I. (1985): Fine structure of the stalk of an isocrinid sea lily (*Metacrinus rotundus*) (Echinodermata, Crinoidea). *Zoomorphology* **105**, 39–50.

Guensburg, T. (1992): Paleoecology of hardground encrusting and commensal crinoids, Middle Ordovician, Tennessee. *J. Paleont.* **66**, 129–147.

Hagdorn, H. (1978): Muschel/Krinoiden-Bioherme im Oberen Muschelkalk (mo1, Anis) von Crailsheim und Schwäbisch Hall (Südwestdeutschland). *N. Jb. Geol. Paläont. Abh.* **156**, 31–86. Stuttgart.

— (1982): *Chelocrinus schlotheimi* (Quenstedt) 1835 aus dem Oberen Muschelkalk (mo1, Anisium) von Nordwestdeutschland. *Veröff. Naturkde. Mus. Bielefeld* **4**, 5–23. Bielefeld.

— (1983): *Holocrinus doreckae* n. sp. from the Upper Muschelkalk and the evolution of preformed rupture points in the stem of Isocrinida. *N. Jb. Geol. Paläont. Mh.* **1983**, 345–368. Stuttgart.

— (1985): Immigration of Muschelkalk crinoids into the German Muschelkalk Basin, in: U. Bayer & A. Seilacher (eds.), *Sedimentary and Evolutionary Cycles*, 237–254. Lecture Notes in Earth Sciences **1**. Springer-Verlag.

— (1991): The Muschelkalk in Germany – An Introduction, in: H. Hagdorn (ed., in cooperation with T. Simon & J. Szulc), *Muschelkalk, A Field Guide*, 7–21. Goldschneck-Verlag.

— (1995): Literaturbericht Triassic crinoids. *Zbl. Geol. Paläont.* Pt. 2 1995/1/2, 1–22. Stuttgart.

— (1996): Palökologie der Trias-Seelilie *Dadocrinus*. *Geol. Paläont. Mitt. Innsbruck* **21**, 18–38.

Hagdorn, H. & Campbell, H. J. (1993): *Paracomatula triadica* sp. nov.: An early comatulid crinoid from the Otapirian (Late Triassic) of New Caledonia. *Alcheringa* **17**, 1–17.

Hagdorn, H. & Gluchowski, E. (1993): Palaeobiogeography and stratigraphy of Muschelkalk echinoderms (Crinoidea, Echinoidea) in Upper Silesia, in: H. Hagdorn & A. Seilacher (eds.), *Muschelkalk. Schöntaler Tagung 1991*, 165–176. Sonderbde. Ges. Naturk. Württ. **2**. Goldschneck-Verlag.

Hagdorn, H., Gluchowski, E. & Boczarowski, A. (1996): The crinoid fauna of the *Diplopora* Dolomite (Middle Muschelkalk, Triassic, Upper Anisian) at Piekary Slaskie in Upper Silesia. *Geol. Paläont. Mitt. Innsbruck* **21**, 38–87.

Hagdorn, H. & Ockert, W.: (1993): *Encrinus liliiformis* im Trochitenkalk Süddeutschlands, in: H. Hagdorn & A. Seilacher (eds.), *Muschelkalk. Schöntaler Tagung 1991*, 245–260. Sonderbde. Ges. Naturk. Württ. **2**. Goldschneck-Verlag.

Hagdorn, H. & Schulz, M. (1996): Echinodermen-Konservatlagerstätten im Unteren Muschelkalk Osthessens. Pt. 1. Die Bimbacher Seelilienbank von Grossenlüder-Bimbach. *Geol. Jb. Hessen* **124**, 97–122. Wiesbaden.

Hagdorn, H. & Simon, T. (1984): Oberer Muschelkalk (Hauptmuschelkalk, mo), in: *Geologische Karte 1:25000 von Baden-Württemberg, Erläuterungen zu Blatt 6921 Grossbottwar*, 14–27. Geologisches Landesamt Baden-Württemberg.

Hall, J. (1882): Descriptions of the species of fossils found in the Niagara Group at Waldron, Indiana. *Indiana Department of Geology and Natural History Annual Report* **11**, 217–345.

Hardie, W. G. (1971): Itinerary I: Wren's Nest Hill Dudley, in: W. G. Hardie, G. M. Bennison, P. A. Garrett, J. D.

Lawson & F. W. Shotton, *The Area around Birmingham*, 7–9. Geol. Ass. Guides **1**.

Harper, D. A. T. (1982): The Late Ordovicain Lady Burn Starfish Beds of Girvan. *Proc. Geol. Soc. Glasgow*, sessions 122/123 for 1980–1981, 28–32.

Harvey, E. W. & Ausich, W. I. (1997): Phylogeny of calceocrinid crinoids (Paleozoic Echinodermata): Biogeography and mosaic evolution. *J. Paleont.* **71**, 299–305.

Haude, R. (1972): Bau und Funktion der *Scyphocrinites*-Lobolithen. *Lethaia* **5**, 95–125.

— (1980): Constructional morphology of the stems of Pentacrinitidae, and mode of life of *Seirocrinus*, in: M. Jangoux (ed.), *Proceedings of the European Colloquium on Echinoderms*, Brussels, 3–8 September 1979, 17–23. Balkema.

— (1992): Scyphocrinoiden, die Bojen-Seelilien im hohen Silur–tiefen Devon. *Palaeontographica* A **222**, 141–187.

Haude, R., Jahnke, H. & Walliser, O. H. (1994): Scyphocrinoiden an der Wende Silur/Devon. *Aufschluss* **45** (March–April), 49–55. Heidelberg.

Hauff, B. & Hauff, R. B. (1981): *Das Holzmadenbuch*. Museum Holzmaden. 136 pp.

Hauff, R. B. (1984): *Pentacrinites quenstedti* (Oppel) aus dem oberen Untertoarcium (Lias Epsilon) von Ohmden bei Holzmaden (SW-Deutschland). *Paläont. Z.* **58**, 255–263.

Hauser, J. (1997): *Die Crinoiden des Mittel-Devon der Eifler-Kalkmulden*. Distributed by Goldschneck-Verlag. 273 pp.

Heath, K. C. & Mullins, H. T. (1984): Open-ocean, off-bank transport of fine-grained carbonate sediment in the Northern Bahamas, in: D. A. V. Stow & D. J. W. Piper (eds.), *Fine-Grained Sediments: Deep-Water Process and Facies*, 199–208. Geol. Soc. Spec. Publ. **15**.

Heinzeller, T., Améziane-Cominardi, N., Fricke, H., & Welsch, U. (1995): Isocrinida (Crinoidea) compared with non-isocrinid forms: Nervous supply of stalk and stem. *Verh. Dtsch. Zool. Ges.* **88**, 34.

Heinzeller, T., Améziane-Cominardi, N. & Welsch, U. (1994): Light and electron microscopic studies on arms and pinnules of the Cyrtocrinid *Gymnocrinus richeri*, in: B. David, A. Guille, J.-P. Féral & M. Roux (eds.), *Echinoderms through Time: Proceedings of the Eighth International Echinoderm Conference*, Dijon, France, 6–10 September 1993, 211–216. Balkema.

Heinzeller, T. & Fechter, H. (1995): Microscopical Anatomy of the Cyrtocrinid *Cyathidium meteorensis* (sive *foresti*) (Echinodermata, Crinoidea). *Acta Zool.* **76**, 25–34. Stockholm.

Heinzeller, T., Fricke, H., Bourseau, J.-P., Améziane-Cominardi, N. & Welsch, U. (1996): *Cyathidium plantei* sp. n., an extant cyrtocrinid (Echinodermata, Crinoidea) – morphologically identical to the fossil *Cyathidium depressum* (Cretaceous, Cenomanian). *Zool. Scr.* **25**, 77–84.

Hess, H. (1950): Ein neuer Crinoide aus dem mittleren Dogger der Nordschweiz (*Paracomatula helvetica* n. gen. n. sp.). *Eclogae Geol. Helv.* **43**, 208–216.

— (1972a): *Eine Echinodermen-Fauna aus dem mittleren Dogger des Aargauer Juras*. Schweiz. Paläont. Abh. **72**. 87 pp.

— (1972b): Planktonic crinoids of Late Jurassic age from Leg 11, Deep Sea Drilling Project. *Initial Reports of the Deep Sea Drilling Project* **11**, 631–643.

— (1975): *Die fossilen Echinodermen des Schweizer Juras*. Veröff. Naturhist. Museum Basel **8**. 130 pp.

Hess, H. & Holenweg, H. (1985): Die Begleitfauna auf den Seelilienbänken im mittleren Dogger des Schweizer Juras. *Tätigkeitsber. Naturf. Ges. Baselland* **33**, 141–177. Liestal.

Hicks, J. F., & Wray, C. G. (in press): *Paleocurrent orientation of crinoids from the Coeymans Limestone of the Helderberg Group at Days Corners, New York*. Yale Postilla.

Holland, S. M. (1993): Sequence stratigraphy of a carbonate-clastic ramp: The Cincinnatian Series (Upper Ordovician) in its type area. *Geol. Soc. Amer. Bull.* **105**, 306–322.

Horowitz, A. S. (1965): *Crinoids from the Glen Dean Limestone (Middle Chester) of Southern Indiana and Kentucky*. Indiana Geol. Survey Bull. **34**, 52 pp.

Hudson, R. G. S., Clarke, M. J. & Sevastopulo, G. D. (1966): The palaeoecology of a Lower Viséan crinoid fauna from Feltrim, County Dublin. *Proc. Roy. Soc. Dublin*, Ser. A, **2**, 273–286.

Hüssner, H. (1933): Rifftypen im Muschelkalk Süddeutschlands, in: H. Hagdorn & A. Seilacher (eds.), *Muschelkalk. Schöntaler Symposium 1991*, 261–269. Sonderbde. Ges. Naturk. Württ. **2**. Goldschneck-Verlag.

Hyman, L. H. (1955): *The Invertebrates*. Vol. 4. Echinodermata – The Coelomate Bilateria. McGraw-Hill. 763 pp.

Ingham, J. K. (1978): Geology of a continental margin. Pt. 2. Middle and late Ordovician transgression, Girvan, in: D. R. Bowes & B. E. Leake (eds.), *Crustal evolution in northwestern Britain and adjacent regions*, 163–176. Geol. J. Spec. Issue **10**.

Jablonski, D., & Bottjer, D. J. (1983): Soft-bottom epifaunal suspension-feeding assemblages in the Late Cretaceous: Implications for the evolution of benthic paleocommunities, in: M. J. S. Tevesz & P. L. Mc Call (eds.), *Biotic Interactions in Recent and Fossil Benthic Communities*, 747–812. Plenum.

Jaekel, O. (1891): Ueber Holopocriniden mit besonderer Berücksichtigung der Stramberger Formen. *Z. Deutsch. Geol. Ges.* **43**, 557–670.

— (1892): Ueber Plicatocriniden, *Hyocrinus* und *Saccocoma*. *Z. Deutsch. geol. Ges.* **44**, 619–696.

— (1894): Platte mit *Encrinus Carnalli* Beyr. *Sber. naturforsch. Freunde Berlin* **6**, 155–162.

— (1895): Beiträge zur Kenntnis der palaeozoischen Crinoiden Deutschlands. *Paläont. Abh.* N.F. **3**, 3–116.

— (1918): Phylogenie und System der Pelmatozoen. *Paläont. Z.* **3**, 1–128.

Jäger, M. (1993): *Das Fossilienmuseum im Werkforum.* Werkforum. 128 pp.

Janicke, V. (1969): Untersuchungen über den Biotop der Solnhofener Plattenkalke. *Mitt. Bayer. Staatssammlg. Paläont. hist. Geol.* **9**, 117–181.

— (1970): Lumbricaria. Ein Cephalopoden-Koprolith. *N. Jb. Geol. Paläont. Mh.* **1970**, 50–60.

Jeanette, D. C. & Pryor, W. A. (1993): Cyclic alternation of proximal and distal storm facies: Kope and Fairview Formations (Upper Ordovician), Ohio and Kentucky. *J. Sediment. Petrol.* **63**, 183–203.

Jefferies, R. P. S. (1989): The arm structure and mode of feeding of the triassic crinoid Encrinus liliiformis. *Palaeontology* **32**, 483–497.

Jeffords, R. M. & Miller, T. H. (1968): *Ontogenetic Development in Late Pennsylvanian Crinoid Columnals and Pluricolumnals.* University of Kansas Paleont. Contrib., Echinodermata, Art. 10. 14 pp.

Jenkyns, H. C. (1971): Speculations on the genesis of crinoidal limestones in the Tethyan Jurassic. *Geol. Rundschau* **60**, 471–488.

Jungheim, H. J. (1995): Das Mitteldevon der Eifel, in: W. K. Weidert (ed.), *Klassische Fundstellen der Paläontologie 3*, 56–621. Goldschneck-Verlag.

Kammer, T. W. (1984): Crinoids from the New Providence Shale Member of the Borden Formation (Mississippian) in Kentucky and Indiana. *J. Paleont.* **58**, 115–130.

— (1985): Basinal and prodeltaic communities of the Early Carboniferous Borden Formation in northern Kentucky and southern Indiana (U.S.A.). *Palaeoclimatology, Palaeogeography, Palaeoecology* **49**, 79–121.

Kelly, S. M. (1982): Classification and evolution of Class Crinoidea. *Abstr. 4th North American Paleontological Convention*, A23.

Kendall, G. W., Johnson, J. G., Brown, J. O. & Klapper, G. (1983): Stratigraphy and facies across Lower Devonian–Middle Devonian boundary, central Nevada. *Amer. Ass. Petrol. Geol. Bull.* **67**, 2199–2207.

Kershaw, S. (1993): Sedimentation control on growth of stromatoporoid reefs in the Silurian of Gotland, Sweden. *J. Geol. Soc. London* **150**, 197–205.

Kesling, R. V. & Chilman, R. B. (1975): *Strata and Megafossils of the Middle Devonian Silica Formation.* Museum of Paleontology Papers on Paleontology 8. Friends of the University of Michigan Museum of Paleontology, Inc. 408 pp.

Keupp, H. (1977): *Ultrafazies und Genese der Solnhofener Plattenkalke (Oberer Malm, Südliche Frankenalb).* Abh. Naturhist. Ges. Nürnberg **37**. 128 pp.

Keyes, C. R. (1897): Memorial of Charles Wachsmuth. *Proc. Iowa Acad. Sci.* **4**, 13–16.

Kirk, E. (1911): The structure and relationships of certain eleutherozoic Pelmatozoa. *Proc. U.S. Nat. Mus.* **41**, 1–137.

Klikushin, V. G. (1982): Taxonomic survey of fossil isocrinids with a list of the species found in the USSR. *Géobios* **15**, 299–325.

— (1987a): Crinoids from the Middle Liassic Rosso ammonitico beds. *N. Jb. Geol. Paläont. Abh.* **175**, 235–260.

— (1987b): Thiollericrinid crinoids from the Lower Cretaceous of Crimea. *Géobios* **20**, 625–665.

— (1996): Late Jurassic crinoids from Sudak environs (Crimea). *Palaeontographica* A **238**, 97–151.

Koenigswald, R. von (1930): Die Arten der Einregelung ins Sediment bei den Seesternen und Seelilien des unterdevonischen Bundenbacher Schiefers. *Senckenbergiana* **12**, 338–360. Frankfurt a.M.

Kozur, H. (1974): Biostratigraphie der germanischen Mitteltrias. *Freiberger Forschh.* C **280** (1), 1–56, **280** (2), 1–71, **280** (3), Tables 1–15. Leipzig.

Kristan-Tollmann, E. (1970): Die Osteocrinusfazies, ein Leithorizont von Schwebcrinoiden im Oberladin – Unterkarn der Tethys. *Erdöl und Kohle, Erdgas, Petrochemie* **23**, 781–789. Hamburg.

— (1977): Zur Gattungsunterscheidung und Rekonstruktion der triadischen Schwebcrinoiden. *Paläont. Z.* **51**, 185–198.

— (1980): Tulipacrinus tulipa n. g. n. sp., eine Mikrocrinoide aus der alpinen Obertrias. *Ann. Naturhist. Mus. Wien* **83**, 215–229.

— (1986): Triassic of the Tethys and its relations with the Triassic of the Pacific Realm, in: K. G. McKenzie (ed.), *Shallow Tethys 2*, 169–182. Balkema.

— (1988): Unexpected communities among the crinoids within the Triassic Tethys and Panthalassa, in: R. R. Burke, P. V. Mladenov, P. Lambert & R. L. Parsley (eds.), *Echinoderm Biology: Proceedings of the Sixth International Echinoderm Conference, Victoria, British Columbia*, 133–142. Balkema.

— (1990): Mikrocrinoiden aus der Obertrias der Tethys. *Geol. Paläont. Mitt. Innsbruck* **17**, 51–100.

Kristan-Tollmann, E. & Tollmann, A. (1983): Ueberregionale Züge der Tethys in Schichtfolge und Fauna am Beispiel der Trias zwischen Europa und Fernost, speziell China. *Schriftenreihe erdwiss. Komm. Österr. Akad. Wiss.* **5**, 177–230.

Landing, E., & Brett, C. E. (1987): Trace fossils and regional significance of a Middle Devonian (Givetian) disconformity in southwestern Ontario. *J. Paleont.* **61**, 205–230.

Lane, H. R. (1978): The Burlington shelf (Mississippian, north-central United States). *Geol. Palaeont.* **12**, 165–176.

Lane, N. G. (1963a): Two new Mississippian camerate (batocrinid) crinoid genera. *J. Paleont.* **37**, 691–702.

(1963b): The Berkeley crinoid collection from Crawfordsville, Indiana. *J. Paleont.* **37**, 1001–1008.

(1971): Crinoids and reefs. *Proceedings of the North American Paleontological Convention*, Pt. J, 1430–1443. Springer-Verlag.

(1973): *Paleontology and Paleoecology of the Crawfordsville Fossil Site (Upper Osagian: Indiana)*. University of California Publ. Geol. Sci. **99**. 141 pp.

(1978): Historical review of classification of Crinoidea, in: R. C. Moore & C. Teichert (eds.), *Treatise on Invertebrate Paleontology.* Pt T. Echinodermata 2, Vol. 1, T348–T359. Geological Society of America and University of Kansas Press.

(1984): Predation and survival among inadunate crinoids. *Paleobiology* **10**, 453–458.

Lane, N. G. & Ausich, W. I. (1995): Interreef crinoid fauna from the Mississinewa Shale Member of the Wabash Formation (Northern Indiana, Silurian). *J. Paleontol.* **69**, 1090–1106.

Lane, N. G. & Sevastopulo, G. D. (1981): Functional morphology of a microcrinoid: *Kallimorphocrinus punctatus* n. sp. *J. Paleont.* **55**, 13–28.

Lane, N. G., Waters, J. A. & Maples, C. G. (1997): Echinoderm faunas of the Hongguleleng Formation, Late Devonian (Famennian), Xinjiang-Uygur Autonomous Region, People's Republic of China. *J. Paleont.* **71**, Suppl. to No. 2, Mem. 47.

Lane, N. G. & Webster, G. D. (1966): *New Permian Crinoid Fauna from Southern Nevada*. University of California Publ. Geol. Sci. **63**. 60 pp.

Laporte, L. F. (1967): Carbonate deposition near mean sea-level and resultant facies mosaic: Manlius Formation of New York State. *Amer. Ass. Petr. Geol. Bull.* **51**, 73–101.

(1968): *Ancient Environments.* Prentice-Hall. 115 pp.

(1975): Carbonate tidal deposits of the Early Devonian Manlius Formation of New York State, in: R. N. Ginsburg (ed.), *Tidal Deposits: Casebook of Recent Examples and Fossil Counterparts*, 243–250. Springer-Verlag.

Laudon, L. R. (1931): The stratigraphy of the Kinderhook Series of Iowa. *Iowa Geol. Survey Ann. Report* **35**, 333–452.

Laudon, L. R. & Beane, B. H. (1937): The crinoid fauna of the Hampton Formation at LeGrand, Iowa. *Iowa University Studies Nat. History* **7**, 227–272.

Laudon, L. R. & Severson, J. L. (1953): New crinoid fauna, Mississippian, Lodgepole Formation, Montana. *J. Paleont.* **27**, 505–536.

Lehmann, D. M., Brett, C. E., Cole, R. M. & Baird, G. C. (1995): Distal sedimentation in a peripheral foreland basin: Ordovician black shales and associated Flysch of the western Taconic foreland, New York State and Ontario. *Geol. Soc. Amer. Bull.* **107**, 708–724.

Lehmann, U. & Weitschat, W. (1973): Zur Anatomie und Oekologie von Ammoniten: Funde von Kropf und Kiemen. *Paläont. Z.* **47**, 69–76.

Lehmann, W. M. (1957): *Die Asterozoen des rheinischen Unterdevons.* Abhandl. hess. Landesamt. f. Bodenforschg. **21**. Hessisches Landesamt für Bodenforschung. 160 pp.

Leuthardt, F. (1904): Die Crinoidenbänke im Dogger der Umgebung von Liestal. *Tätigkeitsber. Naturf. Ges. Baselland* **1902–1903**, 89–115. Liestal.

(1930): Ueber *Pentacrinus Nicoleti* Desor. *Tätigkeitsber. Naturf. Ges. Baselland* 1926–1930, 217–222. Liestal.

Lewis, R. D. (1981): *Archaetaxocrinus*, new genus: The earliest known flexible crinoid (Whiterockian) and its phylogenetic implications. *J. Paleont.* **55**, 227–238.

Liddell, W. D. (1975a): *Ecology and Biostratinomy of a Middle Ordovician Echinoderm Assemblage from Kirkfield, Ontario.* Unpubl. M. A. Thesis, University of Michigan, Ann Arbor. 41 pp.

(1975b): Recent crinoid biostratinomy. *Geol. Soc. Amer. Abstracts with Programs* **4**, 1169.

Linck, O. (1954): Die Muschelkalk-Seelilie *Encrinus liliiformis*, Ergebnisse einer Grabung. *Aus der Heimat* **62**, 225–235. Öhringen.

(1965): Stratigraphische, stratinomische und ökologische Betrachtungen zu *Encrinus liliiformis* Lamarck. *Jh. geol. Landesamt Baden-Württ.* **7**, 123–148. Freiburg i.Br.

Littke, R., Baker, D. R., Leythaeuser, D. & Rullkötter, J. (1991): Keys to the depositional history of the Posidonia Shale (Toarcian) in the Hils Syncline, northern Germany, in: R. V. Tyson & T. H. Pearson (eds.), *Modern and Ancient Shelf Anoxia*, 311–333. Geological Society Spec. Publ. **58**.

Llewellyn, G. & Messing, C. G. (1991): Local variations in modern crinoid-rich carbonate bank-margin sediments. *Geol. Soc. Amer. Abstracts with Programs* **23**, 344.

Loriol, P. de (1877–1879): Monographie des Crinoïdes Fossiles de la Suisse. *Mém. Soc. Paléont. Suisse* **4**, 1–52, **5**, 53–124, **6**, 125–300.

(1882–1889): *Paléontologie française. Sér. 1. Animaux invertébrés. Terrain jurassique:* **11**, *Crinoïdes*. Pt. 1, 627 pp., Pt. 2, 580 pp. Masson.

Lowenstam, H. A. (1948): *Niagaran Inter-reef Formations.* Illinois State Museum Sci. Pap. **4**. 146 pp.

(1957): Niagaran reefs in the Great Lakes area, in: H. S. Ladd (ed.), *Treatise on Marine Ecology and Paleoecology*, 215–248. Geol. Soc. Amer. Mem. **67** (2).

Lyon, S. S. & Casseday, S. A. (1859): Description of nine new species of Crinoidea from the Subcarboniferous rocks of Indiana and Kentucky. *Amer. J. Sci.* Ser. 2, **28**, 233–246.

Macurda, D. B., Jr. & Meyer, D. L. (1974): Feeding posture of modern stalked crinoids. *Nature* **247**, 394–396.

(1976a): The identification and interpretation of stalked crinoids (Echinodermata) from deep-water photographs. *Bull. Marine Sci.* **26**, 205–215.

(1976b): The morphology and life habitats of the abyssal crinoid *Bathycrinus aldrichianus* Wyville Thomson and its paleontological implications. *J. Paleontol.* **50**, 647–667.

(1977): Crinoids of West Indian Coral Reefs. *Studies in Geology* **4**, 231–237. American Association of Petroleum Geologists.

(1983): Sea lilies and feather stars. *Amer. Scient.* **71**, 354–365.

Magnus, D. B. E. (1963): Der Federstern *Heterometra savignyi* im Roten Meer. *Natur und Museum* **93**, 355–394.

Manni, R. & Nicosia, U. (1986): *Saccocoma schwertschlageri* Walther, 1904 – junior synonym of *Saccocoma tenellum* (Goldfuss), 1829: Evidence of autotomy in fossil crinoids. *Boll. Soc. Paleont. Ital.* **24**, 181–183. Rome.

Manten, A. A. (1971): *Silurian Reefs of Gotland*. Developments in Sedimentology **13**. Elsevier. 539 pp.

Maxwell, W. D. (1989): The end Permian mass extinction, in: S. K. Donovan (ed.), *Mass Extinctions: Processes and Evidence*, 152–173. Belhaven Press.

McKerrow, W. S. (ed.) (1978): *The Ecology of Fossils: An Illustrated Guide*. MIT Press. 384 pp.

Melchin, M. J., Brookfield, M. E., Armstrong, D. K. & Coniglio, M. (1994): *Stratigraphy, Sedimentology, and Biostratigraphy of the Ordovician Rocks of the Lake Simcoe Area, South-Central Ontario*. Field Trip A4, Geological Association of Canada/Mineralogical Association of Canada Annual Meeting, Waterloo, Ontario. 106 pp.

Messing, C. G. (1985): Submersible observations of deep-water crinoid assemblages in the tropical western Atlantic Ocean, in: B. F. Keegan & B. D. S. O'Connor (eds.), *Proceedings of the Fifth International Echinoderm Conference, Galway*, 185–193. Balkema.

(1997): Living Comatulids, in: J. A. Waters and C. G. Maples (eds.), *Geobiology of Echinoderms*, 3–30. Paleontological Society Papers **3**.

Messing, C. G. & Dearborn, J. H. (1990): *Marine Flora and Fauna of the Northeastern United States, Echinodermata: Crinoidea*. NOAA Technical Report NMFS **91** (U.S. Department of Commerce). 29 pp.

Messing, C. G., Neumann, A. C. & Lang, J. C. (1990): Biozonation of deep-water lithoherms and associated hardgrounds in the northeastern Straits of Florida. *Palaios* **5**, 15–33.

Meyer, C. A. (1988): Paläoökologie, Biofazies und Sedimentologie von Seeliliengemeinschaften aus dem Unteren Hauptrogenstein des Nordwestschweizer Jura. *Rev. Paléobiol.* **7**, 359–433.

(1990): Depositional environment and paleoecology of crinoid-communities from the Middle Jurassic Burgundy-platform of Western Europe, in: M. Jangoux & C. De Ridder (eds.), *Echinoderm Research*, 25–31. Balkema.

Meyer, D. L. (1971): Post-mortem disarticulation of Recent crinoid and ophiuroids under natural conditions. *Geol. Soc. Amer. Abstracts with Programs* **3**, 645.

(1981): Biostratinomy and paleoecology of echinoderm lagerstatten from the Cincinnatian Series, in: T. G. Roberts (ed.), *Stratigraphy and sedimentology*, 58–62. Geol. Soc. Amer. Cincinnati '81 Field Trip Guidebooks **1**. American Geological Institute.

(1985): Evolutionary implications of predation on Recent comatulid crinoids from the Great Barrier Reef. *Paleobiology* **11**, 154–164.

(1988): Crinoids as renewable resources: Rapid regeneration of the visceral mass in a tropical reef-dwelling crinoid from Australia, in: R. D. Burke, P. V. Mladenov, P. Lambert & R. L. Parsley (eds.), *Echinoderm Biology: Proceedings of the Sixth International Echinoderm Conference, Victoria, British Columbia*, 519–522. Balkema.

Meyer, D. L. & Ausich, W. I. (1983): Biotic interactions among recent and among fossil crinoids, in: M. J. S. Tevesz & P. L. McCall (eds.), *Biotic Interactions in Recent and Fossil Benthic Communities*, 377–427. Plenum.

Meyer, D. L., Ausich, W. I. & Terry, R. E. (1989): Comparative taphonomy of echinoderms in carbonate facies: Fort Payne Formation (Lower Mississippian) of Kentucky and Tennessee. *Palaios* **4**, 533–552.

Meyer, D. L. & Lane, N. G. (1976): The feeding behaviour of some Paleozoic crinoids and recent basketstars. *J. Paleont.* **50**, 472–480.

Meyer, D. L. & Macurda, D. B., Jr. (1977): Adaptive radiation of the comatulid crinoids. *Paleobiology* **3**, 74–82.

(1980): Ecology and distribution of shallow-water crinoids of Palau and Guam. *Micronesica* **16**, 59–99.

Meyer, D. L. & Meyer, K. B. (1986): Biostratinomy of Recent crinoids (Echinodermata) at Lizard Island, Great Barrier Reef, Australia. *Palaios* **1**, 294–301.

Meyer, D. L. & Oji, T. (1993): Eocene crinoids from Seymour Island, Antarctic Peninsula: Paleobiographic and paleoecologic implications. *J. Paleont.* **67**, 250–257.

Meyer, W. (1988): *Geologie der Eifel*. Schweizerbart. 615 pp.

Miller, J. S. (1821): *A Natural History of the Crinoidea or Lily-Shaped Animals, with Observations on the Genera Asteria, Euryale, Comatula and Marsupites*. Bryan & Co. 150 pp.

Milsom, C. V. (1994): *Saccocoma*: A benthic crinoid from the Jurassic Solnhofen Limestone, Germany. *Palaeontology* **37**, 121–129.

Milsom, C. V., Simms, M. J. & Gale, A. S. (1994): Phylogeny and palaeobiology of *Marsupites* and *Uintacrinus*. *Palaeontology* **37**, 595–607.

Milsom, C. V. & Sharpe, T. (1995): Jurassic lagoon: Salt or soup? *Geology Today*, January – February, 22–26.

Mitchell, C. E., Goldman, D., Delano, J. W., Samson, S. D. & Bergström, S. M. (1994): Temporal and spatial distribution of biozones and facies relative to geochemically correlated K-bentonites in the Middle Ordovician Taconic foreland. *Geology* **22**, 715–718.

Mittmeyer, H.-G. (1980): Zur Geologie des Hunsrückschiefers, in: W. Stürmer, F. Schaarschmidt & H. G. Mittmeyer,

Versteinertes Leben im Röntgenlicht, Kleine Senckenberg-Reihe 11, 26–33. Frankfurt a.M.

Moore, R. C. (1962): *Revision of Calceocrinidae.* University of Kansas Paleont. Contrib., Echinodermata, Art. 4. 40 pp.

— (1967): *Unique stalked crinoids from Upper Cretaceous of Mississippi.* University of Kansas Paleont. Contrib., Art. 17. 35 pp.

— (1978): Flexibilia, in: R. C. Moore & C. Teichert (eds.), *Treatise on Invertebrate Paleontology.* Pt. T. Echinodermata 2, Vol. 2, T759–T812. Geological Society of America and University of Kansas Press.

Moore, R. C. & Jeffords, R. M. (1968): *Classification and Nomenclature of Fossil Crinoids Based on Studies of Dissociated Parts of their Columns.* University of Kansas Paleont. Contrib., Echinodermata, Art. 9. 86 pp.

Moore, R. C., Jeffords, R. M. & Miller, T. H. (1968a): *Morphological Features of Crinoid Columns.* University of Kansas Paleont. Contrib., Art. 8. 30 pp.

Moore, R. C., Jeffords, R. M. & Miller, T. H. (1968b): *Supplement to Echinodermata.* University of Kansas Paleont. Contrib., Echinodermata, Arts. 8–10, Suppl. Ser. 45–47. 18 pp.

Moore, R. C. & Teichert, C. (eds.) (1978): *Treatise on Invertebrate paleontology.* Pt. T. Echinodermata 2. Geological Society of America and University of Kansas Press. 1027 pp.

Moore, R. C. & Vokes, H. E. (1953): *Lower Tertiary Crinoids from Northwestern Oregon.* U.S. Geol. Survey Professional Pap. 233-E, 111–147. Washington, D.C.

Morris, R. W. & Felton, S. H. (1993): Symbiotic association of crinoids, platyceratid gastropods, and *Cornulites* in the Upper Ordovician (Cincinnatian) of the Cincinnati, Ohio region. *Palaios* 8, 465–476.

Mu, A. T. (1949): On the discovery of the crown of *Traumatocrinus. Bull. Geol. Soc. China* 24, 85–92.

Müller, A. H. (1956a): Über eine eigenartige Einbettungsform von *Encrinus carnalli* Beyr. aus dem "Schaumkalk" (mu2χ) von Gutendorf bei Weimar (Thür.). *Geologie* 5, 26–29. Berlin.

— (1956b): Weitere Beiträge zur Ichnologie, Stratinomie und Ökologie der germanischen Trias. Pt. 1. *Geologie* 5, 405–423. Berlin.

— (1969): Zum *Lumbricaria*-Problem (Miscellanea), mit einigen Bemerkungen über *Saccocoma* (Crinoidea, Echinodermata). *Mber. Dt. Akad. Wiss. Berlin* 11, 750–758.

Nichols, D. (1994): Reproductive seasonality in the comatulid crinoid *Antedon bifida* (Pennant) from the English Channel. *Phil. Trans. Roy. Soc. London B* 434, 113–134.

Nicosia, U. & Parisi, G. (1979): *Saccocoma tenella* (Goldfuss): Distribuzione stratigrafica e geografica. *Boll. Soc. Paleont. Ital.* 18, 320–326.

O'Brien, N. R., Brett, C. E. & Taylor, W. L. (1994): The significance of microfabric and taphonomic analysis in determining sedimentary processes in marine mudrocks: Examples from the Silurian of New York. *J. Sediment. Res. Sec. A. Sedimentary Biology and Processes* A64, 847–852.

Oji, T. (1990): Miocene Isocrinida (stalked crinoids) from Japan and their biogeographic implication. *Trans. Proc. Palaeont. Soc. Japan, N.S.* 157, 412–429.

— (1996): Is predation intensity reduced with increasing depth? Evidence from the west Atlantic stalked crinoid *Endoxocrinus parrae* (Gervais) and implications for the Mesozoic marine revolution. *Paleobiology* 22, 339–351.

Opitz, R. (1932): *Bilder aus der Erdgeschichte des Nahe-Hunsrück-Landes Birkenfeld.* Birkenfeld. 234 pp.

Oschmann, W. (1995): Black shales models: an actualistic approach. *Europalaeontology* 8, 26–35.

Palmer, T. J. & Fürsich, F. T. (1974): The ecology of a Middle Jurassic hardground and crevice fauna. *Palaeontology* 17, 507–524.

Parsons, K. M., Brett, C. E. & Miller, K. B. (1988): Taphonomy and depositional dynamics of Devonian shell-rich mudstones. *Palaeogeography, Palaeoclimatology, Palaeoecology* 63, 109–141.

Paul, C. R. C. (1992): *Amphorometra* (Crinoidea, Echinodermata) from the London Clay of Aveley, Essex. *Tertiary Res.* 13, 117–124.

Peck, R. E. (1936): Lower Mississippian microcrinoids of the Kinderhook and Osage groups of Missouri. *J. Paleont.* 10, 282–293.

— (1943): Lower Cretaceous crinoids from Texas. *J. Paleont.* 17, 451–475.

— (1948): A Triassic crinoid from Mexico. *J. Paleont.* 22, 81–84.

— (1973): *Applinocrinus*, a new genus of Cretaceous microcrinoids, and its distribution in North America. *J. Paleont.* 47, 94–100.

Pisera, A. & Dzik, J. (1979): Tithonian crinoids from Rogoznik (Pieniny Klippen Belt, Poland) and their evolutionary relationships. *Eclogae Geol. Helv.* 72, 805–849.

Plodoski, G. (1996): Das Exponat des Monats: Eine Platte mit 44 Kronen freischwimmender Seelilien aus Marokko. *Natur und Museum* 126, 165–167.

Prauss, M., Ligouis, B. & Luterbacher, H. (1991): Organic matter and palynomorphs in the 'Posidonienschiefer' (Toarcian, Lower Jurassic) of Southern Germany, in: R. V. Tyson & T. H. Pearson (eds.), *Modern and Ancient Shelf Anoxia,* 335–351. Geological Society Special Publication 58,

Prokop, R. J. & Petr, V. (1986): Revision of superfamily Melocrinitacea d'Orbigny, 1852 (Crinoidea, Camerata) in Silurian and Devonian of Bohemia. *Sborník Národ. Muz. Praze (Acta Musei Nationalis Pragae)* 42, B, 197–219.

— (1987): *Marhoumacrinus legrandi*, gen. et sp. n. (Crinoidea, Camerata), from Upper Silurian–Lowermost Devonian of Algeria. *Sborník Národ. Muz. Praze (Acta Musei Nationalis Pragae)* 43, B, 1–14.

Quenstedt, F. A. (1874–1876): *Petrefactenkunde Deutschlands.* 1. Abt., 4. Band, *Echinodermen (Asteriden und Encriniden).* Fues's Verlag.

Ramsbottom, W. H. C. (1954): *The British Lower Palaeozoic Crinoidea.* Unpublished Ph.D. thesis, University of London. 290 pp.

– (1961): The British Ordovician Crinoidea. *Palaeontogr. Soc. Monogr.* **114** (492), 1–37.

Rasmussen, H. W. (1961): *A Monograph on the Cretaceous Crinoidea.* K. Danske. Vidensk. Selsk., Biol. Skrifter **12**. 428 pp.

– (1972): *Lower Tertiary Crinoidea, Asteroidea and Ophiuroidea from Northern Europe and Greenland.* K. Danske Vidensk. Selsk., Biol. Skrifter **19**. 83 pp.

– (1977): Function and attachment of the stem in Isocrinidae and Pentacrinidae: Review and interpretation. *Lethaia* **10**, 51–57.

– (1978): Articulata, in: R. C. Moore & C. Teichert (eds.), *Treatise on Invertebrate Paleontology.* Pt. T. *Echinodermata* 2, Vol. 3, T813–T928. Geological Society of America and University of Kansas Press.

Rickard, L. V. (1962): *Late Cayugan (Upper Silurian) and Helderbergian (Lower Devonian) Stratigraphy in New York.* New York State Museum and Science Service Bull. **386**. 157 pp.

Riddle, S. W. (1989): Functional morphology and paleoecological implications of the platycrinitid column. *J. Paleont.* **63**, 889–897.

Riddle, S. W., Wulff, J. I. & Ausich, W. I. (1988): Biomechanics and stereomic microstructure of the *Gilbertsocrinus tuberosus* column, in: R. D. Burke, P. V. Mladenov, P. Lambert & R. L. Parsley (eds.), *Echinoderm Biology: Proceedings of the Sixth International Echinoderm Conference, Victoria, British Columbia,* 641–648. Balkema.

Riegraf, W. (1985): *Mikrofauna, Biostratigraphie und Fazies im unteren Toarcium Südwestdeutschlands und Vergleiche mit benachbarten Gebieten.* Tübinger mikropaläontologische Mitteilungen **3**. Tübingen. 232 pp.

Ringueberg, E. N. S. (1890): The Crinoidea of the Lower Niagara Limestone at Lockport, New York. *N.Y. Acad. Sci., Ann. Rept.* **5**, 301–306.

Robison, R. A. (1965): Middle Cambrian eocrinoids from Western North America. *J. Paleont.* **39**, 355–364.

Rollins, H. B. & Brezinski, D. K. (1988): Reinterpretation of the crinoid–platyceratid interaction. *Lethaia* **21**, 207–218.

Rosenkranz, D. (1971): Zur Sedimentologie und Oekologie von Echinodermen-Lagerstätten. *N. Jb. Geol. Paläont. Abh.* **138**, 221–258.

Roux, M. (1980): Les Crinoïdes pédonculés (Echinodermes) photographiés sur les dorsales océanique de l'Atlantique et du Pacifique: Implications biogéographiques. *C.R. Acad. Sci. Paris* **291**, D, 901–904.

Sable, E. G. & Dutro, J. T., Jr. (1961): New Devonian and Mississippian formations in DeLong Mountains northern Alaska. *Amer. Ass. Petrol. Geol. Bull.* **45**, 585–593.

Schmidt, H. (1992): Mikrobohrspuren in Makrobenthonten des Oberen Muschelkalks von SW-Deutschland, in: H. Hagdorn & A. Seilacher (eds.), *Muschelkalk: Schöntaler Symposium 1991,* 271–278. Sonderbde. Ges. Naturk. Württ. **2**. Goldschneck-Verlag.

Schmidt, W. E. (1934): *Die Crinoideen des Rheinischen Devons.* Pt. 1. *Die Crinoideen des Hunsrückschiefers.* Preuss. Geol. Landesanstalt, Abh., N.S., **163**. 147 pp.

– (1942): *Die Crinoideen des Rheinischen Devons.* Pt 2. A. *Nachtrag zu: Die Crinoideen des Hunsrückschiefers,* B. *Die Crinoideen des Unterdevons bis zur Cultrijugatus-Zone (mit Ausschluss des Hunsrückschiefers).* Reichsstelle Bodenforsch. Abh., N.S., **182**. 253 pp.

Schubert, J. K., Bottjer, D. J. & Simms, M. J. (1992): Paleobiology of the oldest known articulate crinoid. *Lethaia* **25**, 97–110.

Schuchert, C. (1904): On siluric and devonic Cystidea and *Camarocrinus. Smithsonian Misc. Coll.* **47**, 201–272.

Schumacher, G. A. (1985): Fossil lag deposits: A key to Upper Cincinnatian crinoid biostratinomy. *Geol. Soc. Amer. Abstracts with Programs* **17**, 325.

Schumacher, G. A. & Ausich, W. I. (1985): Catastrophic sedimentation: Impact on a Late Ordovician crinoid assemblage. *Geol. Soc. Amer. Abstracts with Programs* **17**, 278.

Schumacher, G. A. & Meyer, D. L. (1986): Tempestites and variable crinoid preservation: Examples from the Upper Ordovician of Ohio and Indiana. *Geol. Soc. Amer. Abstracts with Programs* **18**, 323.

Scott, R. W., Root, S. A., Tenery, J. H. & Nestell, M. (1977): Morphology of the Cretaceous microcrinoid *Poecilocrinus* (Roveacrinidae). *J. Paleont.* **51**, 343–349.

Seilacher, A. (1960): Strömungsanzeichen im Hunsrückschiefer. *Notizbl. hess. Landesamt. f. Bodenforschg.* **88**, 88–106. Wiesbaden.

– (1961): Ein Füllhorn aus dem Hunsrückschiefer. *Natur und Volk* **91**, 15–19.

– (1970): Begriff und Bedeutung der Fossil-Lagerstätten. *N. Jb. Geol. Paläont. Mh.* 1970, 34–39. Frankfurt A. M. Stuttgart.

– (1990a): Taphonomy of Fossil-Lagerstätten: Overview, in: D. E. G. Briggs & P. C. Crowther (eds.), *Paleobiology: A Synthesis,* 266–270. Blackwell.

– (1990b): Die Holzmadener Posidonienschiefer: Entstehung der Fossillagerstätte und eines Erdölmuttergesteins, in: W. K. Weidert (ed.), *Klassische Fundstellen der Paläontologie* 2, 107–131. Goldschneck-Verlag.

– (1993): Fossillagerstätten im Muschelkalk, in: H. Hagdorn & A. Seilacher (eds.), *Muschelkalk: Schöntaler Symposium 1991,* 215–222. Sonderbde. Ges. Naturk. Württ. **2**. Goldschneck-Verlag.

Seilacher, A., Drozdzewski, G. & Haude, R. (1968): Form and

function of the stem in a pseudoplanktonic crinoid (*Seirocrinus*). *Palaeontology* **11**, 275–82.

Seilacher, A. & Hemleben, C. (1966): Beiträge zur Sedimentation und Fossilführung des Hunsrückschiefers, 14, Spurenfauna und Bildungstiefe der Hunsrückschiefer (Unterdevon). *Notizbl. Hess. Landesamt. f. Bodenforschg.* **94**, 40–53.

Sevastopulo, G. D., Ausich, W. I. & Franzén-Bengston, C. (1989): Echinoderms, in : C. H. Holland & M. G. Bassett (eds.), *A Global Standard for the Silurian System*, 264–267. National Museum of Wales, Geol. Ser. **9**. Cardiff.

Shaver, R. H., Ault, C. H., Ausich, W. I., Droste, J. B., Horowitz, A. S., James, W. C., Okla, S. M., Rexroad, C. B., Suchomel, D. M. & Welch, J. R. (1978): *The Search for a Silurian Reef Model Great Lakes Area*. Indiana Geol. Survey Spec. Rep. **15**. 36 pp.

Sieverts, H. (1927): Über die Crinoidengattung Marsupites. Abh. Preuss. Geol. Landesanstalt, N. F., **108**. 73 pp.

Signor, P. W. & Brett, C. E. (1984): The mid-Paleozoic precursor to the Mesozoic marine revolution. *Paleobiology* **10**, 229–245.

Simms, M. J. (1986): Contrasting lifestyles in Lower Jurassic crinoids: A comparison of benthic and pseudopelagic Isocrinida. *Palaeontology* **29**, 475–493.

(1988a): The phylogeny of post-Paleozoic crinoids, in: C. R. C. Paul & A. B. Smith (eds.), *Echinoderm Phylogeny and Evolutionary Biology*, 269–284. Clarendon Press.

(1988b): *An intact Comatulid Crinoid from the Toarcian of Southern Germany*. Stuttgarter Beitr. Naturk. Ser. B, **140**. 7 pp.

(1989): British Lower Jurassic crinoids. *Palaeontogr. Soc. Monogr.* **142** (581), 1–103.

(1990): The radiation of post-Palaeozoic echinoderms, in: P. D. Taylor & G. P. Larwood (eds.), *Major Evolutionary Radiations*, 287–304. Clarendon Press.

(1994a): Reinterpretation of thecal plate homology and phylogeny in the class Crinoidea. *Lethaia* **26**, 303–312.

(1994b): A new interpretation of crinoid thecal plate homology and phylogeny, in: B. David, A. Guille, J.-P. Féral & M. Roux (eds.), *Echinoderms through Time*, 257–263. Balkema.

(1995): Phylogenetic relationships of 'aberrant' Ordovician crinoids, in: R. H. Emson, A. B. Smith & A. C. Campbell (eds.), *Echinoderm Research 1995: Proceedings of the 4th European Echinoderm Colloquium, 1995*, 223–228. Balkema.

Simms, M. J., Gale, A. S., Gilliland, P., Rose, E. P. F. & Sevastopulo, G. D. (1993): Echinodermata, in: M. J. Benton (ed.), *The Fossil Record*, 2d ed., 491–528. Chapman & Hall.

Simms, M. J. & Ruffell, A. H. (1990): Climatic and biotic change in the late Triassic. *J. Geol. Soc. London* **147**, 321–327.

Simms, M. J. & Sevastopulo, G. D. (1993): The origin of articulate crinoids. *Palaeontology* **36**, 91–109.

Smith, W. (1816): *Strata identified by organised fossils, containing prints on coloured paper of the most characteristic specimens in each stratum.*

(1817): *Stratigraphical System of Organised Fossils with Reference to the Specimens of the Original Geological Collection in the British Museum, Explaining Their State of Preservation and Their Use in Identifying British Strata*. J. Cary.

Spencer, W. K. & Wright, C. W. (1966): Asterozoans, in: R. C. Moore & C. Teichert (eds.), *Treatise on Invertebrate Paleontology*. Pt. U, Echinodermata 3, U4–U107. Geological Society of America and University of Kansas Press.

Springer, F. (1901): *Uintacrinus: Its Structure and Relations*. Harvard College Museum Comp. Zool., Mem. **25**. 89 pp.

(1917): *On the Crinoid Genus Scyphocrinus and Its Bulbous Root Camarocrinus*. Smithsonian Inst. Publ. **2440**. 74 pp.

(1926): *American Silurian Crinoids*. Smithson. Inst. Publ. **2871**. 239 pp.

Sprinkle, J. (ed.). (1982): *Echinoderm Faunas from the Bromide Formation (Middle Ordovician) of Oklahoma*. University of Kansas Paleont. Contrib. Art. **1**. 369 pp.

Sprinkle, J. & Collins, D. (1995): *Echmatocrinus* revisited: Still an echinoderm and probably the earliest crinoid. *Geol. Soc. Amer. Abstracts with Programs* **27**, A113.

Stilwell, J. D., Fordyce R. E. & Rolfe, P. J. (1994): Paleocene isocrinids (Echinodermata: Crinoidea) from the Kauru Formation, South Island, New Zealand. *J. Paleont.* **68**, 135–141.

Strimple, H. L. (1963): *Crinoids of the Hunton Group (Devonian–Silurian) of Oklahoma*. Oklahoma Geol. Survey Bull. **100**. 169 pp.

Strimple, H. L. & Koenig, J. W. (1956): Mississippian microcrinoids from Oklahoma and New Mexico. *J. Paleont.* **30**, 1225–1247.

Strimple, H. L. & Moore, R. C. (1971): *Crinoids of the La Salle Limestone (Pennsylvanian) of Illinois*. University of Kansas Paleont. Contrib, Echinodermata, Art. **11**. 48 pp.

Struve., W. (1957): Ein Massengrab kreidezeitlicher Seelilien. *Natur und Volk* **87**, 361–373.

Stukalina, G. A. (1988): Studies in Paleozoic crinoid-columnals and -stems. *Palaeontographica* A **204**, 1–66. Stuttgart.

Stürmer, W. & Bergström, J. (1973): New discoveries on trilobites by X-rays. *Paläont. Z.* **47**, 104–141.

Südkamp, W. (1992): Nicht alltäglich: Seelilie auf Kieselschwamm. *Fossilien* **9**, 179–181.

(1995): Echinodermen aus dem Hunsrückschiefer. Pt. 2. Seelilien. *Fossilien* **12**, 113–126.

Talbot, M. (1905): Revision of the New York Helderbergian crinoids. *Amer. J. Sci.* **20**, 17–35.

Taylor, P. D. (1983): *Ailsacrinus* gen. nov., an aberrant millericrinid from the Middle Jurassic of Britain. *Bull. British Museum (Natural History), Geology* **37**, 37–77.

Taylor, W. L. & Brett, C. E. (1996): Taphonomy and paleoecology of echinoderm Lagerstätten from the Silurian (Wenlockian) Rochester Shale. Palaios 11, 118–140.

Teichert, C. (1949): Permian Crinoid Calceolispongia. Mem. Geol. Soc. Amer. 34. 132 pp.

— (1951): The marine permian faunas of Western Australia. Paläont. Z. 24, 76–90.

— (1954): A new Permian crinoid from Western Australia. J. Paleontol. 28, 70–75.

Thompson, T. A. (1990): Architectural Elements and Paleoecology of Carbonate Shoal and Intershoal Deposits in the Salem Limestone (Mississippian) in South-Central Indiana. Indiana Geol. Survey Guidebook 14. 75 pp.

Thompson, T. L. (1986): Paleozoic Succession in Missouri. Pt. 4. Mississippian System. Missouri Department of Natural Resources Division of Geology and Land Survey, Report of Investigations 70. 182 pp.

Tobin, R. C. (1986): An assessment of the lithostratigraphic and interpretative value of the traditional 'biostratigraphy' of the type Upper Ordovician of North America. Amer. J. Sci. 286, 673–701.

Trümpy, R. (1957): Palaeozoische und mesozoische Geschichte der westlichen und zentralen Sahara. Bull. Ver. Schweiz. Petrol-Geol. u. Ing. 24, 55–69.

Ubaghs, G. (1953): Classe des Crinoïdes, in: J. Piveteau (ed.), Traité de Paléontologie 3, 658–773.

— (1956a): Recherches sur les Crinoïdes Camerata du Silurien de Gotland (Suède). Introduction générale et Partie I: Morphologie et paléobiologie de Barrandeocrinus sceptrum Angelin. Arkiv f. Zoologi, Ser. 2, 9 (26), 515–550.

— (1956b): Recherches sur les Crinoïdes Camerata du Silurien de Gotland (Suède). Partie II: Morphologie et position systématique de Polypeltes granulatus Angelin. Arkiv f. Zoologi, Ser. 2, 9 (27), 551–572.

— (1958): Recherches sur les Crinoïdes Camerata du Silurien de Gotland (Suède). Partie III: Melocrinicae. Arkiv f. Zoologi, Ser. 2, 11 (16), 259–306.

— (1969): Aethocrinus moorei Ubaghs, n. gen., n.sp., le plus ancien crinoïde dicyclique connu. University of Kansas Paleont. Contrib. Art. 38. 25 pp.

— (1978): Skeletal morphology of fossil crinoids, in: R. C. Moore & C. Teichert (eds.), Treatise on Invertebrate Paleontology. Pt. T. Echinodermata 2, Vol. 1, T58–T216; Camerata, Vol. 2, T408–T519. Geological Society of America and University of Kansas Press.

Urlichs, M., Wild, R. & Ziegler, B. (1994): Der Posidonien-Schiefer des unteren Juras und seine Fossilien. Stuttgarter Beitr. Naturk. Ser. C, 36. 95 pp.

Van Sant, J. F. & Lane, N. G. (1964): Crawfordsville (Indiana) Crinoid Studies. University of Kansas Paleont. Contrib., Echinodermata, Art. 7. 136 pp.

Verniory, R. (1961): Présence de Saccocoma quenstedti Doreck (in coll.) dans les gorges de la Méouge (Sisteron – Provence). Arch. Sci. Genève 14, 315–320.

— (1962): Une nouvelle forme de Saccocoma (Montbrand, Hautes-Alpes, France). Arch. Sci. Genève 15, 391–397.

Viohl, G. (1994): Fish taphonomy of the Solnhofen Plattenkalk: An approach to the reconstruction of the palaeoenvironment. Géobios M.S. 16, 81–90.

— (1996): The paleoenvironment of the Late Jurassic fishes from the southern Franconian Alb (Bavaria, Germany), in: G. Arratia & G. Viohl (eds.), Mesozoic Fishes: Systematics and Paleoecology, International Meeting, Eichstätt, 1993, 513–528. Verlag Dr. Pfeil.

Vollrath, A. (1958): Beiträge zur Paläogeographie des Trochitenkalks in Baden-Württemberg. Jh. geol. Landesamt Baden-Württemberg 3, 181–194. Freiburg.

Wachsmuth, C. & Springer, F. (1880–1886): Revision of the Palaeocrinoidea. Philadelphia Acad. Nat. Sci. Proc., 1880, 226–378; 1881, 177–422; 1885, 225–364; 1886, 64–226.

— (1897): The North American Crinoidea Camerata. Harvard College Museum Comp. Zool., Mem. 20, 21. 897 pp.

Walker, K. R., Read, J. F. & Hardie, L. A. (1989): Cambro-Ordovician Carbonate Banks and Siliciclastic Basins of the United States Appalachians. 28th Int. Geol. Congress Field Trip Guidebook T161. 88 pp.

Walther, J. (1886): Untersuchungen über den Bau der Crinoiden mit besonderer Berücksichtigung der Formen aus dem Solnhofener Schiefer und dem Kelheimer Diceraskalk. Palaeontographica 32, 155–200.

— (1904): Die Fauna der Solnhofener Plattenkalke, bionomisch betrachtet. Festschrift der Medizinisch-naturwissenschaftlichen Gesellschaft zu Jena 11, 133–214.

Wanner, J. (1920): Ueber armlose Krinoiden aus dem jüngeren Palaeozoikum. Verh. Geol.-mijnb. Genoots. Nederland en Koloniën, Geol. Ser. 5, 21–35.

— (1924): Die Permischen Krinoiden von Timor. Verhand. Mijnw. Ned. Oost-Indië, 1921, Pt. 3. 348 pp.

— (1937): Neue Beiträge zur Kenntnis der permischen Echinodermen von Timor, VIII–XIII. Palaeontographica Suppl. 4 (4), Pt. 2, 59–212.

— (1940): Neue Beiträge zur Kenntnis der permischen Echinodermen von Timor, XIV. Poteriocrinidae, 3. Palaeontographica Suppl. 4, 215–242.

Warn, J. M. & Strimple, H. L. (1977): The Disparid Inadunate Superfamilies Homocrinacea and Cincinnaticrinacea (Echinodermata Crinoidea), Ordovician – Silurian, North America. Bull. Ameri. Paleont. 72 (296). 138 pp.

Waters, J. A. & Maples, G. C. (1991): Mississippian pelmatozoan community reorganization: A predation-mediated faunal change. Paleobiology 17, 400–410.

Watkins, R. (1991): Guild structure and tiering in a high-diversity Silurian community, Milwaukee County, Wisconsin. Palaios 6, 465–478.

Watkins, R. & Hurst, J. M. (1977): Community relations of Silurian crinoids at Dudley, England. Paleobiology 3, 207–217.

Webster, G. D. (1987): Permian crinoids from the type-section of the Callytharra Formation, Callytharra Springs, Western Australia. *Alcheringa* **11**, 95–135.

— (1990): New Permian crinoids from Australia. *Palaeontology* **33**, 49–74.

Webster, G. D. & Lane, N. G. (1967): *Additional Permian Crinoids from Southern Nevada*. University of Kansas Paleont. Contrib., Art. 27. 32 pp.

— (1987): *Crinoids from the Anchor Limestone (Lower Mississippian) of the Monte Cristo Group, Southern Nevada*. University of Kansas Paleont. Contrib., Art. 119. 55 pp.

Weir, G. W. & Peck, J. H. (1968): Lithofacies of Upper Ordovician rocks Exposed beneath Maysville and Stanford, Kentucky. *U.S. Geol. Survey Professional Pap.* **600-D**, 162–168.

Welch, J. R. (1976): Phosphannulus on Paleozoic crinoid stems. *J. Paleont.* **50**, 218–225.

Westhead, S. (1979): Carboniferous crinoids from the Clitheroe area. *Proc. NE Lancashire Group Geol. Ass.* **2**, 465–496.

Wiedenmayer, F. (1979): Modern sponge bioherms of the Great Bahama Bank and their likely ancient analogues, in: C. Levi & N. Boury-Esnault (eds.), *Biologie des Spongiaires*, 289–296. Colloques Int. du C.N.R.S. **291**.

Wignall, P. B. & Simms, M. J. (1990): Pseudoplankton. *Palaeontology* **33**, 359–378.

Wilkie, I. C. & Emson, R. (1988): Mutable collagenous tissues and their significance for echinoderm palaeontology and phylogeny, in: C. R. C. Paul & A. B. Smith (eds.), *Echinoderm Phylogeny and Evolutionary Biology*, 311–330. Clarendon Press.

Willink, R. J. (1979): The crinoid genera *Tribrachyocrinus* McCoy, *Calceolispongia* Etheridge, *Jimbacrinus* Teichert and *Meganotocrinus* n. gen. in the Permian of eastern Australia. *Palaeontographica A* **165**, 137–194.

Witzke, B. J. (1983): Silurian benthic invertebrate associations of eastern Iowa and their paleoenvironmental significance. *Wisconsin Academy Sci. Trans.* **71**, 21–47.

Witzke, B. J. & Strimple, H. L. (1981): Early Silurian camerate crinoids of eastern Iowa. *Proc. Iowa Acad. Sci.* **88**, 101–137.

Wright, J. (1950–1960): *A Monograph of the British Carboniferous Crinoidea*, 2 vols. Palaeontogr. Soc. Monogr. 347 pp.

Yeltisheva, R. S. (1955): Klass Crinoidea – morskie liliy, in: O. I. Nikiforova (ed.), *Field Atlas of Ordovician and Silurian Faunas of the Siberian Platform*, 40–47. VSEGEI (in Russian).

— (1956): Stebli morskikh i ikh klassifikatsija (Stems of crinoids and their classification). *Vest. Leningr. gos. Univ. (Geol. Geog.)* **12**, 40–46 (in Russian).

Yeltisheva, R. S. & Polyarnaya, Zh. A. (1986): Parastratigraphic groups of flora and fauna of the Triassic–Marine organic remains: Finds of traumatocrinids in the Triassic deposits of the Novosibirsk Islands. *Tr. Vses. Nauchno Issled Geol. Inst. A. P. Karpinskogo* **334**, 112–116, 261–263 (in Russian).

Zardini, R. (1976): *Fossili di Cortina*. Cortina d'Ampezzo. 29 pp.

Zenker, J. C. (1833): *Beiträge zur Naturgeschichte der Urwelt*. Friedrich Mauke.

Ziegler, P. A. (1982): *Geological Atlas of Western and Central Europe*. Elsevier and Shell Int. Petrol. Maatschappij B.V. 130 pp.

Zitt, J. (1983): Spoon-like crinoids from Stramberk (Lower Cretaceous, CSSR). *Acta Musei Nationalis Pragae* **39B** (2) 69–114.

General Index

Aalenian, 203
aboral, 4, 5, 7
aboral cup, 3, 17
abyssal 240, 244
Acadian, 46, 134
accretion disc, 13
acrothoracian barnacles, 170, 198
active filter feeder, 29
adoral, 4, 5
Aegean, 165
Agricola, Georgius, 164
algae, 45, 52, 66, 89, 195, 223
algal mats, 104, 195
Algonquin Arch, 105
all or nothing rule, 180
Alnif, 95
ambulacra, 3
ambulacral epidermis, 4
ambulacral furrows, 20
ambulacral groove, 5, 22, 23, 153
ambulacral plates, 29
ambulacrals, 23
American Museum of Natural History, 108, 109
ammonia, 127
ammonites, 47, 86, 177, 184, 221, 226
ammonoids, 114
anal plates, 17, 23, 32, 33, 35, 36, 153
anal pyramid, 20, 23
anal sac, 20, 22, 23, 35, 36, 40, 66, 71, 79, 89, 117, 125, 133, 152, 160
anal tube, 5, 20, 23, 24, 57, 99, 114, 141, 143–145, 151, 227
anal X, 40
anchorage, 15, 125, 174, 179
anchor-like holdfast, 14, 15
Anchor Limestone, 45
Anisian, 165
annelids, 198
anoxic conditions, 53, 179, 195, 223, 231
Antarctic Peninsula, 233, 234

Antelope Spring, xiv
anus, 3, 23, 56, 79, 80, 143
apical cirrus, 16
Appalachian Basin, 92
aragonite, 223, 247
arborescent bryozoan(s), 199
Arbuckle Mountains, 44
archaeogastropods, 56, 79
Arenig, 38
areola, 7, 8
argillaceous rocks, 247
Argovian, 47
Arkona Shale, 53, 129–131
Armorican Massif, 184
arms, 24–30
arms, branching, 27, 28, 159, 181, 189
arthropods, 112–114, 184
articular facet, 8–12, 16, 26, 151
articulation, 4, 5, 7, 9–12, 15, 24–26
Artinskian, 162, 163
Arzo, 41
Ashgill, 42
asteroids, 31, 42, 75, 88, 137, 167, 212, 218
asymmetry, 21, 47, 243
atomous arms, 42, 156, 163
atrypids, 123
attachment, 10, 12–15, 114, 181
attachment disc, 13, 42, 180
Ausable River, 129, 130
Australia, 160–163, 226
Austroalpine faunal province, 166
autecology, 55
autochthonous, 45, 222
autotomy, autotomization, 30, 37, 107, 174, 179, 189, 201, 218
axial canal, 13, 16, 99, 171, 173
axillary, 28

bactritids, 129
Bad Boll, 183
Bahama Bank, 223

Bahamas, 223, 237, 240, 241
Bailey Formation, 97, 98
Bajocian, 203–205
Bakony, 167
Banan Formation, 43
banana plantation, 181
Banff Formation, 52
Bangor Limestone, 46
barnacles, 170, 198
barnacle-like, 241
Bartels Collection, 119, 121
basal, 18
basal circlet, 24, 38
basal disc, 13
basal tubercle, 16
basket star, 27
Basleo Beds, 22, 161
Bassett, D. A., 146
Bassler, R. S., 75
Bath Freestone, 197
Bathonian, 47, 59, 197–199, 205, 212
Bavaria, 54, 216
Bayerische Staatssammlungen für Paläontologie und historische Geologie, 120, 224
Beane, B. H., 136, 138
Beecher, Charles, 103
Beecher slab, 104
Belbek valley, 48
belemnites, 184
Bellevue Formation, 75
Bellman Quarry, 46
Beloit College, 138
benthic, 7, 53, 88, 153, 179, 194, 221, 231
Berriasian, 48
Besleo = Basleo, 22
Beuthen, 164
bicarbonate cements, 127
bifascial, 10, 25
bifrons Zone, 183, 191
bilateral symmetry, 20, 35
bioclastic limestone, 83, 197, 199, 248

bioherms, 45, 48, 81, 169
biomicrite, 248
biosparite, 248
biostrome, 47, 105
bioturbation, bioturbated, 63, 89, 195, 209
Birmenstorf Beds, 47
biserial, 28, 30
biserially pinnulate, 27, 36
Bithynian, 165
bituminous, 93, 183, 194
bivalves, 56, 170, 179, 183, 195, 199, 215, 236
Blackriveran, 42
black shales, 43, 53, 63, 99, 155, 183
Black Ven, Black Ven Marls, 25, 177
blastozoans, 13, 31
blood vessel, 5
Blue Mountain, 64
Blumennester, 174
Bobcaygeon Formation, 69
Bohemia, 97, 166, 217
Bohemian Massif, 184
Bond Formation, 155
Bonifatiuspfennige, 164
Borden Formation, Group, 58, 146
borehole, 56, 58, 174
Boruszowice Beds, 165
bothriocidaroids, 42
boulders, 247
brachials, 18, 24–28
brachioles, 133
brachiopods, 65, 89, 123, 129
brachitaxis, 28
Bradford Clay, 197
Bradford hardground, 198, 200
Bradford-on-Avon, 197
Bradley, F. H., 146
Brassel, G., 121
Braun, Frederick, 87
Brechin, 68
Bridlington, 46
Bristol City Museum, 198
brittle stars, 206
Broccatello, 41
Brockport, 87
Broili, Ferdinand, 120
Bromide Formation, 44
Brownsport Formation, 44
bryozoan, bryozoans, 65, 89, 124
Buckland, William, 177
Buffalo Museum, 92
bulb, 15, 96–102
Bundenbach, 111–114, 117, 119–121
Bundesanstalt für Geowissenschaften und Rohstoffe Berlin, 176
Buntsandstein, 165
buoy, 15, 98
buoyancy, 12, 102, 167, 189, 221
Burgen in Endre, 86
Burgess Shale, 38
Burgundy Gate, 166
Burgundy Platform, 205, 211
Burgundy Strait, 167
Burleigh Hill Member, 87
Burlington Limestone, 41, 52, 139–143

Burlington Shelf, 140
burrowers, 50, 74, 179
burrowing bivalves, 195
burrows, 74, 209, 223
byssus, byssal, 169, 171, 174, 195

calcarenite, 63, 248
calcareous marl, 247
calcareous shale, 63, 87, 248
calcite endoskeleton, 3
Caleb family, 87
callianassid crustacean, 215
Callytharra Springs, 163
calyx, calyces, 4, 17–19
Cambrian, 7, 38
Campanian, 47
Cape Girardeau, 97
Caput Medusae, 183
Caradocian, 63
carbonate sediments, 45, 52, 247
Carboniferous, 23, 31, 39, 40, 43, 45, 57
Carden Quarry, 68
Carlile Shale, 225
Carnian, 43, 47, 167, 176
carpoid, 108
Cedarville Dolomite, 45
Celtic facies, 205
cementation, 53, 63, 72, 107, 127, 171, 200
cemented attachment, 12
Cenomanian, 242
Cenozoic, 42
Centennial Road, 129
central canal, 7, 10, 16
centrodorsal, 15, 16, 37, 48, 212, 233
cephalopods, 114, 221, 225
Chadian, 45
Chalk, chalk, 37, 46, 53, 225, 226, 231, 232, 248
Charmouth, 25, 178, 181
chert, 135, 248
Chesterian, 46
chitinoclastic bacteria, 54
chonetids, 123, 129
ciliated, 3, 4
Cincinnati, 63, 75
Cincinnati Arch, 105
Cincinnatian, 75, 76
circlet, 17, 31, 32, 35–38
circular columnals, 11
circum-Antarctic current, 234
cirral, 16
cirri, 4, 5, 7, 9, 15–17
cirrinodal, 7, 37, 175
cirrus lobolith, 95, 98
cirrus socket, 16
cladistic, 174
classification, 31, 32, 38, 42
clastic sediments, 162, 247
clay, 247
Clinton Group, 87
clionid sponges, 198
Clitheroe, 45, 46
Clitheroe Castle Museum, 46
Coal Measures, 155
coccolithophorids, 221, 223

coccoliths, 184, 216, 221, 223, 248
coelom, 4, 7
coelomic canal, 4
Coeymans Formation, Limestone, 52, 103–105
coiled stem, 12
collagen fibrils, 16
colonies, 122, 135, 181, 209, 212, 231
column, 4
columnals, 5, 7
commensal, commensalism, 55–58, 125
competition, 12, 55
compound radials, 35
cones, 19
connective tissue, 4, 5, 50
contractile tissue, 13
Coplow Quarry, 46
coprophagous, 56, 80, 123
corals, 6, 13, 43, 45, 48, 56, 57
Corey, O. W., 146
Corey's Bluff, 146
Corryville Formation, 78
Cotswolds, 199
Crailsheim, 164, 167, 169
Crailsheim Member, 165, 167, 169
Crawfordsville, 45, 53, 145
crayfish, 217
creeping roots, 13
crenulae, 8, 11
crenularium, 12
Cretaceous, 47, 48, 53, 54, 225
Cretaceous-Tertiary (K/T) boundary, 223
Crimea, 42, 48, 59
crinoid lenses, 52, 177
crinoidal limestone, 41, 52, 73
crown, 4, 12, 27, 42
crustacean, 200, 215, 217
cryptodicyclic, 36
cryptosymplexy, cryptosymplectial, 9, 15, 26, 175, 142
crypto-synarthrial, 218
cup, 3, 4, 17
cyanobacteria, 192, 224
cyclical sedimentation, 46, 155
cyclocrinitids, 66
cyclocystoids, 42, 75
cyclothem, 155, 157
cystoids, 56, 87, 105, 114

Dachschiefer, 111
Dakota Formation, 225
Dakota Sandstone, 227
Danian, 233
Days Corners Quarry, 104
Dayville Member, 104
Deansboro Member, 104
death tracks, 217
decapods, 217
Decew Formation, 88
Deep Run Shale, 122
Deep Sea Drilling Project, 224
dehydration, 223
Denmark Member, 64, 65
density, arm branching, 27
Deutsches Bergbau-Museum Bochum, 121

Develier, 204
Devonian, 13, 15, 38, 45, 93, 103, 111, 122, 129
diagenesis, 42, 127, 223
diastem, 248
dicyclic, 18
digestive system, tract, 3, 17, 79
dioecious, 30
Diplopora Dolomite, 165
diploporans, 31
disarticulation, 23, 50, 179
discoidal holdfast, 71, 117, 134, 174
distal, 5, 7, 8
diversification, 38, 176, 237
division series, 16, 28
Dolomites, 176
Domerian, 41
dorsal, 5
dorsal cups, 233
Dorset, 44, 177
Dotternhausen, 183
Dotternhausen Werkforum, 191, 196
Dra Plain, 93
Drake Passage, 234
driftwood, 44, 51, 177–182, 184, 185, 189, 233
Dudley, 43
Dudley Castle, 43
Dudley Museum and Art Gallery, 43
Dunleith Formation, 42
dysaerobic, 63, 66

East Carpathian Gate, 166
echinoid(s), 44, 48, 52, 55, 161, 196, 233
ecological niche, 55, 86, 153
ectoderm, 24
Edenian, 75, 78
edrioasteroids, 71, 90, 108
Edwardsville Formation, 45, 53, 145
eel grass, 231, 238
Eichstätt, 216, 217, 223, 224
Eifel, 45, 117
Eifelian, 45
Eke Beds, 82
Elkader, 226
Elkhorn Formation, 75
Emsian, 111
encrinite, 41, 46, 52, 140, 145
Encrinus Platten, 169
encrusting algae, 195
endoskeleton, 3
endotomous, 28, 43, 181, 189
Eocene, 233, 234
epidermis, 4, 86
epithelial cells, 5, 16
epizoan, 56–59
Erfoud, 93, 94
Erkerode, 164
Eschenbach-Bocksberg, 119–121
eustatic sea level, 75, 155
evolution, 31, 98, 107, 162
evolutionary history, 31, 38

faecal, 5, 23, 56, 248
faeces, 221

Fairmount Formation, 78
Falang Formation, 43
falciferum Zone, 183, 191
Famennian, 39
Fall Brook, 122
Fassanian, 165
favositid corals, 105
Fäule, 216, 224, 248
feather stars, 237
feeding, 26
feeding strategies, 26, 55
fenestrates, 123
Fennoskandia, 166
Fern Glen Formation, 140
Field Museum of Natural History, 80, 138, 144, 154, 241
Fiji, xi
filament, 16, 17
filtration fan, xii, 25–28, 241
Finger Lakes, 122
firmground, 167, 248
fishes, 221, 226, 238
fists, 20
fistuliporoid bryozoans, 123, 125
fixed (brachials, interbrachials, interradials, pinnulars), 18, 24, 30, 32
Flat Creek Shale, 64
Fleins, 183
flexibility, 10, 15, 24
flexible ligament, 10
Flinz, 216, 223
floating, 22, 47, 98, 178–181, 189, 218, 230
floating log, 180, 189, 192
flying reptiles, 183, 217
Follingbo, 83, 85
food-gathering, 3–5, 24, 26
food groove, 4, 21, 26, 27, 221, 230
food processing, 19
foraminifera, 170, 184, 195, 225, 248
Forest Marble Formation, 197, 203
Fort Hayes Limestone, 227
Fort Payne Formation, 54, 154
Franconian Alb, 183
Frasnian, 38
Freyburg an der Unstrut, 173
Fringelites, 51
fulcral ridge, 9, 11, 15, 25, 151
fulcrum, 10, 25

Gagat, 184
Gaismühle, 167
Galena Group, 42
gall, 55, 58
galleried stereom, 5
gap (pinnules), 28
Gascoyne River, 162
gastropods, 56, 79, 123, 125, 131
Gemünden, 111, 117
Gelbe Basisschichten Member, 173
genital cord, 5
genital pinnule, 29, 30
Geologisch-Paläontologisches Institut und Museum Universität Göttingen, 94, 102, 176
Germanic Basin, 166, 174

Gerolstein, 45
Girvan district, 42
Givetian, 45, 122, 129
glacio-eustatic, 155
Glen Dean Limestone, 46
Glenmark Shale, 88
Gmünd, 52
Gogolin, 165, 174
Gogolin Beds, 165
Golden Cap, 179
gonads, 24, 30
goniatites, 129
Gorazdze Beds, 165
gorgonocephalid ophiuroids, 27
Gotland, 52, 81
grade groups, 37
grainstone, 46, 248
Grand Bahama Island, xii
Grande Comore, 242
Graneros Shale, 225
grapnel-like holdfast, 15
grappling devices, 13
graptolites, 93
graptolitic shales, 93
gravity flow, 223
Great Barrier Reef, 237
Great Lakes, 45
Great Meteor Bank, 242
green algae, 52
Greenhorn Limestone, 225
gregarious, 171, 212, 231
Grenville, 73
growth, growth rate, 7, 24, 46, 171, 181, 189, 206
Guadalupian, 161
Guam, 238
Guanling, 43
Guizhou, 43
Gulf Coast, 47
Gull River Formation, 69
Gundolding, 216
Gutendorf, 173
gypsum, 225
gyre, 182

haemal system, 4, 17
Hainzen, 183
Hall, James, 135
Hallstätter Kalk, 47
Hamilton Group, 53, 122, 129
Hampton Formation, 135
Harbor Branch Oceanographic Institution, 240
hardground, 14, 42, 47, 52, 68, 108, 174, 198, 238, 248
Harenberg, Johann Christophorus, 164
Haroum, 93
Hassmersheim Member, 167, 169
Hauff Museum, 184, 196
Hauptrogenstein, 52, 205
Helderberg Group, 104
Hemse reefs, 81
Herkimer County, 103
heteromorphic stem, 97, 107
heterotomous branching, 27, 189, 209

hexactinellid sponges, 166
Hexengeld, 164
Hicks, Jason, 104
Hiemer, 183
Hierlatz limestone, 41
high-energy environments, 12, 52, 66, 74, 108, 197, 209
Hildesheim, 164
hinged crown, 13
Högklint Beds, 81
Hogsty Reef, 240
Hohlwurzel, 13
holdfast, 13–15
Holocrinus-Bank, 165
holothurians, 114, 167, 218
holothurian sclerites, 47
Holston Formation, 41
Holzmaden, 184, 196
homalozoans, 114
Homerian, 87
Homocrinus Beds, 87–90
Honduras, xi
Hook Head, 43, 45
Hook Head Formation, 43
Hopkinton Dolomite, 45
horseshoe crab, 217
Hottwil, 203, 212
House Range, xiv
Hovey, Horace, 145
Humboldt University Berlin, 120, 176
Hungry Hollow Formation, 130
Hunsrück, 23, 53
Hunsrück Slate, Hunstrückschiefer, 45, 111
Hunterian Museum, 43
Huntington Dolomite, 45
Hybonoticeras hybonotum Zone, 216
hydrocarbons, 184, 209
hydropore, 37
hypersaline, hypersalinic, 54, 195, 223

Iapetus Ocean, 43
iceberg strategy, 231
Illinois Basin, 140, 154
Illinois Geological Survey, 159
Illyrian, 165
Indian Creek, 56, 146, 150
infauna, 117, 153, 167, 169, 174
infrabasal, 18, 38
infranodal, 26, 175, 176
inoceramid clams, 225
Institut und Museum für Geologie und Paläontologie Universität Tübingen, 176, 190, 196
interbrachials, 19, 24, 32
internodals, 7–9
interradial processes, 21
interradials, 8, 19, 30, 32
interradius, 19, 30
Irondequoit, 88
iron oolites, 203
isotomous branching, 28

Jago, 242
Jeddo Creek, 87
jellyfish, 114, 217

Jerusalem Hill, 103
jet, 184
Jew Stones, 164
Jissoumour, 95
Johnson Sea Link submersible, xii
Jura Museum Eichstätt, 224
Jurassic, 21, 36, 37, 40–42, 44, 47, 48, 53, 59, 177, 183, 197, 203, 216
juvenile, 12, 95, 99, 114, 170, 206, 218, 232

Kalkberg Formation, 104
Karkovice Beds, 165
Kashong Creek, 122
Kashong Member, 123
Kaub, 113
Keasey Formation, 233
Kelheim, 216, 222
Kennet and Avon Canal, 197
Keokuk Limestone, 41, 140, 154
kerogen, 194
Keuper, 165
Kimmeridgian, 44
Kinderhookian, 135
Kirkfield, 68, 71
Kirkfield Limestone, 68
klados, 36
Klingnau Formation, 205, 212
Koblenzer, 183
Kogruk Formation, 41
Konglomeratbänke, 165
Kope Formation, 75
Kotel Island, 30
Krumme Lagen, 216, 223
Kueichow, 43
Kutscher, F., 120

labyrinthic stereom, 5
Lachmund, Fridericus, 164
Ladinian, 165
Lady Burn Starfish Beds, 42
Lagerstätten, xv
lagoon, lagoonal, 6, 72, 104
Lake Simcoe, 68
Lambton County, 129
lamellibranchs, 112, 189
La Meseta Formation, 234
Lancashire, 45
Lane, Gary, 147
Langenaltheim, 216
La Pouza, 59
Lapworth Museum, 43
La Rochelle, 44
larvae, 12, 99, 171, 180
larval, 22, 67, 99, 171, 180
LaSalle Member, Limestone, 155
Laurentia, 63
Lausen Quarry, 206
Lebanon, 54
Lebanon Limestone, 42
LeGrand, 51, 56, 135
Lehmann, W. M., 121
Lettenkeuper, 165
Leuthardt, Franz, 203
Lewiston Member, 87
Lias, liassic, 41, 183, 195

Liberty Formation, 78
lichenocrinid, 42, 76, 78
Liesberg, 51
Liesberg Beds, 6, 51
Liestal, 203
ligament, 3, 5, 7, 9–11, 16, 24–26, 50
ligament fossa, 116
ligament pit, 25
lignitized driftwood, 191
limestone, 41, 247
lime mudstone, 63, 72, 216, 248
Lindsay Formation, 64, 69
Litchfield, 103
lithification, 198
lithography, 216
living fossils, 21, 241
Livingstone County, 156
Lizard Island, 237
Llandovery, Llandoverian, 38, 45, 81, 93, 96
lobolith, 15, 95, 98
Lochen Beds, 47
Lochkovian, 104
Lockport, 87
Lodgepole Formation, 45
Logan County, 226
logs, 180, 184
London-Brabant Massif, 166, 184
London Clay, 233
Longobardian, 165
Lower Crinoid Beds, 207
Lower Gogolin Beds, 174
Lower Hauptrogenstein, 208
Lower Oolitic Series, 205
Lower Saxony, 164, 173
Loyauté Islands, 20
Lucas County, 129
Ludlow, 81, 93
lumen, 10, 42, 58
Lyme Regis, 177

Maastrichtian, 44
Madison County, 63
Maghreb, 94
malformation, 59
Malm, 216
Manado, 5, 238
manganese, 223
Manlius Formation, 103
marl, 247
marl-clay, 247
marly limestone, 247
Martin, 226
Maubisse Formation, 161
Maxberg Museum, 224
Maynes Creek Member, 135
Maysvillian, 75
Mazon Creek, 155
McLeansboro Group, 155
McMurdo, 4
Mecsek Mountains, 167
Medusa Cement Company, quarries, 129
micrite, 248
microcrinoids, 22, 30, 46
microstructure, 3
Middleport, 87

GENERAL INDEX

Mid-Llandovery, 93
Mill, 63
Mississippian, 41, 45, 52, 135, 139, 145
mitrate, 108
Model City Limestone, 88
Mohawkian, 42, 44, 63
Mohawk River, 63
molasse, 65, 248
monaxon spicules, 170
monocyclic, 18, 32
Monroe County, 153
Monteagle Limestone, 46
Montgomery County, 146
Montmédy, 44
Moravia, 48
Mörnsheim, 216
Morocco, 51, 93
Morozaki Group, 234
mosasaurs, 226
Moscow Formation, 122
moulds, 42, 45, 226, 234
Mount St Helens, 181, 189
Mountain Lake Member, 44
mouth, 3, 4, 23, 24, 36
Much Wenlock Limestone Formation, 43
mucus, 4
mudstone, 248
Mud Creek, 122
multi-armed, 24, 85, 238
multi-radiate stem, 37
multi-plated bowl, 17
Muschelkalk, 41, 48, 51, 164
Muschelkalkmuseum Hagdorn Ingelfingen, 176
muscles, 5, 24, 50
muscular articulations, 5, 25
muscular fossae, 227
Museum am Besucherbergwerk Bundenbach, 120
Museum für Geologie und Paläontologie Tübingen, 171, 176, 228
Museum für Naturkunde Humboldt-Universität, Berlin, 176
Museum of Comparative Zoology, Harvard University, 139, 144, 232
mutable collagenous tissue, 7, 30
Muttonville, 122
myelodactylids, 15, 142
Myophoria Beds, 165

nannoplankton, 231
När, 83, 86
Nardi, James, 122
National Museum of Natural History, Smithsonian Institution, 80, 87, 92, 138, 144, 154, 232
Naturhistorisches Museum (Natural History Museum) Basel, 102, 215
Naturhistoriska Riksmuseet, Stockholm, 86
Naturmuseum Senckenberg, Frankfurt a. Main, 102, 120, 191
Natural History Museum London, 43, 46, 182, 202
Neckarwestheim, 167
Neckarwestheim Member, 165

neotenous, neoteny, 22, 40, 46
nervous system, 4, 7, 17
Nevada, 41, 45, 163
New Caledonia, 6, 20, 47, 242
New Genstar Cement Company, 129
New Providence Shale, 46, 146, 154
New York State Museum, 92, 110, 128
Niagara, Niagara Escarpment, 87
Nickles, J. M., 75
Niobrara Chalk, Formation, 53, 225
nodals, 7, 8, 15
non-pinnulate arms, 27, 30, 35, 36
Norian, 47
Normandy, 44
Northleach, 199
North Sea Central Swell, 166
Nottawasaga Group, 68
Nouméa, 6
Novosibirsk, 30
nuculids, 129
Nusplingen, 221
nutrients, 24, 86, 99, 168, 195, 209

obrution, 52
obtusum Zone, 177
oesophagus, 3
Ohmden, 183, 185, 191
Olenkian, 165
Oligocene, 233, 234
Olney Member, 104
oncolitic limestone, 248
Onondaga Escarpment, 87
Oolithbanke, 165
oolitic, 167, 173, 197, 205, 248
oomicrite, 248
oosparite, 248
Opalinum Clay, 203
ophiuroids, 5, 27, 56, 57, 88, 218, 236
Opitz collection, 120
opposing spine, 16
oral, 4, 5, 16, 22, 24-26
oral pinnule, 5, 29, 30, 239
orals, 46
Ordovician, 7, 9, 19, 31, 36, 38, 41, 42, 49, 63, 68, 75
Oregon, 234
orientation, 5, 30, 51
orthids, 89
Orthoceras limestone, 94
orthocone nautiloids, 93, 119
Osagean, 45, 48, 140, 146, 153
ossicles, 3, 5, 10, 50
Osteocrinus facies, 47
ostracodes, 104, 130, 184
Oued Ziz, 95
Ougarta chains, 93
Oxfordian, 13, 40, 47, 59, 224
oyster beds, 225

packstone, 248
Painten, 216
Palaeocene, 233
palaeoecology, 55, 138, 147, 176, 197
Palaeogene, 40, 233
Paläontologisches Museum Zürich, 168

Palau, 238
Paleocene = Palaeocene, 233
Paleogene = Palaeogene, 40, 233
palms, 233
Pangaea, 155
papillae, 4
Papua New Guinea, 237
parabolic filtration fan, 25, 238-241
paracrinoid, 42, 69
parasite, parasitism, 21, 47, 55, 56, 58
Parkinsoni Beds, 205
pathological, 47, 58, 242
pectinid bivalve, 6
pelagic, 37, 40, 46-48, 53, 86, 99, 114, 177, 217, 221, 224, 230, 232
pelmatozoans, 3, 5, 7, 31, 65
Pelsonian, 165
Pennington Formation, 46
Pennsylvanian, 46, 155
Pentacrinite Bed, 177
pentacrinoid stage, 13
pentaradiate symmetry, 3, 17, 32
perilumen, 151
peripheral cirrus, 16
perisome, 33
peristaltic gut, 221
Permian, 31, 40, 48, 160
Persian Gulf, 198, 223
petal, petals, 7, 185
Pfalzpaint, 216
phenanthroperylene quinones, 51
phoronids, 176, 198
phylogeny, 31, 37
phytoplankton, 194, 209, 212, 223
Pienid Klippen Belt, 47
Pierre Shale, 225
pigments, 51, 135
pinnular, 29, 171
pinnular muscle, 29
pinnulation, 28
pinnule gap, 28
pinnule socket, 29, 219
pinnules, 28
planar filtration fan, 28
plankton, 107, 172, 189, 212, 221, 238, 248
planktonic, 15, 98, 195, 209, 225, 231, 232, 244
planispirally coiled, 10, 142
plate circlets, 17, 31
plate diagram, 33-37
platelets, 3, 4
plate lobolith, 94, 98
Plattenkalk, 54, 216
platyceratid gastropods, platyceratids, 56, 79, 125, 131
plesiosaurs, 177, 183, 225
Pliensbachian, 41, 44, 49, 179
pluricolumnals, 9, 42
podium, podia, 3, 5, 30
Pooleville Member, 44
Posidonia Shale, Posidonienschiefer, 53, 183, 247
posterior interradius, 19
Pragian, 111
Prairie Bluff Chalk, 44

Precambrian Shield, 73
pre-formed rupture points, 15
predation, 7, 55, 163, 238, 242
Preussisches Geologisches Landesmuseum Berlin, 176
Pridoli, Pridolian, 45, 93, 97
primaxillary, 8
primibrachial, IBr, 8, 21, 28
protective morphology, 21
proximal, 5, 7, 8
proxistele, 152
pseudoplanktonic, 44, 179, 189, 191
pteriniids, 108
ptilodictyids, 69
pyrite, 53, 112, 177
pyritized, pyritization, 53, 112, 129, 134, 184, 233

Racine Dolomite, 45
radial, 3, 5, 12, 18, 27, 32, 38
radial canals, 3
radianal, 35, 36
radicular cirri, 11–15
radius, 8
radix, radices, 11, 13, 15
Rädersteine, 164
Ralingen, 165
Ralingen-Kersch, 175
Ramp Creek Formation, 146
ramulate arms, 27, 107
ramus, 27
raptorial feeding, 242
rays, 15, 28
Recent, 10, 237
Recoaro, 174
rectum, 3
recumbent, 12, 19, 36, 124, 142
red tides, 209
reef forms, 20, 48, 161
reef-knolls, 45
Reef of Wittnau, 206
reefs, 13, 40, 45, 81, 105, 221, 237
regeneration, 21, 24, 168, 175, 235, 238, 242
Reichsstelle für Bodenforschung Berlin, 120
reproductive system, 4, 30
reradiation, 39
respiration, 4, 24, 125
respiratory organs, 55
Rheinisches Schiefergebirge, 111
Rhenish-Bohemian Massif, 217
Rhenish Massif, 184
Rhineland-Palatinate, 175
rhombiferan cystoid, 65, 88, 90, 105
rhombiferans, 31, 42, 45, 69, 75, 108
rhynchonellids, 89, 108
ribbon rock facies, 108
Richmondian, 75, 78
Richter, Rudolf and Emma, 120
Ries, 217
rigidity, 17, 31
ring canal, 3
ring commissure, 243
Ringueberg, Eugene, 87
Rochester Museum and Science Center, 92
Rochester Shale, 87

Rockaway, 88
Rocky Mountain Geosyncline, 226
Rogen, 205
Rogoznik, 21, 47
Rondout Formation, 104
root, 6, 13–15, 51
Rosinus, Michael Reinhold, 164
Rösnäs Formation, 233
Röt, 165
Rosso Ammonitico, 42
Rosso Ammonitico Lombardo, 42
Royal Ontario Museum, 74, 92
Rundle Formation, 52
runners, 13, 73
Rust Member, 63, 66
Rust Quarry, 65

Sadler Ranch Formation, 41
Sakrau, 174
Salem Limestone, 46
salinity, 55, 166, 174, 196, 223
Salthill Quarry, 46
sand, 247, 248
sand waves, 42, 205, 206
Santonian, 46, 226
Sargasso Sea, 182
Saurier-Museum, Aathal, 149
scavengers, 50, 54, 83, 117, 179, 202, 209, 236
Schamhaupten, 216
Schaumkalk, 173
Schernfeld, 218, 222
Schinznach Quarry, 205
Schleberoda, 173
Schlossparkmuseum Bad Kreuznach, 121
Schuchert, Charles, 103
Schwäbisch Hall, 164, 170
Schwimmplatten, 221, 224
sclerites, 47, 114
scuba, 238
sea anemones, 209
sea cucumber, xiii
sea floor, 4, 5, 12, 15, 27, 50, 53, 105, 114, 134
seamount, 20, 41, 48, 242
sea urchin, xiii
secondary stereom, 14
secundibrachial (IIBr), 8, 28
sedimentation rate, 14
Seegrasschiefer, 195
Senckenberg, 95, 189, 191
Senefelder, Alois, 216
sensory cells, 4
serpulids, 59, 174, 215
sexual dimorphs, 99
Seymour Island, 234
Shadow Lake Formation, 69
shale, 247, 248
shallow-water, 12, 43, 52, 197, 209, 233, 237
Sharps Hill Formation, 199
shear stress, 12
shearing, 11
Sheinwoodian, 87
shell-boring parasites, 56
shoal-dwelling species, 12
shrimps, 217
Siegenian, 111

Silesia, 164, 174
Silesian-Moravian Gate, 166
Silica Shale, 129
siliceous sponges, 47
siliciclastic, 247
silt, 247
Silurian, 38, 43–46, 81, 86, 92
Simcoe Group, 64
Sinemurian, 177, 189, 191
skeletal limestone, 248
skeleton, 3–5
slate, 247
Slite Beds, 82
Smith, William, 197
Smithsonian Institution, 80, 92, 138, 144, 154, 232
Smokey Hill Chalk, 227
snowshoe-like, 231
soft parts, 3
Solnhofen, 54, 216
Solnhofen Lithographic Limestone, Plattenkalk, 54, 216
Sonnenräder, 164
Spatkalk, 204
spawning, 30, 230
Spickert Knob Formation, 146
spicules, 98, 170
spines, 22, 23, 43, 114, 125, 131, 143, 157, 215, 224
spiriferids, 89, 108, 123
Spiriferina-Bank, 165
Spirit Lake, 181
sponge-algal mounds, 216, 222
sponges, 26, 41, 47, 59, 114, 117, 166, 198, 217, 237
spoon-like cup, 21, 47
Springer Collection, 87, 92
Springer, Frank, 139
squids, 184, 221
Staatliches Museum für Naturkunde Stuttgart, 196
Staffordshire, 43
stalk, 4
stalked, xiii, xiv, 238, 240
starfish, xiii, 52
St. Boniface's pennies, 164
stem, xiv, 4, 5
stemless, xiv
stemmed = stalked
Stephanian, 155
stereom, 3, 5
St. Cassian Formation, 176
St. Hyacinth, 164
Steuben Formation, Member, 66
stokesi Subzone, 179
Stonesfield Slate, 197
storm deposits (tempestites), 50, 78, 104
Straits of Florida, xii
Stramberg, 48
stromatolitic algal mats, 104
stromatoporoids, 13, 45, 81, 105
strophomenids, 65, 89, 123
Stürmer, W., 113, 120
submersible, xi, 238
subtegminal mouth, 32

suffocation, 231
Sugar Creek, 145
Sundre Beds, 81
stylophoran, 75
suspension feeders, xiii, 26, 56
Swabian facies, 47, 183, 203
Swabian Alb, 183
Swansea University, 199
Swiss Jura, 47, 203, 211
Sylvania, 132
symbiotic, 98
symmetry, 3, 17, 32, 35
symplexy, 9
synarthry, 9, 25, 37
synecology, 55
synostosis, synostoses, 9, 25, 30, 189
syntaxial calcite, 179
systematics, 31
syzygy, 4, 22, 26

tabulate corals, 45, 56, 81, 90, 119, 125
Taconic, 63, 73, 92
Tafilalt, 93
Talbot, Mignon, 103
taphonomy, 50, 123, 179, 197, 223, 231
Tarnowice Beds, 165
Taurus Mountains, 47
taxocrinid, 27
Taylor, Paul, 199
Tazoulet, 95
tegmen, 4, 5, 21–23, 32
tempestite (storm deposit), 50, 53, 65, 76, 205, 248
tensile strength, 30
tension, 11
tentacle, 3, 5
tentaculitoids, 112, 114
tenuicostatum Zone, 183, 191, 196
Terebratelbänke, 165
Terebratula Beds, 165
terebratulids, 6
teredinid bivalves, 189
terminal cirri, 9, 15, 140, 175, 206, 243
terminal claw, 16, 193, 206, 212, 236, 239
terminal hook, 222
terquemiids, 167, 169, 170
Tertiary, 233
tertibrachial (IIIBr), 28
Tethys, Tethys Ocean, 42, 47, 166, 195, 243
teuthoid cephalopods, 221
theca, xiii, 125
Thorold, 87
Threave Glen, 42
Thuringia, 164, 173

tier, tiering, 12, 55, 90, 153, 171
Timor, 47, 160
Tithonian, 47, 216, 224
Toarcian, 183, 195
torsion, 11
Tournaisian, 43, 135, 140
toxins, 209, 212
trace fossils, 112, 195
trauma, 14, 26, 167, 179
Traverse Group, 129
Tremadocian, 38
Trenton, 63
Trenton Falls, 64
Trenton Group, 53, 63, 68
Trentonian, 42, 63
Triassic, 28, 37, 40, 41, 43, 47, 51, 107, 160, 164
trilobites, 63, 65, 87, 89, 114, 123, 129
Trochitenbank 6, 167
Trochitenkalk, 165, 172
tube foot, tube feet, 3–5, 24
Tübingen Museum = Institut und Museum für Geologie und Paläontologie der Universität, 176, 190, 196
turbidite, 50, 53, 65, 78, 223, 233, 248

Uinta Mountains, 226
Uintacrinus Beds, 225
Ulrich, E. O., 75
unidirectional current, 21, 47
uniserial, 27, 28, 33, 36
University of Bonn, 121
University of Glasgow, 43
University of Iowa, 159
University of Michigan, 74, 134
University of Rochester, 92
Upper Oolitic Series, 205
Upper Silesia, 164, 174
urns, 19
Utica Shale, 64
Utsumi, 234

vagile benthos, 117
Valanginian, 48
Varians Member, Beds, 205, 212
Variscan mountain building, 112
Venezuela, xi
ventral, 5
vertebrates, 112, 177, 184, 223
Verulam Formation, 73
Villey-Saint-Etienne, 212
Vincent, 122
Vindelician-Bohemian Massif, 166
Vindelician High, 184

Visby Beds, 81
visceral mass, 24
Visean, 45, 146

Wabash College, 146
Walcott, Charles, 63
Waldron Shale, 44
Wandagee Series, 162
Wannen, 222
Wanner, Johannes, 160
Warsaw Formation, 154
waste disposal, 117
water vessel, 5
water-vascular system, 3, 30
wave ripples, 202
Waynesville Formation, 75, 78
wedge-shaped columnals, 12, 142, 163
Wegman's Plaza, 87
Weissjura Zeta 2, 216
Wellenkalk, 173
Wenlockian, 44, 81, 87
Wenlock limestone, 43
Werksmuseum Lauffen Heidelberger Zementwerke, 176
West Australian geosyncline, 161
West Canada Creek, 63
Western Interior Seaway, 225
Wheeler Shale, xiv
Wiarton shoal, 92
Widder Formation, 130
Wilder Schiefer, 191
Wilkowice Beds, 165
Willibaldsburg, 224
Windom Shale, 122
wing oyster bivalves, 108
wing plates, 23, 46
Winkel Quarry, 212
winnowed beds, 248
wood-boring, 189
worms, 58, 114, 195, 215
worm tubes, 88, 170
Wray, Charles, 104
Wren's Nest, 43

X-rays, 112
xenomorphic stem, 10
Xinpu, 43

Y Zone, 109
Yale Peabody Museum, 66, 103, 232

Zandt, 216, 218, 221
Zoologische Staatssammlung Munich, 242
Zscheiplitz, 173

Taxonomic Index

(Numbers in bold-faced type refer to figures)

Abacocrinus tessellatus Angelin, **83**
Abatocrinus Lane, 140
Abatocrinus grandis (Lyon & Casseday), **145**, 146, 147, 150
Abatocrinus macbridei (Wachsmuth & Springer), **136**
Abludoglyptocrinus Kolata, 42
Abrotocrinus coreyi (Meek & Worthen), 150
Abrotocrinus unicus (Hall), 150
Acanthocrinus Roemer, 12, 125
Acanthocrinus rex Jaekel, 107, **117**
Acanthocrinus spinosus (Hall), 125, **127**
Acrosalenia Agassiz, 209
Actinocrinites J. S. Miller, 46
Actinocrinites gibsoni (Miller & Gurley), 56, 147, **148**, 150
Actinocrinites multiramosus Wachsmuth & Springer, 56, **57**
Actinocrinites triacontadactylus J. S. Miller, **33**
Actinocrinitidae Austin & Austin, actinocrinitids, 45, 140, 143
Aethocrinea Ausich, 38
Aethocrinida Ausich, 38
Aethocrinus Ubaghs, 7, 13, 38
Aethocrinus moorei Ubaghs, **7**
Agaricocrinus americanus (Roemer), 147
Agaricocrinus inflatus (Hall), **141**
Agaricocrinus splendens Miller & Gurley, **148, 149**
Agassizocrinus Owen & Shumard, 46, 173
Ailsacrinus Taylor, 161, 173, 199–202, 212
Ailsacrinus abbreviatus Taylor, 199, **201**
Ailsacrinus prattii (Gray), 202
allagecrinids, 46
Amaltheocrinus Klikushin, 42
Ambicocrinus arborescens (Talbot), 103
Ammonicrinus Springer, 10, 45
Ammonicrinus doliiformis Wolburg, **10**

Amphistrophia Hall & Clarke, 89
Amphoracrinus Austin, 46
Amphorometra inornata Paul, 233
Amygdalocystites Billings, 69, 71
Ancyrocrinus Hall, 15
Ancyrocrinus bulbosus Hall, **14**
Annacrinus wyvillethomsoni (Thomson), 240
Anomalocystites Hall, 108
Antedon bifida de Fréminville, 30
Antedonidae Norman, 238
Antholites Davis, 123
Aorocrinus immaturus (Wachsmuth & Springer), 56, 137
Aphelecrinus Kirk, 46
Aphelecrinus elegantulus (Wachsmuth & Springer), **136**
Apiocrinites J. S. Miller, 10, 17, 53, 198–200
Apiocrinites parkinsoni (v. Schlotheim), 197, **198**
Apodasmocrinus Warn & Strimple, 44
Apographiocrinus typicalis Moore & Plummer, **157**
Applinocrinus texanus Peck, 47
Apsidocrinus Jaekel, 48, 160
Apsidocrinus moeschi (Zittel), **21**
Archaeocrinus Wachsmuth & Springer, 42, 44, 69, **72, 73**
Archaeocrinus pyriformis (Billings), **70**
Archaeopteryx v. Meyer, 216, 217
Arctinurus Castelnau, 88, 89, **92**
Arthroacantha Williams, 131, 134
Arthroacantha carpenteri (Hinde), 56, **132, 133**
Articulata J. S. Miller, articulates, 32, 36, 39, 40, 237
Asaphocrinus Springer, 88, 90
Asaphocrinus ornatus (Hall), **89, 92**
Astartellopsis triasina (Roemer), 172
Asteroidea de Blainville, xiii, 31
Atelestocrinus Wachsmuth & Springer, 125

Austinocrinus erckerti (Dames), **12**
Avicula Bruguière, 215
Azygocrinus Lane, 140
Azygocrinus rotundus (Yandell & Shumard), **141**

Bactrocrinites Schnur, 120
Bactrocrinites jaekeli (Schmidt), **120**
Baculites Lamarck, 225
Bakevellia costata (v. Schlotheim), 172
Balanocrinus Agassiz, 48
Balanocrinus gracilis (Charlesworth), 44, 179
Balancocrinus maritimus Bourseau, David, Roux, Bertrand & Clochard, 44
Balanocrinus subteres (Münster in Goldfuss), 47
Balanoglossites Häntzschel, 167
Barrandeocrinus Angelin, 26, 86
Barrandeocrinus sceptrum Angelin, 83, **85**, 86
Barycrinus Wachsmuth, 42, 46, **146**
Barycrinus rhombiferus (Owen & Shumard), **25**, 147, 150
Barycrinus stellatus (Hall), **153**
Bathycrinus Thomson, 233
Bathycrinus aldrichianus Carpenter, **5**
Batocrinidae Wachsmuth & Springer, batocrinids, 140–142
Batocrinus Casseday, 46
Batostoma Ulrich, **70**
Bicidiocrinus wetherbyi (Wachsmuth & Springer), **23**
Bihaticrinus Kristan-Tollmann, 47
Blastoidea Say, blastoids, 31, 125, 134
blastozoans, 13, 31, 76
Bositra buchi (Roemer), 183
Botryocrinus Angelin, 44
Botryocrinus nycteus (Hall), 125
Bourgueticrinida Sieverts-Doreck, 241
Bourgueticrinidae de Loriol, bourgueticrinids, 11, 13, 24, 233, 239, 241

Bourgueticrinus d'Orbigny, 233
Brabeocrinus christinae Strimple & Moore, **156**, 157
Bumastus Murchison, 88, 89

Cainocrinus Forbes, 233
Calamocrinus Agassiz, 233
Calceocrinidae Meek & Worthen, calceocrinids, 18, **19**, 20, 30
Calceocrinus Hall, 42, 45, 90
Calceolispongia Etheridge, 162, 163
Calceolispongia abundans Teichert, **162**
Calceolispongia robusta Teichert, **162**
Calceolispongia spinosa Teichert, **162**
Calceolispongiidae Teichert, calceolispongiids, 48, 162
Calciroda Mayer, 170
Caledonicrinus Avocat & Roux, 243
Caledonicrinus vaubani Avocat & Roux, 243
Calliocrinus d'Orbigny, 45, 83
Calycanthocrinus Follmann, 114
Calymene Brongniart, 89
Camarocrinus Hall, 15, 97, 98
Camerata Wachsmuth & Springer, camerates, 19, 23, 32, 56, 79, 123, 150
Camptocrinus Wachsmuth & Springer, 10
Camptocrinus compressus Wright, **10**
Camptocrinus multicirrus Springer, **10**
Camptocrinus praenuntius Springer, 141, **142**
Carabocrinus Billings, 44, 73
Cardiola Broderip in Murchison, 93
Cardiolampas Pomel, 48
Carpocrinus Müller, 83
Carpocrinus angelini Franzén, **85**
Carpocrinus petilus Franzén, **85**
Caryocrinites Say, 45, 88–90
Caryocrinites ornatus Say, **92**
Catillocrinidae Wachsmuth & Springer, catillocrinids, 152
Cenocrinus asterius (Linné), 238, **240**
Ceraurus Green, 65, **70**, 73
Chariocrinus Hess, 203
Chariocrinus andreae (Desor), 57, 203–206, **207**, **208**, 237, 240
Chariocrinus leuthardti (de Loriol) = *Hispidocrinus leuthardti* (de Loriol), 203
Cheilotrypa Ulrich, 89, **92**
Cheirocrinus Eichwald, 65, 69
Chelocrinus v. Meyer, 174
Chelocrinus carnalli (Beyrich), 51, 165, **173**, **174**
Chelocrinus schlotheimi (Quenstedt), 165, 173
Chilotrypa Miller = *Cheilotrypa* Ulrich, 89
Chladocrinus Agassiz, 41, 164, 195
Chondrites v. Sternberg, 195
Cincinnaticrinus Warn & Strimple, 65, 73, 76, **77**, 78, 79
Cincinnaticrinus heterodactylus (Wachsmuth & Springer), **67**
Cincinnaticrinus pentagonus (Ulrich), **76**, 78
Cincinnaticrinus varibrachialus Warn & Strimple, 78
Cladida Moore & Laudon, cladids, 32, 33, 35, 36, 39, 150, 156, 160
Cladochonus McCoy, 57, **58**

Clarkeocrinus Goldring, 123, 124, 127
Clarkeocrinus troosti (Hall), **124**, **125**
Clathrocrinus clathratus Strimple & Moore, **156**, 159
Cleiocrinus regius Billings, **34**, 71
Codiacrinus Schultze, 114
Coelocrinidae Bather, coelocrinids, 140, 141, 143
Coenites Eichwald, 83, **85**
Coenothyris Douvillé, 170
Coenothyris vulgaris (v. Schlotheim), **172**
Colpodecrinus Sprinkle & Kolata, 38
Columbicrinus crassus Ulrich, **35**, 42
Comatilia iridometriformis Clark, 239
Comatula Lamarck, 230
Comatulida Clark, comatulids, 15, 16, 32, 37, 192, 237, 238
Compsocrinina Ubaghs, 32, 33
Conocrinus d'Orbigny, 233
Coolinia Bancroft, 89
Cordylocrinus Angelin, 107, 108
Cordylocrinus plumosus (Hall), 103, **109**
Corocrinus nodosus Kier, **133**, 134
Cotylacrinna Brower, 42
Cotyloderma Quenstedt, 42
Cremacrinus Ulrich, 42, 45, 65, 69, 71
Cribanocrinus watersianus (Wachsmuth & Springer), **136**, 138
Crinoidea J. S. Miller, 3, 17, 31, 32, 38
Crinobrachiatus Moore, 88–90
Cromyocrinus simplex Trautschold, **36**
Crotalocrinites Austin & Austin, 26, 45, 83
Ctenocrinus Bronn, 103, 107
Ctenocrinus nobilissimus (Hall), 103, 107, **108**
Ctenocrinus pachydactylus (Conrad), 103, **106**
Cupressocrinites abbreviatus Goldfuss, 45
Cupressocrinites gracilis Goldfuss, 45
Cupulocrinus d'Orbigny, 42, 68, 76
Cupulocrinus jewetti (Billings), **70**, **73**
Cupulocrinus polydactylus (Shumard), **80**
Cyathidium Steenstrup, 21, 27, 241
Cyathidium depressum Sieverts, 242
Cyathidium foresti Cherbonnier & Guille, **242**
Cyathidium holopus Steenstrup, 234
Cyathidium meteorensis Fechter = *Cyathidium foresti* Cherbonnier & Guille, 242
Cyathidium plantei Heinzeller, 242
Cyathocrinina Bather, 32, 36
Cyathocrinites J. S. Miller, 46
Cyathocrinites iowensis (Owen & Shumard), 150
Cyathocrinites multibrachiatus (Lyon & Casseday), **28**, 146, 149, 150
Cyclonema Hall, 56, **80**
Cyrtocrinida Sieverts-Doreck, cyrtocrinids, 20, 21, 27, 32, 47, 48, 241–243
Cyrtocrinus Jaekel, 48
Cyrtocrinus nutans (Goldfuss), 21, 47, **59**

Dactylioceras Hyatt, 184
Dadocrinus v. Meyer, 184
Dadocrinus kunischi Wachsmuth & Springer, 174, **175**
Dalejina Havlicek = *Mendacella* Cooper, 89
Dalmanites Barrande, 88, 89

Decadocrinus Wachsmuth & Springer, 125, 134
Decadocrinus nereus (Hall), 125
Democrinus Perrier, 233, 239, 241
Democrinus rawsoni (Pourtalès), **11**
Dendrocrinina Bather, dendrocrinids, dendrocrines, 33, 36–39
Dendrocrinus Hall, 88–90
Desmidocrinus Angelin, 83
Desmidocrinus pentadactylus Angelin, **85**
Devonoblastus Reimann, 125
Diabolocrinus Wachsmuth & Springer, 44
Diademopsis crinifera (Quenstedt), 196
Dichocrinidae S. A. Miller, dichocrinids, 140, 141
Dichocrinus Münster, 45, 46
Dichocrinus delicatus Wachsmuth & Springer, 138
Dichocrinus hammondi Laudon & Beane, 138
Dictenocrinus semipinnulatus (Schmidt), 117
Dimerocrinites Phillips in Murchison, 44, 45, 83, 88–90
Dimerocrinites cf. *liliiformis* (Hall), **90**
Dinocrinus Wanner = *Calceolispongia* Etheridge, 162
Diplobathrida Moore & Laudon, diplobathrids, 32, 150
Diplocrinus maclareanus (Thomson), 239
Disparida Moore & Laudon, disparids, 32, 33, 35, 38, 78
Dizygocrinus Wachsmuth & Springer, 46, 140, 142
Dizygocrinus indianensis (Lyon & Casseday, 149, 150
dolatocrinids, 123
Dolichocrinus de Loriol, 47
Dorycrinus missouriensis (Shumard), 141, **143**
Dunnicrinus mississippiensis Moore, 44

Echinoidea Leske, xiii, 31
Echmatocrinea, Echmatocrinida Sprinkle & Moore, 32
Echmatocrinus Sprinkle, 38
Echmatocrinus brachiatus Sprinkle, 38
Ectenocrinus J. S. Miller, 42, 65, 76
Ectenocrinus grandis (Meek), **18**
Ectenocrinus simplex (Hall), **67**
Edrioaster bigsbyi Billings, **70**, **72**
Edrioasteroidea, edrioasteroids Billings, 42, 69, 71, 87–90, 108
Edriophus levis (Bather), 69
Eleutherocrinus Shumard & Yandell, 125
Eleutherozoa Bell, 31
Embryocrinus hanieli Wanner, **22**
Enallocrinus d'Orbigny, **83**
Enantiostreon Bittner, 167, **169**
Enantiostreon difforme (v. Schlotheim), 172
Encrinidae Dujardin & Hupé, encrinids, 30, 36, 167–175
Encrinites fossilis Blumenbach = *Encrinus liliiformis* Lamarck, 164
Encrinites moniliferus J. S. Miller = *Encrinus liliiformis* Lamarck, 164
Encrinites trochitiferus v. Schlotheim = *Encrinus liliiformis* Lamarck, 164

Encrinus Lamarck, 164, **170**, 174
Encrinus aculeatus v. Meyer, 165
Enrinus greppini de Loriel, 165
Encrinus liliiformis Lamarck, **29**, 51, 164, **168, 169, 171**
Endelocrinus tumidus spinosus Strimple, **158**, 159
Endoxocrinus parrae (Gervais), 209, 238, **239, 241**
Eometacrinus australis Baumiller & Gazdzicki, **234**
Eoparisocrinus Ausich, 42
Eoparisocrinus siluricus (Springer), **19**
Eospirifer Schuchert, 89
Eretmocrinus Lyon & Casseday, 140
Eretmocrinus remibrachiatus (Hall), 141, **142**
Erisocrinus typus Meek & Worthen, **158**, 159
Eucalyptocrinites Goldfuss, 26, 44, 45, 83, 89
Eucalyptocrinites caelatus (Hall), **90**
Eucatillocrinus bradleyi (Meek & Worthen), **152**
Eucladocrinus pleuroviminus (White), **17**, 141, 143
Eudiplobathrina Ubaghs, 32
Eugeniacrinites J. S. Miller, 21, 47
Eugeniacrinites cariophilites (v. Schlotheim), 47
Eugeniacrinites caryophyllatus Goldfuss = *Eugeniacrinites cariophilites* (v. Schlotheim), 47
Eugeniacrinites zitteli Jaekel, 48
Euptychocrinus Brower, 42
Euspirocrinus spiralis Angelin, **35**
Eutaxocrinus Springer, 114
Eutaxocrinus stürtzii (Follmann), **119**
Eutrochocrinus Wachsmuth & Springer, 140
Eutrochocrinus christyi (Shumard), **24**
Exocrinus wanni Strimple, **156**, 157

Favosites parasiticus Hall, **90**
Fenestella Lonsdale, 89, 129
Flexibilia Zittel, flexibles, 27, 32, 152
Flexicalymene Shirley, 65
Forbesiocrinus de Koninck & le Hon, 17
Forbesiocrinus wortheni Hall, **18**, 152

Gaurocrinus S. A. Miller, 76
Gennaeocrinus Wachsmuth & Springer, 124
Gennaeocrinus eucharis (Hall), **126**
Geocoma carinata (v. Münster), 218
Gervillella mytiloides (Zieten), 174
Gilbertsocrinus Phillips, 46, 125, 131, 134, 141
Gilbertsocrinus ohioensis Stewart, 131
Gilbertsocrinus spinigerus (Hall), 125
Gilbertsocrinus tuberosus (Lyon & Casseday), 56, 131, 146, 149, **150, 151**, 152
Girvanella Nicholson & Etheridge, 195
Gissocrinus Angelin, 43, 45
Glyptocrinina Moore, 32
Glyptocrinus Hall, 56, 65, 76, 79
Glyptocrinus decadactylus Hall, **33**
Glyptocystites multiporus Billings, **70**
Gogia Walcott, xiii, 7, 13
Gogia spiralis Robison, **xiv**
Goniomya Agassiz, 195
Guillecrinus Roux, 242

Guillecrinus neocaledonicus Bourseau, Améziane-Cominardi & Roux, 242
Guillecrinus reunionensis Roux, 242
Gustabilicrinus Guensburg, 42
Gutticrinus Klikushin, 42
Gymnocrinus de Loriol, 14, 20, 27, 48, 242
Gymnocrinus richeri Bourseau, Améziane-Cominardi & Roux, **21**, 51, 242, **243**

Haereticotaxocrinus Franzén, 83
Haereticotaxocrinus asper Franzén, 83, **85**
Haeretocrinus wagneri Strimple & Moore, 157, **158**
Halimeda Lamouroux, 52
Halysiocrinus Ulrich, **146**
Halysiocrinus tunicatus (Hall), 150
Hapalocrinus Jaekel, 114
Hapalocrinus elegans Jaekel, **114**, 119
Hapalocrinus frechi Jaekel, **119**
Harpoceras Waagen, 184
Hemibrachiocrinus Arendt, 48
Hemicidaris langrunensis Cotteau, 44
Hemicrinus d'Orbigny, 21, 48
Hemicrinus astierianus d'Orbigny, **21**
Hemicystites Hall, 90
Hemistreptocrinida Arendt, 31, 32, 36
Herpetocrinus Salter = *Myelodactylus* Hall, 86
Hesperornis regalis Marsh, 226
Heterocrinus Hall = *Cincinnaticrinus* Warn & Strimple, 65, 78
Heterometra savignii (Müller), 238
Heterotrypa Nicholson, 69
Hexacrinites elongatus (Goldfuss), 45
Himerometra bassleri Gislén, **26**
Hispidocrinus Simms, 203
Hispidocrinus leuthardti (de Loriol), 203–205, 212, **215**
Histocrinus coreyi (Worthen), 149, 150
Hoernesia socialis (v. Schlotheim), **172**
Holectypus depressus (Leske), 212
Holocrinidae Jaekel, holocrinids, 40, 167, 171, 175
Holocrinus Wachsmuth & Springer, 175
Holocrinus acutangulus v. Meyer, 165
Holocrinus doreckae Hagdorn, 165
Holocrinus dubius (Goldfuss), 165, 175, **176**
Holocrinus meyeri Hagdorn & Gluchowski, 165
Holopus d'Orbigny, 14, 27, 48
Holopus alidis Bourseau, Améziane-Cominardi, Avocat & Roux, **20**, 242
Holopus rangii d'Orbigny, 242
Holothuroidea de Blainville, xiii, 31
Howellella Kozlowski, 108
Homocrinus Hall, 87, 88–90, 92
Homocrinus parvus Hall, **92**
Hybocrinida Jaekel, hybocrinids, 31, 32, 36
Hybocrinus Billings, 42, 44
Hybocrinus conicus Billings, **37**
Hybocystites Wetherby, 69, 71
Hybocystites eldonensis (Parks), **70**
Hyolithelminthes Fisher, 58
Hylodecrinus gibsoni (White), 150
Hyocrinidae Carpenter, 240
Hyocrinina Rasmussen, 241

Hyperoblastus Fay, 134
Hyperoblastus reimanni Kier, **133**

Icthyocrinus Conrad, 90
Illemocrinus amphiatus Eckert, 69, **71**
Imitatocrinus Schmidt, 114
Inadunata Wachsmuth & Springer, inadunates, 32, 33
Indocrinidae Strimple, indocrinids, 163
Indocrinus Wanner, 160
Inoceramus Sowerby, 184
Inoceramus (Platyceramus) platinus Logan, 231
Inozoa Steinmann, 41
Iocrinus Hall, 42, 65, 76
Iocrinus trentonensis Walcott, **66**
Isocrinida Sieverts-Doreck, isocrinids, 7, 15, 32, 37, 238–240
Isocrinus v. Meyer, 195, 236
Isocrinus basaltiformis (J. S. Miller), **59**
Isocrinus blakei (Carpenter), 209, 239
Isocrinus cingulatus (Münster in Goldfuss), 47
Isocrinus nehalemensis Moore & Vokes = *Raymondicrinus oregonensis* (Moore & Vokes), 236
Isocrinus nicoleti (Thurmann), 44, 204, 205, 209, **211, 212**
Isocrinus (Chladocrinus) tuberculatus (Miller), 41
Isotelus DeKay, 65
Isotomocrinus typus Ulrich, 69, **71**
Isselicrinus Rovcreto, 233
Isselicrinus subbasaltiformis (J. S. Miller), 233

Jimbacrinus Teichert, 162
Jimbacrinus bostocki Teichert, 162, **163**

Kallimorphocrinus Weller, 46
Kallimorphocrinus lasallensis (Strimple & Moore), 156, **157**

Lampterocrinus Roemer, 45
Lanecrinus depressus (Meek & Worthen), 150
Lantenocrinus Kristan-Tollmann, 47
Lasiocrinus Kirk, 107–109
Lasiocrinus scoparius (Hall), **109**
Lecanocrinus Hall, 44, 45, 90
Leocrinus Kristan-Tollmann, 47
Lepocrinites Conrad, 105, 108
Leptaena Dalman, 89
Leptolepides sprattiformis (Agassiz), 217
Lichenalia Hall, 89
Lichenocrinus Hall, 13
Liliocrinus Rollier, 10, 14, 51
Liliocrinus munsterianus (d'Orbigny), **6, 51**
Lingula tenuissima Bronn, **172**
Lithaster Hess, 218
Lobolithus Barrande = scyphocrinitid lobolith, 98
Lonchocrinus Jaekel, 21, 47
Lumbricaria intestinum Goldfuss, 221
Lyriocrinus Hall, 44

Macrocrinus Wachsmuth & Springer, 140, 141

Macrocrinus mundulus (Hall), 56, **149**, 150, **151**
Macrostylocrinus Hall, 43, 45, 88, 89
Macrostylocrinus ornatus Hall, **89**
Marhoumacrinus Prokop & Petr, 97
Marhoumacrinus legrandi Prokop & Petr, **96**, 97
Mariacrinus beecheri Talbot = *Ctenocrinus nobilissimus* (Hall), 103
Marsupiocrinus Morris, 45
Marsupites J. S. Miller, 232
Marsupites testudinarius (v. Schlotheim), **25**, 46, 232
Mediospirifer Bublichenko, 123
Megistocrinus Owen & Shumard, 124
Melocrinites pachydactylus Conrad = *Ctenocrinus pachydactylus* (Conrad), 103
Melocrinus Agassiz = *Melocrinites* Goldfuss, 103
Mendacella Cooper, 89
Mesolimulus Størmer, 217
Metacrinus Carpenter, 233, 234, 236
Metacrinus angulatus Carpenter, **4**, **8**
Metacrinus fossilis Rasmussen, 234
Microcaracrinus conjugulus Strimple & Moore, 157, 158
Millericrinida Sieverts-Doreck, millericrinids, 32, 37, 200
Millericrinus mespiliformis v. Schlotheim, 222
Millericrinus nobilis Walther = *Pomatocrinus nobilis* (Walther), 222
Monobathrida Moore & Laudon, monobathrids, 32
Monobrachiocrinus ficiformis granulatus Wanner, **22**
Monodechenella Stumm, 123
Montlivaltia Lamouroux, **6**
Mucrospirifer Grabau, 123, 129, 134
Myalina de Koninck, **169**, **171**
Myelodactylus Hall, 42, 45, 86
Myelodactylus ammonis (Bather), 43
Myelodactylus fletcheri (Salter), 43, **86**

Nasutocrinus Kristan-Tollmann, 47
Naticonema Perner, 80
Naumachocrinus hawaiiensis Clark, **11**
Nemaster Clark, 237
Nemaster rubiginosus (Pourtalès), **16**
Neocrinus decorus Thomson, **xii**, **29**
Neoschizodus ovatus (Goldfuss), **172**
Nevadacrinus Lane & Webster, 163
Newaagia Hertlein, 167, **169**
Nielsenicrinus Rasmussen, 233
Notocrinus rasmusseni Meyer & Oji, 234
Notocrinus seymourensis Baumiller & Gazdzicki, 234

Ohiocrinus Wachsmuth & Springer, 42, 78, 79
Ohiocrinus brauni Ulrich, **79**
Onychaster Meek & Worthen, 56, **57**
Onychocrinus exsculptus Lyon & Casseday, **35**, 147, 149, 150
Onychocrinus ramulosus (Lyon & Casseday), 146, 150
Ophiopetra lithographica Hess, 224

Ophiuroidea Gray, xiii, 31
Orophocrinus conicus Wachsmuth & Springer, 136
Orophocrinus fusiformis (Wachsmuth & Springer), 136
Orthoceras Bruguière, 93, 94, **114**
Osteocrinus Kristan-Tollmann, 47
Osteocrinus rectus (Frizzell & Exline), 47
Ottawacrinus typus Billings, 70
Oxycomanthus bennetti (Müller), **238**, **239**
Oxytoma Meek, 184

Pachyantedon gracilis (Walther) = *Solanocrinites gracilis* (Walther), 222
Pachylocrinus aequalis (Hall), 149, 150
Palaeocomaster schlumbergeri de Loriol, 44
Palaeocrinus Billings, 44
Paracidaris florigemma (Phillips), **6**
Paracomatula Hess, 212
Paracomatula helvetica Hess, 203–205, **212**, 232, 237
Paracremacrinus Brower, 44
Paradichocrinus polydactylus (Casseday & Lyon), 147
Parapisocrinus quinquelobus (Bather), **27**
Paraspirifer Wedekind in Salomon, 131
Parisangulocrinus Schmidt, 114
Parisangulocrinus zeaeformis (Schultze), **113**, **114**
Parulocrinus pontiacensis Strimple & Moore, **159**
Paucicrura Cooper, 65
Pellecrinus hexadactylus (Lyon & Casseday), 147
Pelmatozoa Leuckart, xiii, 13
Pentacrinites Blumenbach, 15, 25, 51, 191
Pentacrinites briareus württembergicus Quenstedt = *Pentacrinites dichotomus* (McCoy), 191
Pentacrinites dargniesi Terquem & Jourdy, 51, 203, 209, **211**
Pentacrinites dichotomus (McCoy), 44, 181, 191, **193**, **194**
Pentacrinites doreckae Simms, 44
Pentacrinites fossilis—Blumenbach, **25**, 44, 177, **178**, 179, **180**, **181**, 191
Pentacrinites quenstedti Oppel = *Pentacrinites dichotomus* (McCoy), 191
Pentaramicrinus Sutton & Winkler, 46
Pentremites Say, 46
Pentremitidea medusae Jaekel, **117**
Periechocrinus Morris, 44, 45
Petalocrinus Weller & Davidson, 26, 45
Phacops Emmrich, 123, 129
Phanocrinus Kirk, 46
Phosphannulus Müller, Nogami & Lenz, **58**
Phrynocrinidae Clark, 241
Phyllocrinus d'Orbigny, 47, 48, 160
Physodoceras Hyatt, 221
Pilocrinus Jaekel, 47
Pilocrinus moussoni (Desor), 47
Placunopsis Morris & Lycett, 170
Plagiostoma striatum (v. Schlotheim), **172**
Platyceras Conrad, **56**, 114, 149
Platyceras dumosum Hall, 131

Platyceratidae Hall, platyceratids, 56, 79, 80
Platycrinites J. S. Miller, 10, 45, 46, 57, 141, 143, **151**
Platycrinites hemisphaericus (Meek & Worthen), 56, 149–151
Platycrinites regalis (Hall), **11**
Platycrinites saffordi (Hall), 149, 150, 151
Platycrinites symmetricus (Wachsmuth & Springer), **136**, 137
Platycrinitidae Austin & Austin, 140
Platystrophia King, 69, 73
Pleurocrinus Austin & Austin, 46
Pleurocystites Billings, 69, 71
Plicodendrocrinus Brower, 42, 43, 73, 76
Plicodendrocrinus casei (Meek), **11**
Plotocrinus Peck, 47
Poecilocrinus Peck, 47
Pomatocrinus nobilis (Walther), 222
Porocrinus Billings, 43
Porphyrocrinidae Clark, 241
Portheus Cope, 226
Poteriocrinina Jaekel, poteriocrinines, 32, 36, 37
Poteriocrinites J. S. Miller, 46
Praecupulocrinus Brower, 42
Prasopora Nicholson & Etheridge, 65, 69
Proapsidocrinus Wanner, 160
Procomaster pentadactylus Simms, 192, **195**
Proctothylacocrinus esseri Kesling, **133**, 134
Proisocrinidae Rasmussen, 243
Proisocrinus Clark, 242
Proisocrinus ruberrimus Clark, 242
Promachocrinus Carpenter, **29**
Promachocrinus kerguelensis Carpenter, **4**
Prophyllocrinus Wanner, 160
Protodouvillina Harper, Boucot & Johnson, 123
Psalidocrinus Remes, 48
Pseudomytiloides Koschelkina, 184, 191
Pseudomytiloides dubius (Sowerby), 185, **187**, 189, 190, 195
Pseudoperna congesta (Conrad), 225
Pteranodon Marsh, 226
Pterocoma Agassiz, 222
Pterocoma pennata (v. Schlotheim), 218, 221, **222**, 223
Pterotocrinus Lyon & Casseday, 23, 46
Pycnocrinus S. A. Miller, 12, 56, 76, **78**
Pycnocrinus dyeri (Meek), 56, 79, **80**
Pygope Link, 48

Quenstedticrinus Klikushin, 42

Rafinesquina Hall & Clarke, **77**
Rafinesquina deltoidia (Conrad), 65
Raymondicrinus Klikushin, 236
Raymondicrinus oregonensis (Moore & Vokes), 233, **235**, 236
Resserella Bancroft, 89
Reteocrinus Billings, 42
Rhadinocrinus Jaekel, **120**
Rhipidocrinus crenatus (Goldfuss), 45
Rhizocrinus lofotensis Sars, **13**
Rhizostomites Haeckel, 217
Rhodocrinites J. S. Miller, 45

Rhodocrinites kirbyi (Wachsmuth & Springer), 34, **136**, **137**, 138
Rhodocrinites nanus (Meek & Worthen), 138
Rhodocrinites nanus glyptoformis (Laudon & Beane), 138
Rhodocrinitidae Roemer, 140
rhombiferans, 31
Rhynchonella varians v. Schlotheim, 215 = *Rhynchonelloidella alemanica* (Rollier)
Rhynchotetra Weller, 89
Ristnacrinus Öpik, 42
Roveacrinida Sieverts-Doreck, roveacrinids, 22, 32, 37, 46–48
Roveacrinidae Peck, 46, 47
Roveacrinus Douglas, 47

Saccocoma Agassiz, 51, 216–224
Saccocoma cretacea Bather = *Applinocrinus cretaceus* (Bather), 47
Saccocoma pectinata (Goldfuss) = *Saccocoma tenella* (Goldfuss), 218, 221
Saccocoma quenstedti Verniory, 224
Saccocoma schwertschlageri Walther = *Saccocoma tenella* (Goldfuss), 218
Saccocoma tenella (Goldfuss), **218**, **219**, **220**, 224
Saccocrinus Hall, 89, 90
Saccocrinus speciosus Hall, **90**
Sagenocrinida Springer, 32, 33
Sagenocrinites Austin & Austin, 17
Saracrinus Clark, 234, 236
Scaphites Parkinson, 225
Sclerocrinus strambergensis Jaekel, 48
Scyphocrinites Zenker, 93–95, 97, 98
Scyphocrinites elegans Zenker, 97, 98
Scyphocrinitidae Jaekel, scyphocrinitids, **15**, 28, 48, 51, 53, 93, **96**, **97**, 98–101
Scytalocrinus decadactylus (Meek & Worthen), **148**, 149
Scytalocrinus robustus (Hall), 57, 149, 150, 152
Seirocrinus Gislén, 10, 12, 25, 51, 99, 178, 180, 181, 189
Seirocrinus subangularis (J. S. Miller), 44, 183–186, **187**, **188**, **189**, **190**, **191**, **192**
Senariocrinus Schmidt, 23
Senariocrinus maucheri Schmidt, **20**, 117
Shroshaecrinus Klikushin, 42
Silesiacrinus Hagdorn & Gluchowski, 165

Sinosura kelheimensis (Boehm), 224
Siphonocrinus S. A. Miller, 45
Solanocrinites gracilis Walther, 222
Solemya Lamarck, 195
Somphocrinidae Peck, somphocrinids, 22, 37, 40, 46, 47
Somphocrinus mexicanus Peck, 47
Sowerbyella Jones, 65
Springericrinus magniventrus (Springer), 147
Stegerhynchus Foerste, 89
Steinmannia Fischer, 184
Steinmannia radiata (Goldfuss), 183
Steinmannia radiata parva (Quenstedt), 195
Stellarocrinus sp. cf. *S. virgilensis* Strimple, **159**
Stenaster Billings, 73
Stenaster salteri Billings, **70**
Stenopecrinus sp. cf. *S. planus* (Strimple), **156**
Stephanocrinus Conrad, 89
Stiptocrinus Kirk, 45
Striispirifer Cooper & Muir-Wood, 88, **89**, 90, **92**
Strimplecrinus inornatus (Wachsmuth & Springer), **136**, **137**, 138
Strotocrinus glyptus (Hall), 141, **143**
Stylometra spinifera (Carpenter), **xii**
Sulcoretepora d'Orbigny, 123, 129, 130
Sundacrinus Wanner, 160
Synaptocrinus Springer, 134
Synbathocrinus Phillips, 45, 46

Talpina v. Hagenow, 170
Taxocrinida Springer, 32, 33
Taxocrinus Phillips, 46
Taxocrinus colletti White, 149, 150
Taxocrinus intermedius Wachsmuth & Springer, 137
Taxocrinus stürtzii Follmann = *Eutaxocrinus stürtzii* (Follmann), **119**
Taxocrinus ungula Miller & Gurley, **149**, 150, 152, 153
Teliocrinus springeri (Clark), 233, **234**
Tentaculites v. Schlotheim, 104, 108
Terminaster Hess, 218
Tetanocrinus Jaekel = *Dolichocrinus* de Loriol, 47
Tetracrinus moniliformis Goldfuss, 47
Thallocrinus Jaekel, 114
Thallocrinus procerus Schmidt, **119**

Thamnasteria Lesauvage, **6**
Thiolliericrinidae Clark, thiolliericrinids, 48
Thiolliericrinus Etallon, 48
Tholocrinus Kirk, 46
Tholocrinus wetherbyi (Wachsmuth & Springer) = *Bicidiocrinus wetherbyi* (Wachsmuth & Springer), **23**
Thylacocrinus Oehlert, 124
Thylacocrinus clarkei Wachsmuth & Springer, **127**
Thysanocrinus arborescens Talbot = *Ambicocrinus arborescens* (Talbot), 103
Timorechinus Wanner, 160
Timorocidaris Wanner, 161
Timorocidaris sphaeracantha Wanner, 161
Torynocrinus Seeley, 48
Trampidocrinus Lane & Webster, 163
Traumatocrinus Wöhrmann, 28, 44
Traumatocrinus caudex (Dittmar), 43
Triacrinus Münster, 114
Triacrinus koenigswaldi Schmidt, **121**
Tribrachyocrinus McCoy, 160
Troosticrinus Shumard, 45
Trypanites Mägdefrau, 69, **70**, 71, 167, **174**
Tryssocrinus endotomitus Guensburg, 42
Tulipacrinus Kristan-Tollmann, 47
Tylasteria Valette, **6**

Uintacrinida Broili, uintacrinids, 32, 37
Uintacrinus Grinnell, 30, 46, 51, 53, 226
Uintacrinus socialis Grinnell, **17**, 212, 227, **228**, **229**, **230**, **231**
Uperocrinus Meek & Worthen, 140
Uperocrinus nashvillae (Hall), **23**

Volviceramus grandis Conrad, 231
Vostocovacrinus Yeltisheva & Polyarnaya, 28
Vostocovacrinus boreus Yeltisheva & Polyarnaya, **30**

Xenocrinus S. A. Miller, 43, 76
Xenocrinus baeri (Meek), **79**
Xenocrinus penicillus S. A. Miller, **18**
Xiphactinus Leidy, 226

Zeacrinites Hall, 46
Zophocrinus S. A. Miller, 45
Zygodiplobathrina Ubaghs, 32
Zygospira Hall, 69